Promoting Productive Cooperation Between Space Lawyers and Engineers

Anja Nakarada Pecujlic
University of Cologne, Germany

Matteo Tugnoli
European Space Policy Institute, Austria

A volume in the Advances in Public Policy and Administration (APPA) Book Series

IGI Global
DISSEMINATOR OF KNOWLEDGE

Published in the United States of America by
IGI Global
Information Science Reference (an imprint of IGI Global)
701 E. Chocolate Avenue
Hershey PA, USA 17033
Tel: 717-533-8845
Fax: 717-533-8661
E-mail: cust@igi-global.com
Web site: http://www.igi-global.com

Library of Congress Cataloging-in-Publication Data

Names: Nakarada Pecujlic, Anja, 1989- editor. | Tugnoli, Matteo, 1984- editor.
Title: Promoting productive cooperation between space lawyers and engineers / Anja Nakarada Pecujlic and Matteo Tugnoli, editors.
Description: Hershey, PA : Information Science Reference, an imprint of IGI Global, [2019] | Includes bibliographical references and index.
Identifiers: LCCN 2018022058| ISBN 9781522572565 (hardcover) | ISBN 9781522572572 (ebook)
Subjects: LCSH: Astronautics. | Aerospace engineering. | Space flight--Planning. | Astronautics and state. | Space law.
Classification: LCC TL790 .P75 2019 | DDC 341.4/7--dc23 LC record available at https://lccn.loc.gov/2018022058

This book is published in the IGI Global book series Advances in Public Policy and Administration (APPA) (ISSN: 2475-6644; eISSN: 2475-6652)

British Cataloguing in Publication Data
A Cataloguing in Publication record for this book is available from the British Library.

All work contributed to this book is new, previously-unpublished material. The views expressed in this book are those of the authors, but not necessarily of the publisher.

For electronic access to this publication, please contact: eresources@igi-global.com.

Advances in Public Policy and Administration (APPA) Book Series

ISSN:2475-6644
EISSN:2475-6652

Mission

Proper management of the public sphere is necessary in order to maintain order in modern society. Research developments in the field of public policy and administration can assist in uncovering the latest tools, practices, and methodologies for governing societies around the world.

The **Advances in Public Policy and Administration (APPA) Book Series** aims to publish scholarly publications focused on topics pertaining to the governance of the public domain. APPA's focus on timely topics relating to government, public funding, politics, public safety, policy, and law enforcement is particularly relevant to academicians, government officials, and upper-level students seeking the most up-to-date research in their field.

Coverage

- Government
- Law Enforcement
- Political Economy
- Politics
- Public Administration
- Public Funding
- Public Policy
- Resource Allocation
- Urban Planning

IGI Global is currently accepting manuscripts for publication within this series. To submit a proposal for a volume in this series, please contact our Acquisition Editors at Acquisitions@igi-global.com or visit: http://www.igi-global.com/publish/.

The Advances in Public Policy and Administration (APPA) Book Series (ISSN 2475-6644) is published by IGI Global, 701 E. Chocolate Avenue, Hershey, PA 17033-1240, USA, www.igi-global.com. This series is composed of titles available for purchase individually; each title is edited to be contextually exclusive from any other title within the series. For pricing and ordering information please visit http://www.igi-global.com/book-series/advances-public-policy-administration/97862. Postmaster: Send all address changes to above address.

Titles in this Series

For a list of additional titles in this series, please visit: www.igi-global.com/book-series

Social Jurisprudence in the Changing of Social Norms Emerging Research and Opportunities
Karla L. Drenner (Purdue University Global, USA)
Information Science Reference • copyright 2019 • 174pp • H/C (ISBN: 9781522579618) • US $135.00 (our price)

Capital Management and Budgeting in the Public Sector
Arwiphawee Srithongrung (University of Illinois at Springfield, USA) Natalia B. Ermasova (Governors State University, USA) and Juita-Elena (Wie) Yusuf (Old Dominion University, USA)
Business Science Reference • copyright 2019 • 300pp • H/C (ISBN: 9781522573296) • US $225.00 (our price)

Handbook of Research on Advocacy, Promotion, and Public Programming for Memory Institutions
Patrick Ngulube (University of South Africa, South Africa)
Information Science Reference • copyright 2019 • 453pp • H/C (ISBN: 9781522574293) • US $265.00 (our price)

Marketing Peace for Social Transformation and Global Prosperity
Alexandru-Mircea Nedelea (Stefan cel Mare University of Suceava, Romania) and Marilena-Oana Nedelea (Stefan cel Mare University of Suceava, Romania)
Information Science Reference • copyright 2019 • 304pp • H/C (ISBN: 9781522574644) • US $195.00 (our price)

Mathematical Approaches to Understanding Democracy Emerging Research and Opportunities
Alberto Arteta (Troy University, USA)
Information Science Reference • copyright 2019 • 148pp • H/C (ISBN: 9781522575580) • US $165.00 (our price)

Advanced Methodologies and Technologies in Government and Society
Mehdi Khosrow-Pour, D.B.A. (Information Resources Management Association, USA)
Information Science Reference • copyright 2019 • 726pp • H/C (ISBN: 9781522576617) • US $275.00 (our price)

Handbook of Research on Promoting Peace Through Practice, Academia, and the Arts
Mohamed Walid Lutfy (London School of Economics, UK) and Cris Toffolo (Northeastern Illinois University, USA)
Information Science Reference • copyright 2019 • 553pp • H/C (ISBN: 9781522530015) • US $275.00 (our price)

Handbook of Research on Military Expenditure on Economic and Political Resources
Ramesh Chandra Das (Katwa College, India)
Information Science Reference • copyright 2018 • 564pp • H/C (ISBN: 9781522547785) • US $325.00 (our price)

IGI Global
DISSEMINATOR OF KNOWLEDGE

701 East Chocolate Avenue, Hershey, PA 17033, USA
Tel: 717-533-8845 x100 • Fax: 717-533-8661
E-Mail: cust@igi-global.com • www.igi-global.com

For our parents

Editorial Advisory Board

Table of Contents

Detailed Table of Contents

of thousands of kilograms of material built on the ground and transported to space for assembly. It houses an artificial atmosphere to sustain life in outer space and has been continually inhabited for over 15 years. This chapter describes the technical complexity of the ISS, the background of how it was assembled, its major systems, details of crew life onboard, commercial usage of the resource, and examples of mishaps that have occurred during the ISS's operation. The technical details of the ISS provide a glimpse into what future space stations that might orbit the Moon and Mars will resemble.

The International Space Station is certainly one of the most astounding achievements of humankind in space. Especially from a legal point of view, the creation of the Intergovernmental Agreement (IGA) and its sub-instruments specifically for the ISS was a major success: the IGA was the first grand-scale multi-national legally binding space-related treaty drafted, ratified, and implemented by the major space faring nations since the drafting of the five UN Space Treaties roughly 20 years before. And still today, the legal framework of the ISS is a stand-alone legal system which can serve as model for other missions, as it refines and develops in an innovative way the rules laid out in the five UN Space Treaties and at the same time manages to coordinate and organize management, utilization, and financing between all the partners. This chapter intends to shed light on the complex legal system governing the ISS and to point out the novelties of the IGA-Structure in comparison with the conventional body of international space law.

This chapter focuses on satellite mega-constellations, as mega-constellations raise several technical issues from telecommunication to launching these huge sets of satellites. Through an overview of basic signal science (periodic waves, types of modulation) and a discussion about telecommunication systems, actual 4G and future 5G networks, the chapter aims at giving without prior scientific background, sufficient material to better understand mega-constellation from its basic functioning to sophisticated satellites systems on orbit providing wide communication services all around the Earth. Major topics are evoked about such constellations be it on launching, space debris, challenges, and technologies used from a technical point of view.

As a reflection of the fast-paced world we currently live in, a new era in satellite technology begins to emerge. This sector, of vital importance for our day-to-day lives, is in a period of significant change, characterized by the use of large satellite constellations. These mega-constellations aim to provide, among other things, worldwide internet coverage. Since 2015, satellite internet providers such as Space X and OneWeb have been working on the idea of mega-constellations and intend to launch circa 4.000 and 882 satellites respectively, thus forcing satellite manufacturers to lower the costs drastically. However, many legal issues may arise since the vast amounts of satellites will significantly increase the risk of collision with other satellites or spacecraft.

The human exploration of planetary bodies started with the Apollo missions to the Moon, which provided valuable lessons learned and experience for the future human exploration. Based on that, the design of hardware and operations need to further be developed to also overcome the new challenges, which arise when planning crewed missions to Mars and beyond. This chapter provides an overview about the environment and structure of the Red Planet and discusses the challenges on operations and hardware correlated to it. It further provides insights into the considerations regarding the hardware development which need to be investigated and defined before launching a crewed mission to Mars.

This chapter examines the legal implications of the ambitious plans to send manned missions to Mars and establish permanent human presence in outer space announced. It addresses the concept of "colonizing planets" and its potential consequences. It offers an overview of the applicable existing legal norms and presents a number of potential legal scenarios concerning the governance of permanent human habitant in outer space. Even though technology is not completely there yet, we should not wait for the space shuttle to be ready to board in order in the nearest future to start discussing potential legal frameworks required to govern multi-planetary existence.

This chapter explores the concept of space policy and sheds light on its importance in understanding the nature and evolution of space activities and programs. More specifically, this chapter seeks to offer an introduction on how space policy can be defined and understood, followed by an overview of the main current trajectories of space policies of both national actors as well as international and intergovernmental institutions. Finally, this chapter formulates some conclusions regarding the future potential avenues of space policy and describes how these scenarios could impact the future of human activities and behaviour in space.

Foreword

As the Director of the UN Office for Outer Space Affairs (UNOOSA), I lead a talented team responsible for engaging with the wider space community across the policy, legal and technical capacity building aspects of space affairs. We work every day to support international cooperation in the space environment, engaging with stakeholders in some of the most fundamental discussions on space affairs, such as how to best address an increasingly congested and contested space environment, how to legislate space to preserve and offer its benefits to all of humankind.

The aim of this excellent book is to support such endeavours by encouraging cooperation between the legal and technical sides and experts in the space sector. The authors make a valuable contribution towards enhancing mutual understanding between the legal and technical communities that are involved in space activities. Facilitating dialogue between different communities within the space sector has been a goal of the United Nations since the dawn of the space age.

This book presents a selection of example space activities from members of the international space community diligently dissected from the legal and technical perspectives. It is great to see contributors drawn from a range of organisations, from public sector entities such as space agencies to the private sector and academia. Bringing in such a plethora of perspectives is fundamental to ensuring a robust dialogue on what the future of international cooperation in space activities should look like.

The UN Outer Space Treaty, which celebrated its 50th anniversary in 2017, serves as the foundation of international space law. The Treaty confirms, inter alia, that States Parties to the Treaty desire 'to contribute to broad international cooperation in the scientific as well as the legal aspects of the exploration and use of outer space for peaceful purposes. Since the UN Outer Space Treaty, the international community, through the UN's Committee on the Peaceful Uses of Outer Space, of which UNOOSA is the Secretariat, has established a growing list of treaties, principals and guidelines that, collectively, constitute the normative legal framework governing activities in outer space. This framework has been instrumental in sustaining over half a century of exponential growth and successes in space technology, exploration activities and the global space economy. As this book testifies, along with the Outer Space Treaty itself, the teamwork between the legal and technical aspects of space affairs is one of the fundamental bricks over which this unprecedented period of development in the space sector has been built.

This year, the world will mark the 50th anniversary of the Apollo 11 Moon landing. Just like the normative framework governing outer space activities, a lot has changed, over these past decades, on the technical side of space activities. For example, the world is launching record numbers of new space objects (in 2017, 553 new space objects were registered with the United Nations, with just under half of those launches conducted by private sector entities). These activities are conducted against the backdrop of a growing list of national and regional space agencies, reflecting an ever-increasing interest in the frontier field of space exploration and innovation from every part of the world.

The time, therefore, is now to pause and reflect on how we have come this far and what must be done to ensure the future of space exploration is just as bright as its past. The importance of maintaining meaningful, results-based cooperation between the legal and technical space communities will be a fundamental part of this process.

That is why I encourage you to read this book and take time to consider the perspectives it raises and how you can be part of the process of ensuring international cooperation in the peaceful uses of outer space for the benefit of everyone, everywhere.

Simonetta Di Pippo
United Nations Office for Outer Space Affairs

Preface

One of the major non-technical challenges in the space sector today is the need for establishing a better mutual understanding between different professionals that work together in developing space missions, particularly in the early stages. This is often translated into the need to ensure shared views between engineers who design the mission, and space lawyers who have the task to cover the relevant legal aspects pertaining to these missions. Naturally, problems in communication and cooperation arise mostly due to the fact that typically neither engineering nor space law education curricula cover relevant aspects of the other field, for obvious reasons. In the case of space engineering, this situation often leads to a lack of understanding as to why certain technologies although they might be available could be difficult to implement and employ from a legal perspective. On the other hand, space lawyers may have difficulties in grasping key technical aspects that in turn could have relevant implications for the legal status of the mission.

Members of the two communities usually reach some level of understanding only after many years of experience in the space industry, through repeated contact with the topics relevant to their projects. This, however, means that the shared knowledge is limited to very specific areas of space missions, and that a significant amount of time is needed in order to attain just some basic knowledge. This obstacle inevitably slows down the development of space projects. Therefore, we believe that a basic understanding of the most important legal and technical aspects acquired early on in a space lawyer and engineer's career will facilitate cooperation in future space missions and contribute to the advancement of the space sector in general.

This is especially important and even more so, necessary, nowadays as there is a proliferation of space actors and activities. Investments in space sector are growing, each year there is an increasing number of countries engaging in the space domain and developing their national space programmes, as well as the rise of private actors becoming interested in the commercial side of space exploration and activities. These private actors are now ranging from the "billionaires club" (e.g. Elon Musk, Jeff Bezos, Richard Branson), to a plethora of start-ups currently developing ground-breaking space technologies or applications. In order to have a better understanding of all of these developments, the technology involved, the legal regime governing space activities, and possible global political implications that the usage and exploration of outer space may cause, this book offers a unique cross-educational manual to help the reader navigate through this realm. Outer Space is *res communis,* it belongs to all human beings, and as such, it is in the interest of all of us to acquire basic information on what is technologically feasible, what is planned, what the law allows and prohibits, and how can all of this in the end affect our daily life.

Therefore, we compiled a handbook that provides detailed insights into legal and technical topics that are related to a selection of space activities, described in an accessible way. The two sides are elaborated using the examples of seven high-profile space missions, three of which are ongoing and the remaining four are being planned for the future. More specifically, the book describes technical and legal aspects of: space launchers, suborbital spaceflight, the International Space Station, mega-constellations, on-orbit servicing and space debris removal, asteroid mining and in-situ resource utilization, and finally human planetary exploration. The structure is presented with an overarching narrative that starts from leaving the Earth's surface and ultimately touches down a planetary body, through intermediate and interlinked steps.

The introductory chapter of the book is devoted to the Space Generation Advisory Council (SGAC), an organisation and network that aims to represent university students and young space professionals ages 18-35 to the United Nations, space agencies, industry, and academia. The SGAC has been supporting this book from the outset as it perceived it as an important tool to express the views of the future generation of space leaders as regards their long-term visions on space policy, law, engineering, science and their implementation. The SGAC introductory chapter explains the relevancy of this publication as the authors demonstrate the benefits and importance of having a multi-disciplinary approach to space activities. Furthermore, it emphasizes its importance for the organisation itself because it shows the perspective of the next generation on cross-sectorial activities, which is put forward by a number of actors in/from the industry. It showcases that technical and non-technical students and young professionals can learn from each other, which is in line with SGAC's mandate.

Following the introductory chapter about SGAC, the book details seven space missions:

- Space launchers
- Suborbital spaceflight
- The International Space Station
- Mega-Constellations
- On-orbit servicing and active debris removal
- Asteroid mining and in-situ resource utilization
- Human planetary exploration

Chapter 2 and 3 describe space launchers and access to space in general. Specifically, chapter 2 covers the technical side explains how space travel counts as one of the greatest technological efforts and requires careful planning and systematic approach in every technical and human aspect. In this chapter, a space mission example is divided into four consecutive phases, starting with payload and launcher definition and ending with the launch of the rocket. The discussion of the first phase provides general considerations of manned and unmanned commercial missions, orbits and trade-offs for small satellites. Subsequent phase covers transportation issues of the payload to the launch site. The discussion of the payload integration process and launch proceeding complete the technical contribution of this chapter. Lastly, a broad market overview is presented in the conclusion. In the chapter covering the legal side, certain core legal topics that arise in connection with the delivery of a payload into or beyond Earth orbit are outlined. The first part deals with some of the established approaches to procuring launch services, as well as some of the common features of launch service agreements that balance the interests of the launch service provider and its customer. The second part of the chapter looks at governmental authorisation required to carry out a launch. While safety standards and success rates continually improve, launching a space object is still the riskiest part of most space missions and is therefore a carefully regulated aspect of space activity, with participants having to obtain prior authorisation from a competent national authority. Finally, the third part explores some of the legal consequences in international law of launching a space object, including the maintenance of a register of space objects launched, and the burden of liability that is placed on 'launching States'.

The second space mission discussed is suborbital flights. The technical contribution depicts how one of the major historical milestones of spaceflight, suborbital flights played a vital role in the development of technologies and procedures for the operation of spacecraft. Today suborbital flights still play an important part in scientific and commercial fields and in this chapter the basic terms and values are defined and also explained in the historical context. Small launching system for scientific applications called sounding rockets have a significant importance due to an unbeatable flexibility and cost effectiveness and the new and emerging field of commercial human flights to the edge of space is growing as an economically sound business model for private participants. Both of these aspects are detailed in the appropriate sub-chapters, followed by an outlook on the future possibilities of suborbital flights. In the second part, the legal aspects of suborbital spaceflight under international law are outlined with the aim of clarifying upcoming legal challenges to the existing space treaties. One of major problems lies in the applicability of air law to the spaceflight. In the case of using aircraft-type reusable launch vehicle (RLV), it fits into the definition of aircraft in air law. The other problems are also addressed which include: the different legal status of airspace and outer space; the lack of specific concepts and terms related the flights in the existing space treaties; and a need for international regulation to ensure safety in the flights. The overall chapter concludes with recommendations regarding Space Traffic Management.

The third space mission that is explored is the International Space Station (ISS). In the technical chapter, the new era of spaceflight that dawned following the conclusion of the United States and Russian space race is depicted. This new era has been marked by the design, assembly, and operation of one of the greatest engineering feats mankind has accomplished, the ISS. The Station is comprised of hundreds of thousands of kilograms of material built on the ground and transported to space for assembly. It houses an artificial atmosphere to sustain life in outer space and has been continually inhabited for over 15 years. This chapter describes the technical complexity of the ISS, the background of how it was assembled, its major systems, details of crew life onboard, commercial usage of the resource, and examples of mishaps that have occurred during the ISS's operation. The technical details of the ISS provide a glimpse into what future space stations that might orbit the Moon and Mars will resemble. The legal chapter of the ISS mission, emphasizes how the ISS is the most astounding achievement of humankind in space. Especially from a legal point of view, the creation of the Intergovernmental Agreement (IGA) and its sub-instruments specifically for the ISS was a major success: the IGA was the first grand-scale multi-national legally binding space-related treaty drafted, ratified and implemented by the major space faring nations since the drafting of the five UN Space Treaties roughly 20 years before. And still today, the legal framework of the ISS is a stand-alone legal system which can serve as model for other missions, as it refines and develops in an innovative way the rules laid out in the five UN Space Treaties and at the same time manages to coordinate and organize management, utilization and financing between all the partners. This chapter sheds light on the complex legal system governing the ISS and points out the novelties of the IGA-Structure in comparison with the conventional body of international space law.

The fourth space mission selected is represented by upcoming broadband mega-constellations projects. In the technical chapter, the major technical topics around mega constellations are described, ranging from raise, ranging from telecommunications to the unprecedented launch of thousands of satellites. Through an overview of basic signal theory (periodic waves, types of modulation) and a discussion about telecommunication systems, actual 4G and future 5G networks, the chapter aims at giving -- without a necessary prior scientific background - sufficient material to better understand mega-constellations from its basic functioning to sophisticated satellites systems on orbit providing wide communication services all around the Earth. Major topics are elaborated regarding such constellation from launching, space debris, challenges and technologies used, all from a technical point of view. The legal chapter of the mega-constellation mission reflects on the fast-paced world, we currently live in, and a new era in satellite technology that is beginning to emerge. This sector, of vital importance for our day-to-day lives, is in a period of significant change, characterised by the use of large satellite constellations. These mega-constellations aim to provide, among other things, worldwide Internet coverage. Since 2015, satellite Internet providers such as Space X and OneWeb have been working on the idea of mega-constellations and intend to launch circa 4.000 and 882 satellites respectively, thus forcing satellite manufacturers to lower the costs drastically. However, many legal issues may arise since the vast amounts of satellites will significantly increase the risk of collision with other satellites or spacecraft and this chapter offers an overview what these legal questions are.

The fifth space mission that is considered is On-orbit Servicing and Active Debris Removal (OOS/ADR). The challenge of space debris and the concepts of on-orbit servicing and active debris removal are introduced in the chapter devoted to the technical aspect. The evolution of the space debris population is put into a historical perspective, followed by a description of observation and modelling methods, and a brief introduction of internationally-agreed mitigation strategies. Proposed mission types and the required technologies for servicing and removal activities are detailed and their challenges explained. Where applicable, past, current, and proposed activities are summarized to illustrate the concepts. In the legal chapter devoted to the on-orbit servicing and space debris removal mission, the legal issues associated with this kind of activities are examined. This chapter addresses these legal issues from the perspective of military, civil, and commercial space activities. It argues that the major international legal challenges for these technologies are avoiding their weaponization in order to use them to provide for the sustainability of the space environment. Further, it argues that the development of domestic authorization regimes will be necessary precursor to the deployment of OOS and ADR.

The sixth space mission in focus is Asteroid Mining and In-situ Resource Utilization. The technical chapter explains how the human spaceflight is an expensive endeavour, as every kilogram that needs to be transported to low Earth orbit or beyond costs tens of thousands of dollars, with the cost increasing exponentially the farther humanity extends its reach into the solar system and beyond. It is therefore prudent, if not necessary, to consider the use of resources that are available at the destination of a given exploration mission. This concept is called in-situ resource utilization (ISRU). The processes that are required to extract useful materials from the local environment can not only be used to support a human crew, but also to obtain resources that are of value on Earth and can thus be returned there for commercial gain. This chapter provides background information on ISRU in general and discusses the most important technologies and processes that are currently employed or under development. The legal chapter discusses asteroid mining and in-situ resource utilization, whether and how the existing rules of international space law can be applied to regulate the activity of space resource exploitation. Special focus on in-situ resource utilization is given. As such, the chapter aims to establish how natural resources should be defined in the context of space law, how the activity of exploitation compares to other types of activities in space, in particular the prohibited activity of national appropriation, and the extent to which the type and supply of space resource may impact on its regulation. The chapter concludes with a brief overview of the most recent developments in national and international forums attempting to regulate space resource exploitation.

The seventh and final space mission considered is human planetary exploration. The technical chapter starts with a reflection on the Apollo missions to the Moon, which provided valuable lessons learned and experience for the future human exploration. Based on that, the design of hardware and operations need to further be developed to also overcome the new challenges, which arise when planning crewed missions to Mars and beyond. This chapter provides an overview about the environment and structure of the Red Planet and discusses the challenges on operations and hardware correlated to it. It further provides insights into the considerations regarding the hardware development which need to be investigated and defined before launching a crewed mission to Mars. The part devoted to legal aspect explains the human planetary exploration, the legal implications of the ambitious plans to send manned missions to Mars and establish permanent human presence in outer space. This chapter addresses the concept of "colonizing planets" and its potential consequences, it offers an overview of the applicable existing legal norms and presents a number of potential legal scenarios concerning the governance of permanent human habitant in outer space. Even though technology is not completely there yet, this chapter advises us not to wait for the space shuttle to be ready to board in order in the nearest future to start discussing potential legal frameworks required to govern multi-planetary existence.

Finally, this book concludes with a chapter introducing space policy in a broad context. The aim of this chapter is to provide the reader with fairly comprehensive basic insights into space activities and programmes through the lenses of space policy. More specifically, this chapter seeks to offer an introduction on how space policy can be understood, followed by an overview of the main current trajectories of space policies of both national actors as well as international and intergovernmental institutions. Finally, this contribution formulates some conclusions regarding the future potential avenues of space policy and describes how these scenarios could impact the way in which space activities are conducted.

This book is intended for early-stage professionals, graduates and students in (but not limited to) space technology, space law, engineering, and international law, as well as young professionals in the public and private space sector and academia. This book could also serve as lecture material for any basic and interdisciplinary course on space law or space technology. Lastly, as the book lays out the basics of space engineering and space law, it is also suitable for anyone that wants to become acquainted with the space domain in general.

The idea and motivation for such an interdisciplinary guide for space practice originally stemmed from the feedback received across the international space community. By observing the current status of the space industry and discussing with professionals in the space sector at various levels, the growing necessity for a wider common understanding of issues from a legal, practical, programmatic, and technical side become evident. It is vital, for the success of any space endeavour, that the various communities are allowed to cooperate to the fullest, appreciate the expertise each has to offer, and maintain a results-oriented relationship. This publication aims at helping enhance these capabilities at an early stage and promote the comprehension and cooperation between experts in different fields that is necessary to drive space utilization and exploration for the benefit of all mankind. The expected synergies from heightened understanding will nurture and enhance the entire space industry, already consisting of professionals who are passionate about their respective fields, by promoting understanding of other related domains.

Anja Nakarada Pecujlic
University of Cologne, Germany

Matteo Tugnoli
European Space Policy Institute, Austria

Acknowledgment

The Editors would like to sincerely thank all the Authors for their high motivation, constant and tireless work, without whom this project would not have been possible. Furthermore, our gratitude goes to our Editorial Advisory Board members, for their invaluable comments and great support demonstrated from the very beginning, and to UNOOSA Director Simonetta Di Pippo for the foreword of this book. Also, we would like to thank the former SGAC Executive Director Minoo Rathnasabaphaty, and current SGAC Executive Director Clémentine Decoopman, for having encouraged this project and offered the dedicated support of the organization; and all the SGAC members for their help. Lastly, we want to thank all our colleagues, family and friends that followed progress of this project over the years, and supported us throughout the whole process.

Acknowledgment

Chapter 1
The Space Generation Advisory Council

Clémentine Decoopman
Space Generation Advisory Council, Austria

Jennifer Lauren Napier
Space Generation Advisory Council, Austria

ABSTRACT

The Space Generation Advisory Council (SGAC) in support of the United Nations Programme on Space Applications is a global non-governmental, non-profit organization and network which aims to represent university students and young space professionals ages 18-35 to the United Nations, space agencies, industry, and academia. The organisation was created in 1999 at the United Nations Conference on Exploration and Peaceful Uses of Outer Space (UNISPACE III), and its members include university students and young professionals with very diverse educational and professional backgrounds, collaborating together on various projects and activities. This chapter provides a detailed description of the organization, including its mandate and scope. In particular, it outlines the benefits and importance of cross-sectorial and multi-disciplinary cooperation in the space sector, and beyond.

INTRODUCTION

The Space Generation Advisory Council in Support of the United Nations Programme on Space Applications (Space Generation Advisory Council [SGAC or SpaceGen], 2018) is a "global non-governmental, non-profit (US 501(c)3) organisation and network which aims to represent university students and young space professionals ages 18-35 to the United Nations, space agencies, industry, and academia." The organisation was created in 1999 at the United Nations Conference on Exploration and Peaceful Uses of Outer Space (UNISPACE III) and according to The Space Millennium: Vienna Declaration on Space and Human Development (1999) the aim was to "create, within the framework of the Committee on the Peaceful Uses of Outer Space (COPUOS), a consultative mechanism to facilitate the continued participation of young people from all over the world, especially young people from developing countries and young women, in cooperative space-related activities."

DOI: 10.4018/978-1-5225-7256-5.ch001

SGAC members are university students and young professionals with very diverse educational and professional backgrounds from engineering to management, law and policy, international relations, economics, sciences, mission operations and more. The members also come from different countries but yet work remotely on various projects and activities together ranging from the planning and organisation of one of our numerous events to the publication of a report, the organisation of a competition or the production of a video. This diversity and global reach includes geographic, ethnic, and gender equality which only strengthens the credibility and vision of SGAC and what its members can bring to the international space community.

Building on this sustained growth of the past few years and having succession planning and long-term strategies in place, SGAC (2018), "continues to achieve its mission through these five pillars", as also seen in Figure 1:

- Hosting events at the international, regional and local levels, bringing our members together to discuss and engage with current leaders from space agencies, industry and academia;
- Maintaining SGAC project groups year-round in order to enable our members to shape key topics on space through technical papers, policy briefs and recommendations;
- Providing scholarships that enable our members to participate in space-related events around the world;
- Empowering our members with the skills and experiences to become the space leaders of tomorrow;
- Partnering with leading space agencies, corporations and organizations from around the world to nurture the next generation of space leaders.

In light of these different initiatives, SpaceGen has a strong commitment to demonstrating how further involvement of the next generation can strengthen the global space community.

Figure 1. Five pillars of SGAC
Source: SGAC, 2018

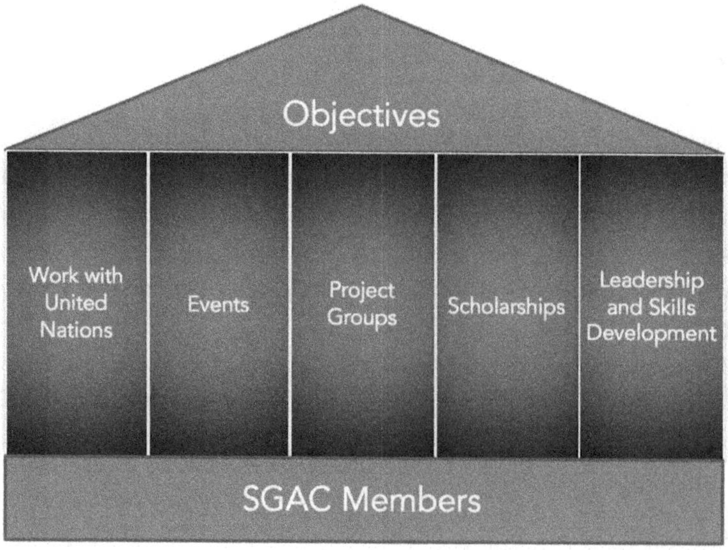

This book, "*Promoting Productive Cooperation Between Space Lawyers and Engineers*" is a perfect example of a project put in place by 21 members to contribute to the future of space activities, bringing into the process the views of the future generation of space leaders as regards to their long-term visions for space policy, law, engineering, science and the tools with which to act. Through this book, the authors demonstrate the benefits and importance of having a multi-disciplinary approach in space activities. This book has a particular importance for the organisation because it shows the perspective of the next generation on cross-sectorial activities which is put forward by many actors in the industry. It showcases that technical and non-technical students and young professionals can learn from each other which is in line with SGAC's mandate.

The organisation's success over the recent years has secured its position as one of the largest international network of students and young professionals in the space sector. SGAC (2018), "continues to fulfil its primary goal set 19 years ago: to enable the next generation of international space sector leaders to share their perspectives with policy makers, and develop their professional skills in the process."

THE SPACE GENERATION ADVISORY COUNCIL AND THE UNITED NATIONS AND THE UNITED NATIONS

The Space Generation Advisory Council was created out of the Space Generation Forum at UNISPACE III in 1999. From that moment SGAC has actively engaged with the United Nations. In 2001 the organisation gained Permanent Observer Status at the Committee on the Peaceful Uses of Outer Space (COPUOS) and its subcommittees and in 2005 gained consultative status within the United Nations Economic and Social Council (UN ECOSOC). Additionally, SpaceGen added the title "The Space Generation Advisory Council in Support of the United Nations Programme on Space Applications" in order to further connect the organisation with the work of UNOOSA. With all of these United Nations connections, SGAC works to give regular input to COPUOS and its subcommittees. No other space organisation run by young adults has a Permanent Observer status within COPUOS and is as active or as engaged in the work of the UN on space topics as SGAC. Currently SGAC is also high committed to the UN Sustainable Development Goals and the Space for Women initiative.

THE SPACE GENERATION ADVISORY COUNCIL NETWORK

Headquartered in Vienna, Austria, the SGAC network of members, volunteers and alumni has grown to more than 15,000 members representing more than 150 countries as seen here in Figure 2.

SGAC's extensive network is not limited to it membership. There are a large number of professionals and experts that contribute to the success of the organisation as well. They contribute through mentorship, speaking, acting as subject matter experts, and as inspiring role models for the students and young professionals. These senior members, traditionally alumni and founders, can also be part of SGAC's Advisory and Honorary Boards which reflect diversity (regionally, academically, and professionally), knowledge and the heritage of the organisation.

Figure 2. SGAC representation
Source: SGAC, 2018

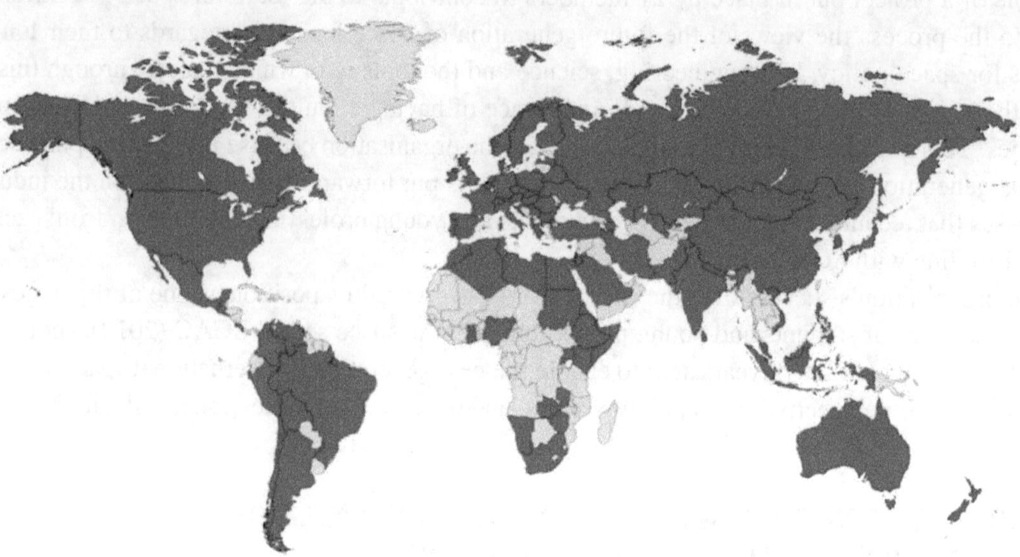

THE SPACE GENERATION IMPACT

SGAC's impact reaches as far as our member base and as broad as our involvement in the space sector. The products of SGAC events and project groups provide inputs at every level and has been referenced by other delegations at UN COPUOS. SGAC's participation and contribution at the international level (including at the UN) is of utmost importance for the youth-led organization because it allows the members an important platform for collecting and gaining insights from the international space community. Having SpaceGen members connected to the UN and the international space community also allows them a chance to present and discuss critical and key issues from the next generation perspective. Additionally, the outcomes created from the eight SGAC project groups; SGAC's local, regional, and global events; as well as activities managed at the regional and national levels can be turned into reports, presentations and recommendations that are then presented to SpaceGen's partners and sponsors as well the international space community during UN COPUOS subcommittee and plenary sessions.

The SpaceGen membership of students and young professionals from over 150 countries allows this organisation to provide a comprehensive and global viewpoint of the space industry. With this global viewpoint at hand, the organisation is able to discuss important and politically sensitive topics based on the fact that it can draw upon various member experiences, backgrounds and nationalities to provide fair, well-rounded assessments to the global space industry. This 'umbrella view' allows SGAC to provide non-biased feedback on space activity, space mission plans, space policies, and more. This allows the SGAC members to really argue for plans that they deem fair for future space activity. With this credibility the organisation has seen an increase in recommendations taken by space stakeholders that were originally put forward by SGAC members and the organization as a whole. The network is continuing in this way to be a 'sounding board' for the space community as the next generation with fresh ideas on topics such as future exploration, space resources governance, and continued global partnerships, just to name a few.

THE IMPORTANCE OF HAVING A MULTI-DISCIPLINARY APPROACH IN SPACE ACTIVITIES

The Space Generation Advisory Council is international and diverse, therefore the organisation promotes multi-disciplinary approaches to space discussions and activities such as this Handbook dedicated tor Space Lawyers and Engineers.

SGAC members are university students and young professionals with very diverse educational and professional backgrounds from engineering to management, law and policy, international relations, economics, sciences, mission operations and more. The members also come from different countries but yet work remotely on various projects and activities together ranging from the planning and organisation of one of our numerous events to the publication of a report, the organisation of a competition or the production of a video.

SGAC believes in empowering its members; "As a volunteer-run organisation, we rely on the skills of various members from around the world to operate. Various roles are available within SGAC not only to sustain the organisation, but also to provide an opportunity for our members to gain valuable skills to help promote personal and professional growth." For instance, SpaceGen members with an engineering background who might not have the opportunity to approach the industry or the commercial side of the space sector through their professional environment can be given the opportunity through SGAC. They can join one of our event organising teams and be in charge of the communication, the sponsorships, the scholarships or the logistical aspects of the event. They learn to work with a team of international volunteers, to put together a budget, to look for sponsors and much more. It provides them with a unique opportunity to learn new skills, which might prove to be useful for their future careers.

Diversity in the members' backgrounds has enabled SGAC to grow and has proved to benefit the activities they are involved in. It has also enabled its members to expand their understanding by exposing them to different fields, disciplines, ideas, and cultures. SGAC's strength is its cross-cultural communication and collaboration, as the organization continues to strive to put in place pathways for the recognition of exemplary individuals and team behaviours that promote diversity and inclusion. As such, SGAC plays an important role in global capacity building, providing fora for networking, using unique project groups to engage with young professionals all around the world, and providing professional development opportunities to their members.

The same applies to the space industry at large. There is an African proverb which says *"if you want to go quickly, go alone. If you want to go far, go together."* Diversification of the space community has been proven essential in today's space industry. There will also be great value if we could liaise with non-space organizations, active in those domains where the use of space products and data is envisaged. Cross-sectoral cooperation is twofold and should be addressed outside of the space industry as part of agenda items, either by continuing to propose more dedicated joint workshops aimed at informing these communities of the benefits offered by the use of space services and applications in their own industrial sector or more structured cooperation to support specific projects. Additionally, cross-sectorial cooperation should continue to grow and deepen within the space sector as more academic backgrounds are included in key space topic debates. It is no longer just a matter of astronomy, engineering or space law but the industry also includes economics, business, political science and more. All of these students, young professionals and experts should work together to better understand the bigger picture of space application and exploration as it takes everyone working together from various backgrounds to promote and grow the space industry. This is why this Handbook for Space Lawyers and Engineers is so critical. It is important to see all sides of a key issues before making decisions and recommendations on the way forward. Having the tools and the opportunity to see the technical and non-technical sides of a challenge are of great value especially in an ever changing and unique sector such as this.

FOSTERING COMMUNICATION BETWEEN LAWYERS AND ENGINEERS

Encouraging the development of platforms for increased knowledge and information-sharing between lawyers and engineers is extremely important as both areas are strongly linked. By providing students and young professionals from both sides of the space industry with an appropriate network, platform or forum for discussion, they will be able to collaborate amongst themselves on major projects which may be infeasible or impractical for them to conduct alone. By working together, and gaining advice from each other's, they will have a far higher likelihood of success.

SpaceGen facilitates this goal through encouraging the sharing of best practices, the diversification of the space sector and education through projects such as this book. Through this project, the organisation supported young engineers and lawyers to get together and educated each other on the basics of both disciplines.

Also, by highlighting the benefits of cross-sectoral analysis it brings a deeper layer to already researched discussions and raises awareness of other academic backgrounds in the sector. This also encourages conducting capacity building activities which empowers the space leaders of tomorrow. These may include fostering collaborations across the SGAC Project Groups providing opportunities to participate in a variety of projects and encouraging the exchanges between disciplines.

THE IMPORTANCE OF RAISING AWARENESS ABOUT SPACE TO THE GENERAL PUBLIC

Although modern societies are heavily reliant on space systems, the vast majority of people are still not aware of how, and how much, outer space impacts our daily lives. Public outreach and diversification of the space community have been discussed during the past few years and been proven essential. Public outreach will prove beneficial to raise awareness amongst governmental and commercial actors, and the general public alike, of the benefits and opportunities that space brings. It is important to show the general public the support that space technologies give us in everyday life, which in turn can also motivate the future generations to study STEM and other backgrounds with a focus on space.

Space services are enablers of change, playing a substantial role in implementing the Sustainable Development Goals (SDGs). But space is not only technology it is also cooperation. Cooperation not exclusively among institutions but also with non-governmental organisations as well as private sector as supporting actors for the development of new technologies.

THE IMPORTANCE OF INTERNATIONAL COOPERATION AND CAPACITY BUILDING

We are at a critical and amazing point in humankind's interaction with space: we are moving from exploration toward utilisation. The global effort to develop space will continue to attract new actors, many from the private sector. Partnerships that cross national borders, sectors, and disciplines are necessary and will increase as activity in space grows.

Space-based technologies are instrumental in every country as a crucial aspect of a country's infrastructure. With the increasing number of countries involved in space activities, the need for effective laws and policies on space activities, not just on an international level but also on the national and regional levels, is becoming more and more apparent. One of the pillars that support the development of legal and policy frameworks at the national level is the availability of professionals to be able to provide services in that field. Capacity building, training and education helps to promote international development and cooperation in space activities, and helps build national expertise and capacity in countries with emerging space capabilities. It also provides the means for a better understanding of the interdependent roles of science, technology and law in space activities

SGAC encourages new generations to take the lead and address global challenges in the domain of space and its applications to tackle global challenges. Younger generations will inherit the decisions made today by leaders and influencers, for this reason SGAC wants the next generation to actively take part in the decision-making process, take responsibility for tomorrow space policies from science and technology, to commercial to generate a sustainable sector. Space includes different players from government, academia, to industry. International cooperation is necessary to allow a societal understanding of space and its role in our everyday life and its long term impact on it.

CONCLUSION

The Space Generation Advisory Council aims to continue to facilitate and encourage the fostering of collaboration across disciplines by providing capacity building opportunities to our members and the wider SpaceGen network. SGAC prides itself on being the platform for information sharing on a variety of current and up and coming space topics. What makes all of this possible, is the incredible diversity of the SGAC network based on geography, gender, generations, and cultural and academic backgrounds.

To conclude, through this book, SGAC sees a valuable instrument to foster and apply its goals, an important educational resource for both young space lawyers and engineers but more generally as an accessible resource for any other young space professional who does not have a background in law or engineering but has an interest in learning more about these sub-sectors of the space industry. This handbook will give the readers an understanding of technical and non-technical aspects of space which helps them strengthen their voice for tomorrow.

REFERENCES

Space Generation Advisory Council. (2018). Retrieved from https://spacegeneration.org/

United Nations. (1999). *The Space Millennium: Vienna Declaration on Space and Human Development.* Retrieved from: http://www.unoosa.org/pdf/reports/unispace/viennadeclE.pdf

Chapter 2
Space Launchers:
Technical Aspects

Yuriy Metsker
Technische Universität München, Germany

ABSTRACT

Space travel is one of the largest technological efforts and requires careful planning and a systematic approach in every technical and human aspect. In this chapter, a space mission example will be divided into four consecutive phases, starting with payload and launcher definition and ending with the launch of the rocket. The discussion of the first phase provides general considerations of manned and unmanned commercial missions, orbits, and trade-offs for small satellites. The subsequent phase covers transportation issues of the payload to the launch site. The discussion of the payload integration process and launch proceeding complete the technical contribution of the chapter. A rough market overview is presented in the conclusion.

INTRODUCTION

The year is 1957 and the Soviet Union has launched for the very first time in human history a vehicle capable of reaching and putting into Earth orbit an artificial satellite. Once again, by overcoming unknown hazards and inspired by the human yearning for distant, unexplored regions, a new mean of travel was created. As an old Chinese proverb says: ‘a journey of a thousand miles begins with a single step’, so the launch of a spacecraft is the human species’ first step toward exploring far-flung celestial bodies, deep space, and above all, its own planet.

At the beginning of the Space Age, in the midst of the Cold War era, only two nations—the US and the Soviet Union—developed and mastered the capability to access outer space. Since then, over 5770 rockets have been launched into Earth orbit until end of 2018 (Kyle, 2018), and the club of space-faring nations has notably grown. Launchers make possible the very existence of any kind of space activity. From Sputnik to Rosetta, through Apollo and the modules of the International Space Station, space programmes worldwide have invariably relied on the availability of launch vehicles to reach and exploit outer space. However, launching into space is still today an extremely challenging endeavour, which

DOI: 10.4018/978-1-5225-7256-5.ch002

entails a great deal of complexity not just from an engineering and technical perspective, but as well as from a legal and regulatory point of view.

BACKGROUND

From a classical viewpoint, a rocket consists of three main parts: the payload; the rocket engine; and the structure, which includes propellant, tanks, and avionics.

The payload is the raison d'être of any launch, representing the rocket's most valuable component. For institutional activities, it might be an Earth observation satellite for weather forecast or a navigation satellite for providing positioning signals (which our smartphones all utilize). For commercial activities, it might be a telecommunication satellite or even a human being in the case of tourism flights. Similar payloads can also be carried for military applications focusing on Earth surveillance, secure communication, and high-precision positioning.

If we look at ground transportation, it can be observed that every vehicle needs contact with the ground's surface, represented for example by static friction between tires and asphalt, to push itself along. The absence of contact with firm-ground compels reliance on the reaction principle to move matter through air and outer space. In this connection, the vehicle ejects a previously stored substance in the opposite direction to that of the intended movement. Every rocket engine is thus designed to expel gasses at high velocities to generate a force, known as thrust, counter to the flight direction. The process involves converting propellants' chemically bounded energy via combustion into thermal and ultimately the kinetic energy of hot exhaust gasses. To ensure efficient rocket-engine operation, combustion occurs at very high pressures and temperatures, burning up to several tons of propellant per second. For example, one of the biggest rocket engines ever build, the Rocketdyne F-1 – engine burns approx. 2605 kg of liquid oxygen and kerosene per second (Oefelein & Yang, 1993). Liquid-fuel rocket engines usually employ turbo pumps to feed propellants from tanks into a combustion chamber, where they burn and generate hot gases. A nozzle accelerates the hot gasses into the external environment at supersonic speed. Due to physical correlations according to the Tsiolkovsky equation, it is useful to strive for very light, reliable, and powerful engines to increase the lift capabilities of the entire rocket.

The rocket engine, propellant tanks, and the payload are integrated into the rocket structure. Regarding requirements imposed on mass, the rocket engineer's challenge is to design very light structures, providing the necessary stiffness and maximal robustness for the construction. The vehicle's mass decreases during flight due to the propellant consumption and ejection of hot gasses out through the nozzle. On the other hand, the mass of the tank containing the propellants remains constant. The tank's mass is thus rendered superfluous after the rocket's engine has consumed all of the propellants to accelerate during flight. At a certain point, it becomes expedient to jettison the depleted tanks to decrease the system's mass. This procedure is better known as staging and is necessary for all space launchers to achieve Earth orbit.

A typical launcher consists of at least two stages, which have to be separated after burn-out. Depending on the destination of the payload—dedicated orbit or planet—the rocket has to burn stored propellants to reach a minimum velocity. Therefore, it is usual to classify particular missions by their required velocity increments. To get into the low Earth orbit, the closest destination for long-time space operations, the payload has to be accelerated at least up to 8 km/s, without consideration of atmospheric and other losses. In order to place a payload in the geostationary orbit, the necessary velocity rises up to 10,7 km/s (Schmidt, 1999).

The scope of this chapter is to take a closer look at the launch of a rocket bearing a satellite as payload. This process can be divided into sequential mission phases, which will be dealt in more detail in the following paragraphs.

PAYLOAD AND LAUNCHER DEFINITION

During early space exploration, development of the payload—usually an unmanned satellite—accompanied the design and construction of the associated launcher. The adaption of both systems took place on drawing boards taking into consideration not just physical dimensions, but also each component's interfaces and functional characteristics. With the transition of necessary technical abilities onto the private sector, a great number of launchers and payloads were designed within decades. Due to wider launcher selection options for the customer today, payload interfaces are designed more flexibly to enable the transportation vehicle to be changed during the development process to accommodate unforeseen difficulties.

Present-day development of typical commercial payloads such as communication satellites requires prior research into available launch options involving lift capabilities in terms of mass and dimensions for the desired orbit. The customer begins negotiation with launch providers after payload development is nearly completed. Thus, the concrete launch date may be defined only within the last development phases of the payload—several months up to a few years before the actual lift-off (European Space Agency, 1993). Besides the necessary lift capabilities of the launcher, the customer imposes further requirements on the costs and reliability of the launch system. The customer normally prefers the most reliable and reasonable launch options in terms of commercial efficiency.

ORBIT CONSIDERATIONS

The main task of every launcher is to insert the payload into a certain Earth/solar orbit or lift it to the specified altitude for suborbital flights. Orbit selection for the payload depends on the given mission objectives and requires specific aspects to provide optimal operation conditions. Commercial satellites might require aspects such as fixed position above the Earth's surface (satellite TV), altitude above ground necessary for signal coverage and resolution (meteorological satellites), or a particular flyover frequency above certain regions (Earth observation and reconnaissance). A suitable launcher is selected depending on the desired orbit. In general, for the same payload mass, the higher the desired orbit, the more powerful the launcher is required to be. The following are the most relevant orbits for this study:

- Low Earth orbits (LEOs) have characteristic maximal altitudes of up to 5000 km, depending on literature reference (Larson & Wertz, 1999), (Kramer, 1996). These orbits are well suited for Earth observation satellites requiring short flyover periods. Small propagation delays and low transmission losses allow efficient design of communication systems and thus more compact constructions. However, residual atmosphere causes higher drag and limited satellite lifespan (Braun, 2015), (Maini & Agarwal, Satellite Technology: Principles and Applications, 2014).
- Sun-synchronous orbits (SSOs), which can be considered a subset of LEO orbits, provide nearly constant solar angle, meaning that a payload passes over any given region on Earth every day at

the same local solar time (Braun, 2015), (Fortescue, Swinerd, & Stark, 2011). These orbits are preferred for Earth observation and vary between 200 km and 1000 km altitude aboveground (Sutton, 2001).

- Medium Earth orbits (MEOs) feature typical altitudes between 10 000 km and 20 000 km and provide good operation conditions for navigation and communication satellites (Braun, 2015), (Aguirre, 2013). The advantage of these orbits over LEO is the satellite's longer visibility duration over a certain Earth region resulting from higher flight altitudes (Maini & Agarwal, Satellite Technology: Principles and Applications, 2014). Nevertheless, MEOs traverse the Van Allen belt, which is characterised by high-intensity particle radiation. This might cause ionisation effects in the satellite's electronic subsystems, which limits their lifetime.
- Geostationary transfer orbit (GTO) is usually the designated orbit for all launchers carrying payloads destined for Geostationary orbit (GEO) (Larson & Wertz, 1999). Here, the upper stage of a rocket boosts the payload to a velocity necessary to reach the desired altitude, releases it, and decelerates to re-enter the Earth's atmosphere. After arriving at GEO altitude, the payload accelerates using apogee engines to remain there. The latter manoeuvre is known as circularisation of the GTO.
- Geostationary orbit (GEO) is attractive for telecommunication satellites due to relatively stationary positioning of the payload over Earth (Larson & Wertz, 1999). Satellites in this orbit fly at an altitude of about 36 000 km above the Earth's surface and have the same angular velocity as the Earth. Thus, once the satellite is brought onto this orbit, it remains over the same spot over the Equator without readjusting it ever again to receive and send communication signals. Due to long transmission delays, GEO satellites are unsuitable for real-time communication (Kramer, 1996), (Braun, 2015).
- Earth-escape orbits are used for scientific missions to other celestial bodies or into deep space and require greater velocity increments and thus high-performance launchers.

General Considerations for Unmanned Commercial Missions

The cost and reliability of the launch mission is a trade-off product of several technically interlinked considerations. The customer has to take them into account during mission planning to create a solid basis for selection of the most suitable launcher option. The following are the main considerations for unmanned commercial missions:

- **Punctuality of the Launch Event**: From the mission objectives point of view, it's important to start operating the payload at the initially specified date. To increase return on investment, a commercial customer needs to start operating the payload as soon as possible. For scientific purposes, the launch window is calculated taking into account particular planetary configurations. Missing the window may result in a delay before the next launch date providing optimal conditions arrives. In the case, where the launch conditions require a certain celestial constellation, a delay prior the next launch window might last several weeks or even months. Thus, the interval between completion of the payload's development and the time when actual operation begins in orbit should be as brief as possible. This requires high technological maturity of the launch system regarding possible long-term time delays. Furthermore, a less predictable phenomenon—the weather—should be considered a possible short-term obstacle.

- **Position and Nature of the Launch Site**: Depending on the desired orbit, the initial launch condition of the rocket can be optimised by selecting the launch site with a certain position on Earth. Considering the angular speed of the Earth, every spacecraft leaves our planet with an initial tangential velocity that depends on distance to the Earth's rotational axis. This can be compared to the shepherd's sling: the longer the cord, the higher is the velocity of a thrown stone. Tangential velocity increases from the North Pole toward the south, reaching its maximum at the Equatorial plane, and decreases moving from there toward the South Pole, e.g. Table 1. For communication satellites in GEO located directly over the Equator, the gain of tangential velocity is quite profitable. Hence the launch pad should be situated near the Equatorial plane as it is at the Guiana Space Centre near Kourou. In case of MEO and SSO, where weather or other Earth-observation satellites are ordinarily placed, great tangential velocity is quite disadvantageous, so near-pole launch sites will be preferred. Well situated spaceports for the latter orbits are for example the Baikonur Cosmodrome in Kazakhstan or the newest Russian Vostochny Cosmodrome in Siberia. The characteristics of a suitable launch site include a generous availability of land with low population density, a deep-water harbour and airfield with long-carrier runway, and acceptable meteorological and geological risks (Roviera, Bertrand, & Lardot, 2014).
- **Precision of Orbit Injection:** Satellite design is normally adjusted to enable the satellite to perform in the most efficient way under certain operational conditions. Any deviations from the designed orbital position reduce the operational lifetime in a best-case scenario, even if they don't actually cause a total fail out. The launch provider shall ensure a high accuracy of transfer manoeuvre/orbit injection considering the velocity vector, trajectory, and attitude conditions during decoupling of the payload from the last rocket stage.
- **Mechanical, Thermal, and Electromagnetic Stresses:** During ascent, the payload is exposed to several environmental effects, which it must withstand. Mechanical stress is the first and major interference factor. It results from engine vibrations, structure-propulsion coupling, atmospheric pressure drop, separation shocks, and fuselage flows at supersonic velocities. The second effect comes from aerodynamic friction and especially solar radiation causing thermal heating of the system. The third environmental stress source—electromagnetic radiation—can be produced by electronic devices and must not diminish the functionality of launcher and payload communication systems. All of the environmental requirements imposed on each launcher considered shall be simulated on the payload in qualification tests at special facilities. This process takes place during the payload's final development phase. Its duration depends on the number and complexity of the integrated subsystems.

Table 1. Latitudes and tangential velocities at major launch sites

Launch site	Latitude	Tangential velocity
Plesetsk (Russia)	62,7°	213 m/s
Baikonur (Kazakhstan)	51,6°	288 m/s
Canaveral (USA)	28,5°	408 m/s
Kourou (French Guiana)	5,2°	462 m/s

(Haidn O. J., 2014)

- **Centre of Gravity of the Payload:** The spacecraft's transportation capacities have to accommodate the payloads mass specification. Highly unsymmetrical payloads may need trimming weights or even launchers with higher performance to handle the load.
- **Launch-Vehicle Reusability:** This may become an issue, because typical rocket-component safety factors range between 1.1 and 1.5 times the design limits (Sutton, 2001). This means that in case of a malfunction such as the turbo-pump system's causing combustion-chamber pressure 1.5 times that of the operational value, the rocket engine could fail. Compared to the automotive sector, where safety factors are typically 2 and higher, the design of spacecraft systems has to be very well matched with actual loads. In other words, space related components are more susceptible to atypical operating conditions and malfunction more easily.
- **Reliability and Safety Considerations:** These low safety factors are driven by the requirement for very lightweight structures needed to increase the payload's capacities and they cannot be significantly increased. All components of a reusable launcher have to be checked after every launch for the appearance of fatigue or creep to provide the same reliability as that prevailing at commissioning. Depending on the complexity of the system, the development of more reliable launchers together with the necessary post-flight checks might be very expensive, which tends to favour the manufacture of single-use vehicles. A sophisticated launcher design can nevertheless aspire to higher reliability and certainly to lower complexity of overall systems (Futron Corporation, November 2004), (Sauvageanu & Allen, 1998).

The given examples of major technical trade-offs during launcher selection exhibit high overlap with the payload. To reduce development costs, the customer certainly prefers a transportation vehicle imposing the least stringent requirements on its payload contingent concerning the environmental and dimensional restrictions mentioned.

Specific Trade-Offs for Small Satellites

The small-satellites category can be characterised by a total-mass specification of less than 500 kg. Such satellites typically represent secondary payloads on launchers that carry heavier, more complex primary-payload satellites. Due to their secondary status, smallsats have to meet strict requirements for permission to fly with the primary payload.

They must not damage or even affect the primary payload or the spacecraft during launch. Hence, these systems have standardised safety features and are more autonomous with only a few interfaces to the launcher. Depending on payload dimensions, smallsats can be divided into further mass categories, which determine corresponding special regulations for the design and launch procedures. Customers of smallsats also have little say in matters of mission planning such as time schedule or trajectory shaping.

CubeSats, which are typically built for educational purposes at universities and scientific institutions, represent the most popular category of small satellites. They reproduce almost all of the functions of primary-payload satellites, but at lower performance levels with more favourable components, which is what makes this category so attractive.

A very important attribute is the absence of any pyrotechnic devices or propulsion systems, which may be too hazardous to the primary payload or spacecraft. Aside from possible special limitations on CubeSats, their general requirements are standardised by predefined design specifications (Mehpravar, 2014) including such parameters as dimension, weight, and interfaces. These were established to 'reduce the cost and development time, increase accessibility to space, and sustain frequent launchers' (Mehpravar, 2014).

The design restrictions imposed on the design of small satellites may be avoided only by their classification as primary payload (Schulte, 2000). Thus, the customer either has to negotiate with all of the other customers who booked the launcher or fly alone, which is very cost intensive. Due to the continually increasing number of small satellite manufacturers, several ventures are working on the design of small satellite launchers.

Specific Trade-Offs for Manned Missions

The requirements imposed on the reliability of manned spacecraft are generally far more stringent than those for purely hardware flights. Due to the value of human life, no accidents can be accepted.

A variety of additional systems that enable life support and safe return to Earth have to be taken into account when discussing manned campaigns. When classifying different kinds of manned missions, we can distinguish between journeys to orbital space stations (Mir in the past, the ISS nowadays, and the Chinese Outposts and CSS to be early next decade) and suborbital flights for the purpose of space tourism. The physical difference between them is the velocity increment that the spacecraft should attain. Flights into LEO ideally require speeds of at least around 7.9 km/s (Sutton, 2001), where suborbital flights are situated below.

The distinction of both kinds of mission defines the minimum residence duration in outer space and the complexity of the respective life support systems. Several interplanetary missions to the Moon and Mars are planned in the near future. Hence, higher ideal launcher velocity increments in the range of 13.1 km/s to the Moon and 17.5 km/s to Mars are required (Sutton, 2001). The given ideal velocity specification doesn't take into consideration additional margins to overcome aerodynamic drag during flight through the atmosphere or for corrective manoeuvres.

Besides the necessary velocities, which the launcher should provide, the customer's decision to pick a certain carrier is based on two major requirements:

- **Vehicle Acceleration:** During the propelled flight, the previously mentioned mechanical stresses will be transferred to the crew, directly affecting their wellbeing. The spacecraft has to be functionally capable of transporting human beings so as not to exceed certain acceleration limits. A typical solution is to thrust throttle the rocket engine.
- **Transport-Vehicle Reliability:** Here, the reputation of a launch provider relating to the launch history and failure statistics confers the best advertising effects and determines the insurance costs.

Additional requirements such as re-entry functionalities or duration of the space travel have design influences on the launcher, increasing the overall mass by adding necessary subsystems.

TRANSPORTATION

After successful negotiations between the customer and the launch provider, and completion of the payload, the components have to be delivered from manufacturers to the launch site to meet the schedule. The launch window will normally be coordinated between the customer and the launch provider several months before the actual launch. For example, with Arianespace being the launch provider, the possible launch window will be settled at the Preliminary Mission Analysis Review and verified at the Final Mission Analysis Review (Arianespace, 2011). The time interval of approx. 14 months between both these analyses provides enough time for the determination of the exact launch window. After the latter review, no changes in the schedule will be expected.

Affected by previous development phase, several aspects must be considered for transportation of components to the launch site to ensure punctual lift-off. The first and very important aspect is the containment of ITAR (International Traffic in Arms Regulations)-protected components, which are produced in the US, within the payload or the launcher. If these are listed on the USML (United States Munitions List), US authorities must authorise foreign personnel coming into contact with the components or final product.

The second aspect emerges from export restrictions on the final product concerning the launch pad. For example, if a European launcher is to be built in Italy and will incorporate ITAR-restricted inertial measurement components from the US, then the US authorities have to approve the European personnel and the launch campaign. Hence, to reduce bureaucratic proceedings and make the launch date more predictable, less mature or effective technical solutions from non-US suppliers may be preferred during the system's design.

Due to the distance between the two locations and dimensions of the components, different forms of transport can prove expedient. For instance, bigger cargos such as complete spacecraft stages will be normally transported by ship between Europe and the Guiana Space Centre in Kourou. This is the only European spaceport able to launch all Ariane-family rockets as well as Vega and Soyuz vehicles. The launch site, located 5° North of the Equator, has been operating since 1968 and is ideally placed for launches into GEO. A journey from Europe to Guiana may take from 15 to 20 days depending on weather conditions. Rail is preferred for overland transport, for example from Europe to Baikonur, although comparable travel time is expected. Payload and crew, which are usually smaller, can be transported by plane within a day unless intermediate stops are made.

A subcontractor specialised in payload transportation has to meet specific requirements—especially environmental conditions involving humidity, temperature, pressure, and shocks. In some design-driven cases, electronic connections to the payload may be needed to provide power and communications capabilities for functions such as system-status checks or mission programming. The latter requires the manufacturer's full access to the payload during transport. Further supply units with corresponding interfaces must be provided when integrated components with permanent cooling requirements are present.

Additional safety measures during transport to the launch pad should be implemented for satellites equipped with propulsion systems for attitude adjustments and path corrections. These apply primarily to loaded, solid-fuel engines and sealed tanks filled with hypergolic propellants. The payload will normally be fuelled shortly before launch, although there are exceptions. Hypergolic propellants consist of separated oxidizer and fuel, which react in direct contact with each other in the combustion chamber. In contrast to hypergolic propellants, solid-fuel rocket engines use oxidizer and fuel already premixed in one combustion chamber.

Fire suppression in the propulsion system is very difficult if not impossible in case of an accident, such as one caused by an inconvenient lightning strike. Thus, the transport company has to be prepared for the worst-case scenario in which the payload's propellants ignites unintentionally, managing the force majeure to minimise further damaging to the transport vehicle, personnel injury, or complete carriage loss. To meet these special transportation requirements, the carrying company should possess country-specific licenses with corresponding hardware and knowledge about how to handle hazardous, highly flammable, and explosive materials.

More severe restrictions exist for radioactive devices that can be installed on satellites such as scientific probes or radioisotope thermoelectric generators (RTGs), which provide the payload with electric energy in the absence of sunlight or batteries. Besides the necessary licenses and permissions to transport the payload, additional armed escort may be deployed here to prevent any kind abduction or robbery of radioactive materials.

During transport, the carrier is normally equipped with sensors measuring shocks, temperature, pressure, and humidity within the payload container. These record all environmental data for review of influence on the payload after arrival. This might be needed for take-off clearance or for analysis in case of subsequent malfunction resulting from rough handling during transport. The requirements imposed on the transport company for transporting small satellites to the launch pad are proportionally less stringent due to absence of propulsion systems, pyrotechnics, or need for thermal treatment or permanent communication.

After all of the transportation aspects have been clarified, the readiness of the facilities at the port of entry, the personnel, and all of the formal issues have to be prepared to ensure smooth campaign workflow.

INTEGRATION

Several system checks are done after the payload and the launch vehicle arrive at the integration facility. Environmental impacts on the carriage are examined first. Permission to fuel the payload, for example the satellite, is subsequently granted if no anomalies or damage are detected, and it transitions into integration and pre-flight status.

The payload and the launch vehicle have to remain in clean rooms throughout the integration phase. The payload's functional capabilities are checked and monitored during preparations, which requires power supply and communication. All of this necessitates appropriate assembly buildings and infrastructure to provide filtered air and proper thermal conditions.

The propellants' toxicity renders satellite fuelling at the launch site one of the most hazardous processes for the operating personnel. The duration of the integration facilities' preparation depends on payload complexity. For larger systems, it can last up to several months before device arrival. In the case of launch vehicles, the qualification procedures for liquid propellants, especially kerosene-based fuels, are necessary to ensure the rocket's performance, unless the former are supplied by the engine manufacturer.

Propellant compositions are extremely important for engine performance. This is due to specific cooling characteristics, sediments in the tanks and filters, and soot formation in the combustion chamber. Special areas with restricted access might be required for manufacturing of solid rocket engines. The number of involved workers inside the integration building must be limited during and after the integration of solid propellant grains due to the heightened danger of accidental ignition (Roviera, Bertrand, & Lardot, 2014).

The assembly of single stages can be preceded in the horizontal or vertical position depending on launcher design. In the case of Ariane-launch vehicles, the payload is attached to the rocket body in the vertical position few days before the lift-off. Pre-flight checks are performed after rocket assembly.

The launch site's design has to consider possible accidents during the preparations, including inflammation of the spacecraft in its fuelled state. Sufficient distance between the failed rocket and mission-critical ground equipment has to be ensured in these cases. The facilities, which contain command and communication centres, are also reinforced to withstand possible impact of rocket parts or are located below ground.

Government armed forces normally secure the launch sites to prevent sabotage attempts such as those that presumably occurred at facilities in French Guiana (Leitenberger, 2009). Authorities generally grant personnel access to the launch facility only after special clearance.

LAUNCH

The rocket lift-off, carrying its payload is the culmination of the launch campaign, although this is the briefest of all previous phases. Aspects such as launch approach, rocket performance characteristics, and launch proceedings are the most important issues here.

Propellant Considerations and Rocket Performance

Several types of rocket engines qualify for the launch taking into consideration the physical and technical requirements necessary for lift-off. These are distinguished by the phase state of propellants used: liquid, solid, and hybrid (a combination of each). Furthermore, propellants comprising oxidizer and fuel can have a huge range of variations for possible design combinations.

From the early days of space exploration to the present, the central character was represented by liquid propellants due their high performance. Large launchers with complex propulsion systems and extensive payload capacities normally use these.

Notwithstanding this, the oldest rocket-engine concept was based on solid fuels and was employed for military purposes (Sutton, 2001). These rocket engines are characterised by low performance but excellent handling qualities, which are necessary on the battlefield. To this day, solid propellants are the preferred option for missile design and are normally used only in auxiliary boosters on liquid propelled core stages when needed.

However, the Vega launcher represents an exception. It's a four-stage rocket with solid propellant in the first three stages and is designed to deliver small payloads into LEO (Arianespace, April 2014).

The hybrid engine is the last feasible rocket-engine concept for reaching outer space. Here, the solid propellant grain is already located in the combustion chamber. The reaction starts when liquid oxidizer from the tanks is injected into the combustion chamber. The major advantage of the hybrid-propellant concept over that of liquid and solid propellants is handling and operational safety. This is what makes it so attractive for space tourism and education. Nonetheless, hybrid propellants never achieved the performance of their liquid, or even of their solid counterparts, although calculations predict greater theoretical performance parameters for them. The most familiar examples for hybrid propelled rockets are members of the spaceplane family build by The Spaceship Company (The Spaceship Company, 2018).

One of the most important performance parameters to benchmark any rocket-engine technology is the specific impulse. It represents the total impulse per unit mass of propellant (Sutton, 2001). The parameter that results from multiplying specific impulse with the Earth's gravity represents the velocity of the exhaust gasses leaving the nozzle. Using this definition, a rocket engine's thrust can be expressed as:

$$F_{Thrust} = \dot{m} \cdot I_{SP} \cdot g_0$$

where $\dot{m}$ is the mass flow rate representing the mass of the propellant flowing through the rocket engine per unit time in kg / s, g_0 is Earth's gravitational acceleration at sea level in m / s^2 and I_{SP} is the specific impulse in s.

From the designers' perspective, a large thrust value is desirable and can be achieved with large mass flows and specific impulse values. For the purpose of highest efficiency, the least propellant mass should produce a maximum specific impulse. Typical theoretical values for specific impulse of common propellants are roughly estimated in Table 2 (Sutton, 2001), (Haidn O. J., 2008), (Schmidt, 1999), (Davenas, 1993), (Kniffen, McKinney, & Estey, 1990).

The benefit from greater specific impulse entails greater technological complexity. Thus, only very mature launchers operate with liquid cryogenic propellants, where hybrid and solid propellant solutions can be implemented with less effort (Table 2).

Furthermore, rocket-design quality can be measured using two key values: payload mass fraction and structural mass ratio (Sutton, 2001). Similar to the striving for higher specific impulses, tendencies of these two values can be observed. The payload mass fraction is the ratio of the payload mass to overall rocket mass. This factor should be as high as possible. In contrast to this, the structural mass ratio should be as low as possible, due to the relation of the final mass of the rocket after burn out divided by the mass of entirely fuelled rocket before launch.

Types of Launch

In total there are four applicable and already established methods to launch a rocket into space. The most common and oldest approach is a vertical lift-off from the launch pad. Other more exotic methods exploit secondary transport vehicles as launch platforms to provide the rocket with more advantageous initial conditions. These include air launch from a carrier aircraft, sea launch from a maritime platform, and submarine launch. The same criteria as those mentioned in the first phase qualify for selection of the most suitable launch system: system assets and drawbacks, overall complexity, and launch costs.

Launch From the Ground

This is the usual method to start sophisticated, high performance launchers with large payloads and cryogenic propellants. It is most prevalent due to minimal infrastructure-maintenance costs and, more importantly, the possibilities of continual access to and control of the rocket up to lift-off. These possibilities are necessary to monitor the complex fuelling system and engine performance even during first seconds of operation, which enables termination in case of system malfunction. Figure 1 shows some major launch sites on the globe.

Table 2. Typical theoretical specific impulse for major propellant classes

Propellant combination	Classification	$I_{SP}\left[s\right]$
Liquid oxygen (LOX) / Liquid hydrogen (cryogenic)	Liquid fuel	389–455
Liquid oxygen / Rubber, e.g. HTPB	Hybrid rocket fuel	280–302
Ammonium perchlorate / Rubber, e.g. HTPB, mixtures	Solid propellant	262–266

Figure 1. Exemplary selection of major launch sites

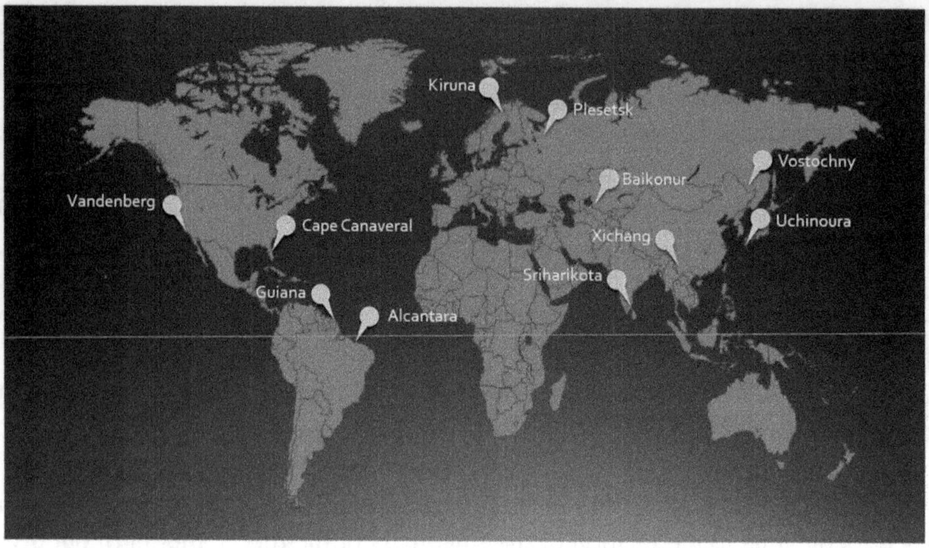

The main drawback for fixed launch sites on Earth is the inability to adapt location for different mission and orbit profiles. The task of correcting initial orbit deviations is transferred to the rocket. Therefore, launch sites such as Plesetsk located toward the north are not preferred for payloads destined for GTO or GEO orbits.

Sea and Submarine Launch

Launching from a floating platform is attractive, because the launch site can be freely positioned thereby providing optimal rocket-performance conditions (Buhrig, Cox, Alexander, Krenitsky, & Huebner, 1996). Private joint venture Sea Launch (Maini & Agrawal, Compendium of Satellites and Satellite Launch Vehicles, 2007), which involves companies from the U.S., Russia, Ukraine, and Norway, conducts most of the launches from floating platforms.

Several restrictions must be taken into account to launch using this method. The main one is the available space on the carrier, which restricts the fuelling of and access to the launcher and payload. Additional facilities for producing cryogenic propellants in the necessary amounts as well as the command control centre and personnel shelter have to be located on supplementary ships.

The time schedule, which is affected by weather conditions prevailing while the rocket is being transported to destined launch location and thus has to be taken into account, represents another limitation. From the provider's side, a highly automated launcher and all-weather start ability are required (Maini & Agrawal, Compendium of Satellites and Satellite Launch Vehicles, 2007). Considering these restrictions, the sea launch could not prevail over the conventional, cost-effective ground launch.

Submarine launches represent a less common method. In the absence of threats, the cold war legacy of intercontinental ballistic missiles offered new possibilities by replacing the warheads with civilian payloads. Only two launches of small satellites were thus conducted using adapted Shtil´-missiles from a Russian submarine in 1998 and 2006 (space.com, 2006). In this case, the navy of a sovereign country assumes the tasks of launch provider. Still, this method cannot support market expansion due to the small number of expired strategic missiles, shifting political relationships among involved parties, and rigorous security restrictions imposed on the customer.

Air Launch

Another often-discussed approach is launching from a flying platform. In this case, an aircraft carries the spacecraft into a defined launch corridor, which provides beneficial orbit conditions for the payload. After achieving the necessary launch position, orientation, and velocity, the rocket is released from the aircraft and falls unpropelled for a few seconds to establish the necessary distance from the carrier. The rocket's engines subsequently start, and the launcher begins to pursue the preassigned trajectory. The aircraft initiates an evasion manoeuvre during this interval to prevent possible collision with the ascending rocket (Schade C. C., 1990).

Besides this method's successful use with the Pegasus missile for 38 successful starts between 1990 and 2016 (Northrop Grumman, 2018), several further approach studies are ongoing. Among them Stratolaunch Systems (Schade & Moiser, 1994), XCOR Lynx (Jones, 2004), Skylon (Mehta, Aftosims, Bowles, & Pandya, 2016), SOAR (Swiss Space Systems, 2016), and possibly others.

A number of advantages in operational flexibility and cost effectiveness of this method are expected (Schade & Moiser, 1994), although only few launch providers have established and pursued the approach to date due to additional risks and low real benefit compared to conventional vertical launch (Van Kesteren, 2013). This launch method's greatest risk is the launcher's separation procedure from the aircraft. Insufficient analysis, design, or slightly deviating operating conditions may lead to collisions between the two vehicles.

Among the main advantages are such factors: as potential and kinetic energy gained from the carrier aircraft, higher propulsion efficiency resulting from higher expansion ratio of the first stage's engine, less aerodynamic drag and smaller gravitational losses, and smaller thrust-direction losses (Schade C. C., 1990), (Sarigul-Kijin & Sarigul-Klijn, 2001), (Salt, 2014). Reduction of energy-consumption losses mentioned leads to increased payload mass.

Still, this approach is limited to the lifting capabilities of the carrier aircraft and therefore appropriate for launching small, highly automated rockets with storable (preferably solid) propellants. In recent decades, since space tourism has become more feasible, requirements concerning launchers' greater safety and controllable thrust have become more important for this certain application. Hybrid engines are best suited for such purposes. Although more complex, LOX/kerosene prototypes have already been tested (Jones, 2004).

The adoption of launchers with sophisticated propulsion systems and therefore greater payload capacities requires more complex and larger aircraft carriers, and severe safety regulations. The certification of air-launched vehicles as well as modified carrier aircraft is also essential to ensuring proper operation modes, especially during the separation procedure. Running costs, including those incurred for aircraft maintenance, must not be neglected either. That the air-launch approach will prevail on the small-satellite market over the more common vertical launches can generally be assumed only under the conditions of very frequent launch events and standardised payload interfaces.

Launch Proceedings and Flight Operations

Lift-off authorisation requires various system checks of the launcher and the payload, proper weather conditions, and a clear flight corridor in the air and on the ground. The proceedings are controlled remotely after personnel are evacuated from the launch site. The command computer takes over the control of the launch sequence shortly before lift-off; however, the launch manager still has the option to stop the countdown.

After lift-off, sensor data measuring the engine's performance, trajectory evolution, and overall system status is transmitted to ground stations. Several radar stations on Earth track the launch vehicle to monitor the rocket's flight performance and the re-entry trajectories of burnt-out stages.

New calibration and correction commands are uploaded to the launch vehicle if deviations are detected between the trajectory measured by radar and data transmitted from the rocket. The trajectory of used and separated components such as tanks, turbo pumps, engines, and the like from the first stages, which always fall back to Earth, is calculated in the manner not to harm populated or conservation areas on the ground. In the case of failed launch and possible contamination of the immediate environment with chemical and radioactive remains, these ideally have to be removed from the impact site. Special cases may arise if the propulsion system in the upper stages fails and the remaining, fully loaded rocket crashes on the territory of otherwise uninvolved countries.

Several analyses considering the spacecraft separation and collision avoidance are conducted during the development phase (Arianespace, 2011), (Arianespace, April 2014), (Arianespace, 2012). These are important for launchers that carry multiple payloads into different orbits.

The launcher's main task ends when the payload is released into its designated position, having imparted the correct velocity vector and altitude conditions to enable stability and functionality in the chosen orbit. The payload's command control is fully transferred to the customer after this phase. Post-launch analysis represents the final interaction between the launch provider and the customer. It involves evaluating the mission's success, especially orbit-injection accuracy.

CONCLUSION

From a logistics perspective, the launch event is a very complex and prolonged process, which must not be underestimated during space campaign planning. From the beginning, it involves all participating parties and requires continuous communication and willingness to negotiate and compromise.

From the beginning of space exploration, the launch process was perfected in order to enhance the safety of the operating personnel and monitor all possible condition data of the carrier rocket and the launch pad. The latter was necessary to analyse the flight, especially in case of failure or minor malfunction. Still, the fundamental procedure of the launch industry hasn't changed since the 1940s.

The market situation nowadays has tended upward since about 2005, with approximately 90 launch attempts per year, including manned and unmanned flights (Kyle, 2018), (see Figure 2). The decreased number of launches from 1990 to the beginning of the 21st century can be explained by the collapse of the Soviet Union and the improved reliability of satellites' electronic components. Since the political motivation is no longer the driver of the scientific space exploration in Europe and the US like it was during the Cold War, the commercial background dominates future trends. Due to the increasing number of CubeSats (The Tauri Group, 2015) as educational projects at universities, it can be assumed that the number of launches will increase slightly in this segment during the coming years. Nevertheless, the increasing number of small satellites on the market can be partially counterbalanced by clustering payloads on single capable launchers, as was done during the Orbital-2 Mission (NASA Media Press Kit, 2014) or in the planned QB50 programme (Von Karman Institute, 2016).

The major change in the evolution of the current market and launcher technology can be only expected with the realisation of satellite mega-constellations, where several thousands of payloads should be deployed in LEO (Grush, 2018). Still, many technical and legal aspects for this venture should be clarified previously. The manned missions to the Moon and Mars planned for the near future awaken very little or no commercial interest and therefore depended on the political and economic situation in participating countries. The best example of sustainable and systematic engineering effort supported by tremendous investments based on political and economic situation can be observed these days in China. The unwillingness of the Western World to let Chinese state companies participate at international space exploration endeavours in the last decades conducted a huge political motivation for the development of their own space strategy (Xiang & Tongyu, 2014).

The demand on information collecting and transmission by satellites is balanced with current financial expenditure and can be only changed by the implementation of completely new space-access technologies. The economic success of space mining, for example, cannot be estimated for the near future with any reliable figures given other than profit-oriented projects related to the permanent launch needs.

Still, the changing market for small satellites and space tourism is profitable at best for small launcher manufacturers such as SpaceX, Blue Origin, Scaled Composites, and the like. The game changer for such companies is positioning the design office, manufacturing plant, integration hall, and, in part, the testing area at the same location. Doing so enables very efficient development methods, reducing transport time and communication losses, and thus, costs. Further features, such as the reusability of

Figure 2. Worldwide number of launches per year

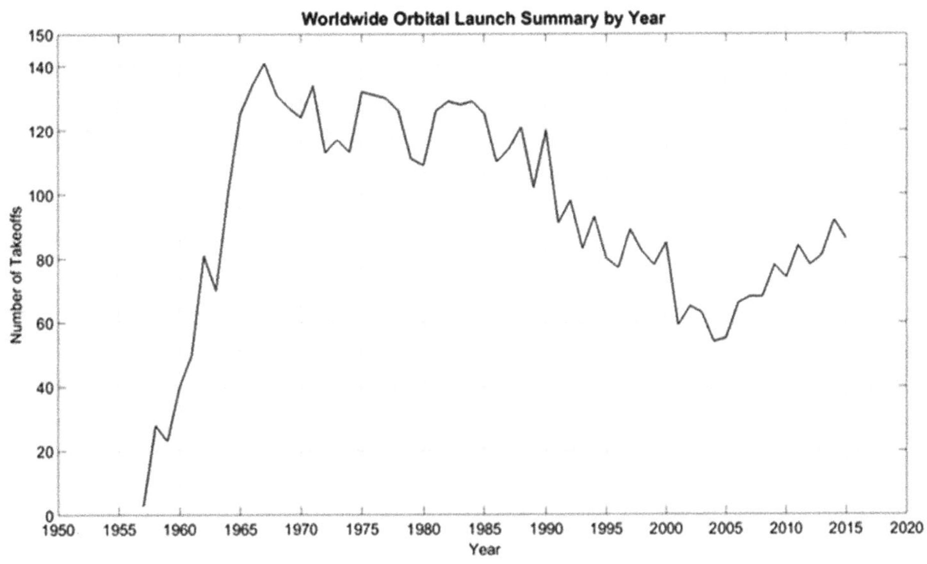

returned components have still to prove themselves from the commercial point of view on longer time scales, since several attempts with the Space Shuttle and Buran programme have failed. Therefore, the newcomers gain market share without achieving the best performances in the hardware technologies opposite to the current developments of key player companies like Airbus, Boeing, or Lockheed Martin.

Nevertheless, the space industry has not lost the courage to conduct research and discover new frontiers beyond our planet over the years. Within the main objective – to facilitate our life on Earth, it is destined to search for new technologies enabling us to transport higher loads with lower costs and extend the human presence in outer space.

REFERENCES

Aguirre. (2013). *Introduction to Space Systems: Design and Synthesis*. New York: Springer.

Arianespace. (2012). *Soyuz User's Manual*. Author.

Arianespace. (2011). *Ariane 5 User's Manual* (5th ed.). Author.

Arianespace. (2014). *Vega User's Manual. Issue 4, Revision 0*. Author.

Braun, J. (2015). *Influence Analysis for Liquid Satellite Attitude Control Thruster Design* (Bachelor's thesis). Technische Universität München, Institute of Flight Propulsion.

Buhrig, A., Cox, S., Alexander, M., Krenitsky, M., & Huebner, C. (1996). Sea Launch Program Overview. In *16th International Communications Satellite Systems Conference* (pp. 58-67). Washington, DC: AIAA.

Davenas, A. (1993). *Solid Rocket Propulsion Technology*. Pergamon Press.

European Space Agency. (1993). *Meteosat Second Generation. The Satellite Development*. ESA BR-153.

Fortescue, B., Swinerd, G., & Stark, J. (2011). *Spacecraft Systems Engineering*. John Wiley & Sons Ltd. doi:10.1002/9781119971009

Futron Corporation. (2004). *Design Reliability Comparison for SpaceX Falcon Vehicles*. Research Report sponsored by SpaceX.

Grush, L. (2018). *The Verge*. Retrieved from SpaceX just launched two of its space internet satellites — the first of nearly 12,000: https://www.theverge.com/2018/2/15/17016208/spacex-falcon-9-launch-starlink-microsat-2a-2b-paz-watch-live

Haidn, O. J. (2008). *Advanced Rocket Engines*. Advances on Populsion Technology for High-Speed Aircraft.

Haidn, O. J. (2014). *Space Propulsion I, Lecture Script*. Garching: Technische Universität München, Institute of Turbomachinery and Flight Propulsion.

Jones, D. (2004). *Reusable Rocket Propulsion for Space Tourism Vehicles. Joint Propulsion Conference and Exhib*. Fort Lauderdale, FL: AIAA.

Kniffen, McKinney, & Estey. (1990). Hybrid Rocket Development at the Ameriacan Rocket Company. In *26th Joint Propulsion Conference*. Orlando, FL: AIAA.

Kramer, H. J. (1996). *Observation of the Earth and Its Environment: Survey of Missions and Sensors*. Berlin: Springer-Verlag. doi:10.1007/978-3-642-97678-0

Kyle, E. (2018, November 21). *Space Launch Report*. Retrieved from http://www.spacelaunchreport.com/

Larson, W. J., & Wertz, J. R. (1999). *Space Mission Analysis and Design*. Mcgraw-Hill.

Leitenberger, B. (2009). *Europäische Trägerraketen 1: von der Diamant zur Ariane 4*. Books on Demand.

Maini, A. K., & Agarwal, V. (2014). *Satellite Technology: Principles and Applications*. John Wiley & Sons Ltd. doi:10.1002/9781118636459

Maini, A. K., & Agrawal, V. (2007). *Compendium of Satellites and Satellite Launch Vehicles. In Satellite Technology: Principles and Applications*. John Wiley & Sons, Ltd.

Mehpravar, A. (2014). *CubeSat Design Specification. Rev 13*. California Polytechnic State University.

Mehta, U., Aftosims, M., Bowles, J., & Pandya, S. (2016). Skylon Aerospace Plane and Its Aerodynamics and Plumes. *Journal of Spacecraft and Rockets, 53*(2), 340–353. doi:10.2514/1.A33408

NASA Media Press Kit. (2014). *Orbital-2 Mission to the International Space Station*. NASA.

Northrop Grumman. (2018). Retrieved from Pegasus: http://www.northropgrumman.com/Capabilities/Pegasus/Documents/Pegasus_Mission_History.pdf

Oefelein, C. J., & Yang, V. (1993). Comprehensive Review of Liquid-Propellant Combustion Instabilities in F-1 Engines. *Journal of Propulsion and Power, 9*(5), 657–677. doi:10.2514/3.23674

Roviera, P. M., Bertrand, J., & Lardot, C. (2014). *ESA Launchers Ground Facilities: Background operational Phase and Future Developments. SpaceOps Conference*. Pasadena, CA: AIAA.

Salt, D. J. (2014). *Could a subsonic air-launched RLV enable a paradigm shift in space operations?* SpaceOps 2014 Conference, AIAA 2014-1897, Pasadena, CA.

Sarigul-Kijin, M., & Sarigul-Klijn, N. (2001). *A Study of Air Launch Methods for RLVs. Space Conference and Exposition*. Albuquerque, NM: AIAA.

Sauvageanu, D., & Allen, B. (1998). Launch Vehicle historical Reliability. In AIAA 1988-3979, Cleveland, OH.

Schade, C., & Moiser, M. (1994). The Pegasus Launch Vehicle new Capabilities and the Future. *15th International Communications Satellite Systems Conference and Exhibit*. 10.2514/6.1994-1172

Schade, C. C. (1990). Pegasus, Taurus and glimpses of the future. *Space Programs and Technologies Conference, AIAA 90-3573-CP*. 10.2514/6.1990-3573

Schmidt, G. (1999). *Technik der Flüssigkeits-Raketentriebwerke*. Taufkirchen bei München: Daimler Chrysler Aerospace.

Schulte, R. (2000). TUBSAT-N, an ultra low cost global communication nanosatellite system. *Air & Space Europe*, *2*(5), 80–83. doi:10.1016/S1290-0958(00)80094-5

space.com. (2006, May 29). Retrieved from Russian Submarine Launches Russian Microsatellite: https://www.space.com/2440-russian-submarine-launches-russian-microsatellite.html

Sutton, G. P. (2001). Rocket Propulsion Elements (7th ed.). John Wiley & Sons, Inc.

Swiss Space Systems. (2016, Jul. 1). Retrieved from official company Link: http://www.s-3.ch

The Spaceship Company. (2018). Retrieved from Making Dreams take Flight: http://thespaceshipcompany.com/

The Tauri Group. (2015). *State of the Satellite Industry Report.* Satellite Industry Association.

Van Kesteren, M. W. (2013). *Air Launch versus Ground Launch: a Multidisciplinary Design Optimization Study of Expendable Launch Vehicles on Cost Performance.* Delft: Master of Science Thesis AE5810.

Von Karman Institute. (2016, Jul. 1). *Official project link.* Retrieved from https://www.qb50.eu

Xiang, M., & Tongyu, L. (2014). *The new Generation Launch Vehicles in China.* Toronto, Canada: IAC.

Chapter 3
Space Launchers:
Legal Aspects

Edmond Boulle
Orbital Witness, UK

ABSTRACT

This chapter outlines certain core legal topics that arise in connection with the delivery of a separated payload into or beyond Earth orbit. The first part deals with some of the established approaches to procuring launch services, as well as some of the common features of launch service agreements that balance the interests of the launch service provider and its customer. The second part of the chapter looks at governmental authorization required to carry out a launch. While safety standards and success rates continually improve, launching a space object is still the riskiest part of most space missions and is therefore a carefully regulated aspect of space activity, with participants having to obtain prior authorization from a competent national authority. Finally, the third part explores some of the legal consequences in international law of launching a space object, including the maintenance of a register of space objects launched, and the burden of liability that is placed on "launching states."

INTRODUCTION

From the outset of the space age, it was clear that a regulatory framework was urgently needed in order to ensure the peaceful development of outer space activities. Both Cold War superpowers were willing to cooperate with one another and with other States, under the auspices of the United Nations, and sought to create a legal framework which could be capable of broad acceptance by both the US and Soviet Union and the wider international community.

In 1967, the United Nations Committee on Peaceful Uses of Outer Space (UN COPUOS) adopted the very first space treaty, the 'Treaty on Principles Governing the Activities of States in the Exploration and Use of Outer Space, Including the Moon and Other Celestial Bodies'; commonly known as Outer Space Treaty (OST). Under Article I of the OST the exploration and use of outer space, including celestial bodies, are fundamental freedoms guaranteed to all States. To be free to use and explore outer space it follows that States must first be free to access outer space. Mankind has struggled bravely and

DOI: 10.4018/978-1-5225-7256-5.ch003

creatively to exercise this freedom: in the 1960s and 1970s scientists experimented with ballistics in an ultimately unsuccessful attempt to develop a supergun (Park, 2016); high altitude balloons have reached near space altitudes in the Earth's atmosphere (Yamagami, 2003); and there are those who are proponents of the space elevator and other non-rocket concepts (ISEC, 2018). However, launch (in the sense of *rocket launch*) remains the only proven means of access to space. Launch is thus critical to exercising the freedoms accorded under the Outer Space Treaty.

Yet, launching a space object is still the riskiest part of most space missions. In 2015, five out of a total of 87 orbital launches were reported as a partial or total failure, about 6%. (Skyrocket & Spaceflight, 2015) A launch failure can result in the destruction of property, damage to the environment, and tragic loss of life, and can precipitate the loss of confidence between the launch service provider and the customer, thus impairing future contracts.

It should therefore come as no surprise that launching a space object is an activity in which law and regulation have a significant role to play. This chapter does not address the legal issues of specific relevance to the launch of manned suborbital vehicles. The scope of this chapter is instead confined to an examination of certain legal issues that arise in connection with the delivery of a separated payload into or beyond Earth orbit. These issues can be divided into the legal topics detailed in each of the subsequent paragraphs.

THE LAUNCH SERVICE AGREEMENT

A launch service agreement (LSA) is a contract for the provision of certain services directed towards the launch of a payload into or beyond Earth orbit. Two important observations should be made at the outset:

- A contract is an agreement between two or more natural or legal persons (the 'parties') intended to be legally binding and enforceable. Contracts are instruments of private law, not international space law (as explained below). An LSA is no different: typically, the parties involved are a launch service provider (LSP) and a customer who may be a commercial satellite owner/operator, a space agency, a research body, or other private or public entity. LSAs are enforceable in national courts in accordance with the national law chosen by the parties.
- An LSA is a contract for services, not for goods. The customer does not acquire any ownership rights in the launch vehicle. Instead the customer pays for certain associated services including the assembly, integration, and testing of the payload and launch vehicle; transport of the assembled launch vehicle to the launch pad; fuelling; launch; and delivery and separation of the spacecraft to the proper orbit.

Government Launch Procurement

In the case of the National Aeronautics and Space Administration (NASA) and the European Space Agency (ESA), a baseline or framework agreement is negotiated setting out the major terms and conditions under which launch services shall be provided. Thereafter mission-specific launch requests can be made (replicating the typical repeat procurement model of Contract plus Purchase Orders).

The NASA Launch Services II (NLS II) contract is the primary mechanism used by NASA to procure a launch service for its own and NASA-sponsored payloads.(NASA Launch Vehicle Performance) Through NLS II, NASA awards Indefinite Delivery/Indefinite Quantity (IDIQ) contracts to domestic LSPs (e.g. SpaceX, Orbital ATK, United Launch Alliance) whereby NASA reserves the unilateral right to order launch services on the LSP's qualifying launch vehicles (e.g. SpaceX's Falcon 9, ULA's Atlas V, or Orbital's Pegasus) for a given period from the contract award.(NASA, 2018) These are called Launch Service Task Orders (LSTOs) and are competed among the NLS II contract holders. The NLS II launch service providers are required to submit responses to each request for launch service proposal or they must request a waiver if the provider is unable to compete. The potential total contract value thus depends on the number of missions ultimately ordered by NASA. Each LSTO procures the entire capacity of the launch service, and NASA may utilize the excess capacity for secondary or auxiliary payloads. Twice a year, there is an opportunity for existing and emerging domestic launch service providers to tender proposals if their vehicles meet the minimum contract requirements. (NASA, 2008) This process is referred to as an *on-ramp*. The NLS II contract assumes award of LSTOs at L-30 months plus/minus 3 months such that procurement normally begins approximately 3 years prior to the launch date since each LSTO under the contract may only take 3-6 months to negotiate.

Similarly, ESA negotiates Framework Contracts with Arianespace for the procurement of launch services. The main purpose of the Framework Contract is to maximise the use of European launchers Ariane 5 and Vega, whose development was funded by ESA, while ensuring competitive launch service prices for ESA missions and defining a set of standard contractual provisions for all launch services procured by ESA from Arianespace. (ESA Press Release, 2009).

Commercial Launch Procurement

LSPs market their services in terms of the capability of a fixed range of launch vehicles. (Arianespace, 2018) Therefore, the basic technical parameters of the mission (such as payload weight, size, environmental tolerance, and orbital trajectory) will determine the range of potential launch services capable of servicing a customer's mission requirements. Additional constraints may be presented by national export control regulations that prevent sensitive technologies from being sent to an overseas launch site.

Customers either approach a LSP directly or through a launch services broker. The customer/broker may procure the entire capacity of the launch vehicle (usually at a premium to the customer) or a part of the capacity ('a place') on the vehicle alongside other third-party payloads. CubeSat missions, for example, exclusively fall into the latter category, as it is not currently economically viable to launch a CubeSat as a primary payload. Generally, concrete negotiations with one or more LSPs do not take place until the payload has successfully passed in-orbit qualification procedures.

Currently, commercial LSA are negotiated for a single or fixed number of launch events, each involving one or multiple payloads. However, as we enter the dawn of large satellite constellations, it is conceivable that new business models and methods of contracting will develop. For instance, where a single entity plans to launch hundreds or even thousands of satellites, it may prefer (and have sufficient commercial leverage) to negotiate an umbrella terms with an LSP akin to NASA's and ESA's framework agreements to ensure standard terms going forward. However, for now the 'initial basics imposed by the nature and specifications of launch activities have been unchanged by the newcomers'. (Alain du Parquet, 2011).

Commonalities in LSAs

The provisions of LSAs are understood to be generally standardised across the small but growing number of commercial entities that offer launch services. (Wheeler & White, 2012). This section shall examine some key provisions of LSAs in further detail.

Performance

Performance in contract law refers to the point at which a party's obligations under the contract are deemed to have been discharged. Though it might at first seem intuitive that an LSP should have discharged its obligations under an LSA only once the payload has been successfully delivered into orbit (in line with the stated purpose of the contract), in fact an LSP is generally deemed to have performed under the contract at the point at which launch operations can no longer be interrupted or reversed. (Wheeler & White, 2012). Depending upon the launch vehicle, this 'point of no return' may be the point of lift-off from the launch pad or it may be some earlier point at which some irreversible command is given to the on-board computers. If the launch is aborted prior to this point having been reached the LSP has not performed or discharged its duties to the customer under the contract. In such cases the LSA may make provision for a re-flight or a refund.

The LSP is therefore under an obligation to use its best efforts to deliver a payload into orbit, without a guarantee of success. (Soucek, 2015). The reason for this somewhat counterintuitive notion of performance in commercial LSAs[1] is due to the inherent risk associated with launch activities. (Skyrocket & Spaceflight, 2015). Given the cost of a re-launch, no LSP can presently guarantee its customers a successful injection into orbit. Instead, launch risk passes to the customer who should seek appropriate insurance for its payload. It is therefore important that insurance policy provisions are consistent with LSA definitions. However, looking ahead, it is conceivable that as launch vehicle technology develops, reliability continues to improve and the risk decreases, performance may come to be defined as successful delivery into the proper orbit.

The customer also has obligations that it must perform under the LSA. In addition to those concerning payment (see below) the customer must also make the payload available for testing and integration in accordance with the launch schedule. The customer is also responsible for obtaining necessary licences or authorisations (see section below) and for ensuring compliance with national export and import laws and regulations.

Launch Scheduling/ Launch Schedule Adjustment

An LSP will advertise its launch manifest to potential customers. This information is crucial to enable customers to align the manufacture and delivery of their payload with the launch opportunity and to ensure that all necessary governmental authorisations have been obtained in good time as well as any filings for the purposes of coordinating and registering frequency assignments under the procedures of the International Telecommunication Union.

Normally, the LSP and the customer will not agree on a specific date for the launch to take place in the LSA. Instead, the parties agree to zero-in on a launch window from an initially identified potential launch period of three to six months. The day of the launch may only be ascertained and agreed a matter of a few months prior to the start of the agreed launch period. (Arianespace, 2011). One exception might be in the case of interplanetary missions that necessarily require a particular launch date and window to remain fixed to coincide with the alignment of planets in the solar system. (Thompson, 2015). As commercial entities seek to conduct activities beyond Earth orbit, for example to extract resources from passing asteroids, fixed launch windows may become a more common feature of LSA.

However, even once a launch date has been set, the customer generally retains the right under the LSA to request a postponement to the launch for a variety of reasons. For instance, a postponement may be requested in the event of a delay in the manufacture or delivery of the payload. The risk to the customer in such cases is that if the LSP has a full launch schedule, it will be 'bumped' to the end of the queue. The LSP may also request or have to accept a postponement in certain circumstances such as adverse weather conditions or where concerns are raised about the reliability of a launch vehicle. In such a scenario the customer will generally retain its place in the queue. (Alain du Parquet, 2011).

Contract Price, Payments and Termination

Unlike satellite procurement contracts, which may include performance-related incentive payments (e.g. staged payments only released upon successful operation of a satellite for a given period), fixed price contracts are the norm in LSAs. (Wheeler & White, 2012). The price determined is of course the outcome of negotiations and will generally reflect the level of difficulty, non-standardisation and risk associated with the launch. Interim payments are likely to be staggered over the period building up to the launch date to enable the LSP to recoup some of its costs incurred during manufacturing and other pre-launch activities.[2]

If the customer terminates the contract prematurely the LSP may stipulate that a sum be paid to the LSP reflecting the stage of the contract that had been reached so that the LSP does not end up in a worse position that it would have been had it not entered into the contract.

Responsibilities Between the Parties

The article in the Agreement titled responsibilities between the Parties refers to the relevant undertakings of each Party to the Contract. This is usually presented as a list of all engagements the Party is taking under the Agreement.

Liability

The LSA will likely include a cross-waiver of liability.[3] A cross-waiver of liability shields parties taking part in launch services from an inter-party claim for compensation for injury to personnel or for damage to or loss of property that was not the result of a wilful or deliberate act or gross negligence on the part of their employees or contractors. Consequently, each party accepts the risk of injury or damage to its own personnel or property during the launch activities. Third party liability is discussed later in this chapter. However, this is especially significant as international law introduces unlimited or uncapped liability on a zero-fault basis for damage sustained anywhere on the Earth's surface, that could result from a launch event. States may not wish to bear the full burden of this risk and therefore may require the launching parties to carry some of the risk themselves through an obligation to carry launch insurance and/or indemnify the State against future losses above a certain threshold. The LSA will reflect the parties' intentions as to how these obligations and the concurrent risk are to be apportioned between the parties.

Licences and Authorisation

Each Party shall ensure it is acting in accordance with national legal requirements before proceeding with the launch, meaning having obtained the necessary licences and authorisation of the responsible State. The contract sets a deadline (*e.g.* up to three months before the launch) by which time each party must have fulfilled the requirements related to licence and authorisation.

Public Relations

Finally, yet importantly, the question of public relations around the launch has to be well defined in the contract. The launch provider and the satellite owner may wish to lay down rules about and communicating to the outside world about the launch and the related satellite mission.

GOVERNMENTAL AUTHORISATION

This section shall look at the relationship between the State and non-governmental entities involved in launch activities. We have therefore moved from the realm of private law (governing legal relations between entities), as in the case of LSAs, to the realm of public law and regulation.

Non-governmental entities require governmental authorisation (typically a licence) to carry out launch activities. The genesis of this requirement entails a very brief insight into international law. International law creates rights and obligations between States. It does not address the natural or legal persons of those States. An international treaty, for example, binds the States that are parties to the treaty; it does not impose obligations directly on the citizenry of those States.

The fundamental instrument of international space law is the Outer Space Treaty, which entered into force in 1967 (OST). As of 2016, the OST has been ratified by 103 States (including all major space-faring nations). Article VI makes State Parties responsible for their national space activities, whether carried out by governmental agencies or non-governmental (private) entities. As a corollary to this responsibility, Article VI further imposes the obligation upon State Parties *to authorise and continuously supervise the activities of non-governmental entities in outer space*. As a consequence, private entities that are not the subject of international law will nonetheless have to demonstrate to the satisfaction of the competent national authority that their activities will be carried out in conformity with international law in order to obtain authorisation.

State Parties to the OST implement their aforementioned international obligation to authorise and supervise non-governmental space activities through the enactment and enforcement of national laws and regulations, often referred to collectively as 'national space laws'. (Dempsey, 2015). To be clear, most State Parties to the OST have yet to pass any such laws–a State whose national space activity is restricted to small-scale governmental activity may not perceive the need to do so[4]however, the Legal Subcommittee of the United Nations Committee for the Peaceful Uses of Outer Space maintains a schematic overview of 26 States (as of 2016) that have some form of national space legislation. (UNCOPUOS, 2018).

The content and scope of national space laws vary considerably reflecting the nature and extent of national non-governmental space activities as well as the wide discretion accorded to State Parties in the OST on how to authorise and supervise. (Tennen, 2010). However, some insight may be gained by examining the relevant provisions of three examples of national space laws from the US, France, and the UK. This section shall also consider a Recommendation issued by the United Nations General Assembly (UNGA) in 2013 to States seeking to enact regulatory frameworks for national space activities. Recommendations of the UNGA are not binding upon the States to whom they are addressed, however they provide a useful insight into the consistent and predictable approach advocated by the international community.

These sources shall be considered, in outline, against the following lines of enquiry:

1. Which activities related to launch services require a licence?
2. How to determine which State is responsible for issuing a licence?
3. What are the procedures and requirements to obtain a licence?

The answers in any given case are necessarily interrelated. For instance, the determination of the particular State responsible for issuing the licence will provide the conditions under which the licence is issued.

Which Activities Related to Launch Services Require a License?

The OST simply obliges States to authorise and continuously supervise 'the activities' of non-governmental entities in outer space without offering any further guidance on the specific activities that require authorisation. (OST, 1967). States are left to interpret this obligation by defining the scope of application of their relevant national laws. States that have promulgated space laws regulate the provision of launch services. Some regulate the operation of launch sites.

US law requires a licence for various activities including the 'launch of a launch vehicle and operation of a launch or re-entry site'. (US Code, 51 U.S.C. § 50904-50906).

French law requires a licence for 'any activity consisting in launching, attempting to launch or procuring the launch of a space object into outer space'. (French Space Operation Act, 2008).

UK law requires a licence for space activities including 'launching or procuring the launch of a space object'. (Outer Space Act, 1986).

The United Nations General Assembly recommends that 'the scope of national regulatory frameworks may include, as appropriate, the launch of objects into space and the operation of a launch or re-entry site'. (UNGA Resolution 68/74, 2013).

This is a snapshot of authorisation requirements in respect of launch activities only. Further licences may have to be obtained in advance of a mission to operate a satellite, handle certain types of data and utilise orbital positions and radio frequencies depending on the purpose of the mission and the regulatory requirements of the relevant national authority.

How to Determine Which State Is Responsible for Issuing a License?

The issue here is that of *jurisdiction*, a term used to describe the territory, persons or sphere of activity over which the legal competence of a sovereign State extends. (Lotus case, 1927) (Cheng, 1997).

National space laws tend to define their jurisdiction on the basis of nationality and territoriality principles.(Freeland, 2010). The UNGA recommends that 'the State should ascertain national jurisdiction over space activities carried out from the territory under its jurisdiction and/or control; likewise it should issue authorisations for space activities carried out elsewhere by its citizens and/or legal persons established, registered or seated in territory under its jurisdictions and/or control'. (UNGA Resolution 68/74, 2013).

The US claims jurisdiction over any legal person of any nationality carrying out a launch activity (as defined above) within its territory as well as any US citizen or entity organised under US law conducting a launch activity outside the US. The US further claims jurisdiction over foreign entities, where a controlling stake is held by a US citizen or entity, acting outside the US, in certain prescribed circumstances. (US Code, 51 U.S.C. § 50904-50906).

France claims jurisdiction over launches carried out by any person from the national territory of France or from facilities falling under French jurisdiction including the Space Centre in French Guiana (CSG), as well as any French operator intending to launch a space object from abroad. France additionally claims jurisdiction over any French citizen or legal entity headquartered in France seeking to procure a launch service. (French Space Operation Act, 2008).

The UK claims jurisdiction over UK nationals, Scottish firms and bodies incorporated under UK law conducting launch activities anywhere in the world. UK law appears 'not to apply to non-nationals carrying out launching activities in the UK'. If the UK Government follows through on its stated ambition to build a commercial spaceport in the UK, the Outer Space Act 1986 may have to be amended. (Outer Space Act, 1986) (Dempsey, 2015).

In instances where more than one State may assert jurisdiction, the UNGA recommends that the State exercising jurisdiction on the grounds of nationality (rather than territory) should consider forbearing from duplicative requirements to avoid unnecessary regulatory burdens.

What Are the Procedures and Requirements to Obtain a Launch License?

In the absence of any detailed requirements or guidelines in the treaties, procedures and requirements in national regulations vary according to national interests and policies. Even so, they tend to be motivated by the need to protect the health and safety of citizens, their property, and the environment.

In the US, The Federal Aviation Administration's (FAA) Office of Commercial Space Transportation is given powers to issue licences, permits and approvals for commercial launches and re-entries.[5] 'Launch-specific' licences authorise the licensee to conduct one or more launches of one type of launch vehicle at one launch site with each launch activity identified in advance. 'Launch-operator' licences authorise the licensee to conduct multiple launches from one launch site within a range of operational parameters of launch vehicles from the same family of vehicles over 2-5 years. (FAA Regulation, 1998). The FAA reviews a licence application within 180 days to determine whether it presents any issues affecting safety of people and property, protection of public health and the environment, space debris mitigation, US national security or foreign policy interests, or international obligations of the United States.

French space law introduced an authorisation regime administered by the French Space Agency (CNES). (French Space Operation Act, 2008) .Among the conditions required to obtain a licence are the provision of moral, and financial and professional guarantees by the applicant together with proof that the systems and procedures to be put in place comply with all relevant technical regulations proposed by CNES in the interest of safety, both of persons and property, as well as the protection of the environment. Licences for a specified duration can also be obtained for launch systems and operations that are used on a recurrent basis and have been certified as compliant. Consequently, Arianespace does not need to continuously reapply for a licence for the standard (non-customised) part of its launch operations from the CSG. CNES provide their technical assessment within a period of 2 weeks to 2 months depending on the type of authorisation sought. A separate administrative power is granted to the President of CNES to provide security and safety regulations dealing with the CSG launch facility.

In the UK licensing powers are conferred on the Secretary of State for Business, Innovation and Skills, who carries out these powers through the UK Space Agency (UKSA). Applications should be submitted at least six months in advance of the proposed launch activity. Before a licence is granted the UKSA must be satisfied that the launch activity will not jeopardise public health, the safety of persons or property, national security, or compromise the UK's ability to comply with its obligations under international space law. An iterative assessment of the application will be made taking into account relevant international industry standards and debris mitigation guidelines. Licensing requirements are flexible, and conditions may be waived, or additional conditions added depending on the circumstances of the application. (UKSA, 2015).

The licencing of a space activity carries an implication for the private entity to provide proof of its insurance for the activity in question. The precise rules may vary from one state to another, but the objective is always the same; shielding the State from the risk of international liability.

Through bilateral agreements between the French Space Agency (CNES), the launch provider (Arianespace), the owner of the launch centre (ESA) and the concerned States, insurance conditions are determined for the development, production and launching of an Ariane rocket. In most cases, France is considered as the Launching State and the French insurance scheme for space activities applies. The ceiling up to which the company requesting the licence needs to demonstrate insurance has been fixed at 60 M€ in 1980 and has been maintained at that level since then and included in the French Space Operations Act. (French Space Operation Act, 2008). For damages above this limit, the French State's liability takes over.

The question of insurance is becoming more technical and complicated for damages occurring late into the mission a satellite or after the functional lifetime of the satellite (i.e. when it considered space debris). Indeed, insurance usually covers a space activity only for a year or two. After the end of the coverage term of the policy, the State's unlimited liability under international law is at risk once more. In the UK, the UK Space Agency now requests proof of insurance coverage for the whole lifetime of the satellite and covering any recourse against the State.

The US' commercial launch insurance requirement can be found at 51 U.S.C. § 50914. It sets out a procedure for determining the 'Maximum Probable Loss' for which a licensee shall demonstrate financial responsibility or obtain insurance to protect the State, but sets out upper limits beyond which the licensee shall not be required to demonstrate financial responsibility or obtain insurance.

REGISTRATION AND LIABILITY

Whereas the previous two sections considered some of the legal and regulatory prerequisites to launch a space object, this final section shall turn to consider some of the consequences triggered by such activities under international law. In particular, this section shall outline the obligation at State level to maintain a register of objects launched into outer space, as well as the generation of 'launching State' status central to the special liability regime in international space law.

Registration

The launch of a space object triggers an obligation to register that space object under international law. The OST establishes the link between registration of a space object and the assertion by a State of *jurisdiction and control* over it (Article VIII). (Soucek, 2015). Registration is therefore the basis on which a State may assert that its national laws apply to a space object, and any personnel thereof, while in outer space.

The OST is however silent when it comes to the mechanism for registering a space object or its proximity and relationship to a launch event. This detail was introduced in a separate treaty, the Registration Convention that came into force in 1976. (Registration Convention, 1972). The Registration Convention requires each State Party to establish and maintain an appropriate national registry for space objects and to make an entry in this register for each space object that is launched into or beyond Earth orbit for which it is the launching State (Article II, para 1). Registration is not required for suborbital launches or for space objects or launcher stages that do not succeed in reaching Earth's orbit. Where there are two or more launching States (see 2.3.3 below) they shall jointly determine which of them shall become the 'State of registry' in respect of the space object (Article II, para 2).

The Registration Convention specifies neither the timeframe for registration nor the information that each entry should contain. This is left to the discretion of the State Parties. However, as soon as practicable following the launch of space object (usually interpreted as a matter of months rather than a matter of days or hours), the State of registry must furnish the UN Secretary General with the following minimum information to input into the UN's own register (Article IV):

1. Name of launching State or States
2. An appropriate designator or registration number
3. Date and location of launch
4. Basic orbital parameters
5. General function of the space object

It is the State of registry, through its competent national authority, and not the LSP, that is obliged to update the register and inform the UN Secretary General. However, it follows that the State will require the relevant LSP or customer to assist the State to comply with its international obligations.

Liability

Including *liability* among the various legal consequences that flow from launch activities merits some explanation. Indeed, it is true that for liability to arise, whether in national or international law, there must be some identifiable loss, damage or injury suffered. Launching a space object *per se* is not an activity or event for which a person or entity may be said to be liable.

In other fields of human activity, the law has generally evolved rules whereby liability is derived from establishing a causal nexus between some element of loss or damage and identifiable act(s) or omission(s). However, in international space law an alternative basis–the concept of the 'launching State'–broadens the scope of liability. How and why the concept of 'launching State' operates to broaden the scope of liability shall be explored further below. However, the key point that follows is that launch activities are intrinsic to the creation of 'launching States' and hence to potential liability exposure.

The Launching State

The OST sets out the 'liability principle' (Article VII): 'State Parties are internationally liable for damage to another State Party or to its natural or juridical persons caused by a space object or its component parts for which the State Party is the "launching State"'. The term 'launching State' carries a four-fold categorisation:

- The State that launches an object; or
- The State that procures the launch of an object; or
- The State from whose territory an object is launched; or
- The State from whose facility an object is launched.

Consequently, all launch activities will have as their direct legal consequence the creation of one or more launching States, even in the scenario where a launch is procured and provided by purely commercial entities. 'Launching State' status is important because it denotes the State or States that will be treated as internationally liable for damage caused by a space object. In other words, launching States have *liability exposure* under international law. Moreover, 'launching State' is a status that sticks: liability exposure does not disappear upon completion of the launch event but subsists for the entire existence of the space object, even beyond its functional lifetime. Nor can it be assigned to another State even if the space object is sold or transferred to a third party and the registration of the space object amended accordingly. (IISL-ECSL Symposium, 2012).

The OST's liability principle has been elaborated through a separate and subsequent international treaty, the Liability Convention, which came into force in 1972. (Liability Convention, 1971). The Liability Convention expands the scope of liability exposure for launching States, adding:

- Joint and several liability of launching States (Article V). This means that where a space object has more than one launching State, a claim may be brought against any or all of the launching States for the full amount of loss or damage. A claimant would not be required to show which particular launching State caused the damage.
- Absolute liability for damage caused by a space object on the surface of the Earth or to aircraft in flight (Article II). Absolute liability refers to a situation in which liability attaches without the need to demonstrate fault and irrespective of whether due care and skill was taken to prevent the damage that occurred. Surface damage is particularly relevant to launch and re-entry activities. By contrast, liability for damage caused to other space objects in outer space requires an element of fault.
- Unlimited liability to restore the damaged individual, entity or State to the condition that would have existed if the damage had not occurred (Article XII).

Thus, the consequences of a launch event in terms of liability are potentially wide-ranging. A hypothetical example may help to illustrate: suppose a private Canadian meteorological research organisation procures the launch of its weather satellite from SpaceX's launch facility in Cape Canaveral, Florida, US. The launch proceeds perfectly and the space object is placed into LEO where it operates faultlessly for 15 years before making a controlled re-entry into Earth's atmosphere. Upon re-entry into the Atlantic Ocean it hits and sinks a Dutch company's container ship resulting in the loss of all crew and high value cargo. The Dutch government could pursue a claim on the company's behalf against the US government for the full amount of the damage 15 years on without the need to prove any fault or causal connection with some act or omission of the US government. The US would be liable purely and simply because it is a launching State (i.e. because the launch took place from US territory). Changing the facts slightly, had the weather satellite's launch been procured by the Canadian Space Agency (an organ of the State) instead of a private entity, then the Dutch government could go after both launching States, Canada and the US, who would be jointly and severally liable.

Space law, through its notion of 'launching State', uses the launch event as the predicate for allocating liability exposure. There is no equivalent 'operating State' liability regime. The routes of this launch-centric approach to liability reside in the historic perception that space activities, especially launch, are inherently risky. By allowing launch activities to take place some States are effectively creating or condoning the risk they pose to third parties who have no say in the matter.

Although launching States cannot divest themselves of 'launching State' status, governments can take steps to re-apportion the burden of liability in the event of damage. Where there are two or more launching States, a State that has paid compensation for damage has the right to present a claim for indemnification to other launching States. (Liability Convention, 1971). States that are not 'launching States' but are benefiting from a space programme or mission sometimes enter into bilateral or multilateral agreements with launching States to share the burden of liability in the event of an incident.[6]

Some 'launching States' seek to mitigate the risk they are exposed to by their national non-governmental entities. They may achieve this by making the grant of a licence contingent upon the private entity demonstrating to the State's satisfaction that its mission specifications pose an acceptably low risk. However, since risk cannot be wholly excluded, the launching State may further require an LSP or its customer to hold third party liability insurance or to be able to otherwise demonstrate its ability to indemnify the State against claims successfully brought against it. (FAA regulation, 1998) (Outer space Act 1986) (French Space Operation Act, 2008).

CONCLUSION

As explained in this chapter, launch activities are bounded by legal rights and obligations, arising in private law (LSA, insurance etc.) as well as public law and regulation (authorisation, export control etc.) Launch events give substance and effect to the freedom of States to explore and use outer space, but also trigger certain obligations of the State in international law (registration) and provide the predicate for allocating liability for harm done to third parties during the launch or anytime thereafter by a space object. There is a sense in which LSA have remained quite static since commercial satellites began to be launched in volume in the 1970s. However, the private launch sector is just beginning to spread its wings and grow independently from the drip-feed of government contracts. New reusable technologies are promising to deliver more efficient and reliable access to space and sustainable business models. Concurrently, demand for launch services is increasing with companies competing to launch large satellite constellation architectures and pursue commercial activities beyond Earth orbit. It remains to be seen whether these developments will impact the way customers procure commercial launch services (see 2.1.2 above), or the formulation of provisions that commonly feature in LSA (see, for example, 2.1.3 - 'Launch scheduling' and 'Performance').

Access to space is no longer the sole preserve of the major space-faring nations. A number of developing space nations including Brazil, Argentina, South Africa, and Australia have announced ambitions to develop domestic launch capability before the end of the decade. Many more countries have launched their own satellites on foreign launch systems. Through the UN COPUOS, States are urged to promulgate national space laws to comply with their international obligations to authorise and supervise national space activities. A growing number of States have established regulatory agencies for this purpose and this is likely to be the most fertile area of legislative growth and evolution in the near term.

International law reinforces the responsibility of States for the acts of their non-governmental entities with strict apportionment of international, unlimited liability to the launching State(s) for third party damage, loss or injury. There are as yet no historical precedents of claims prosecuted under the Liability Convention so it is not yet possible to conduct an empirical analysis of how well the current international legal framework, unchanged since 1972, functions to balance the competing interests of States launching space objects with those of States that are exposed to the risk of loss or damage as a result. However, the robustness of the current legal framework is increasingly likely to be put to the test as we move towards a new era of space infrastructure, enabled by reusable launch vehicles, and featuring low-cost satellite systems that can be deployed in large numbers into constellation architectures. The relative ease with which these systems can deliver access to space for a variety of new entrants, some of whom may have limited heritage or experience in space missions, will compound the existing problem of orbital congestion and space debris. Launching state liability remains highly relevant and has prompted a number of States to enact provisions within their national space laws protecting their national treasuries with insurance or indemnification requirements.

REFERENCES

Alain du Parquet, C., & the Specific Clauses of Launch Services Agreements in Ingo Baumann and Lesley Jane Smith (Ed.). Contracting for Space: Contract Practice in the European Space Sector (Farnham; UK: Ashgate Publishing, 2011) ['Contracting for Space'] at 385

Alain du Parquet, C., Contracting for Space, supra note 11 at 390

Ariane 5 User's Manual Issue 5. (2011). *Process for launch window definition, s.2.7.2*. Available at: http://www.arianespace.com/wp-content/uploads/2015/09/Ariane5_users_manual_Issue5_July2011.pdf

Bender, R. (1998). *Launching and Operating Satellites: Legal Issues*. Kluwer Law Int.

Convention on International Liability for Damage Caused by Space Objects, 961 UNTS 187 (1971): As of 2016, the Liability Convention has been ratified by 92 States (including all major space-faring nations)

Convention on the Registration of Objects Launched into Outer Space, Mar. 29 1972, 1023 U.N.T.S. 15

Dempsey, P. S. (2015). *The Emergence of National Space Law*. Available at SSRN: http://ssrn.com/abstract=2692639

ESA and Arianespace sign a Frame Contract for the Procurement of Launch Services. (2009, June 15). ESA Press Release. Available at: http://www.esa.int/Our_Activities/Launchers/ESA_and_Arianespace_sign_a_Frame_Contract_for_the_procurement_of_launch_services

FAA Regulation 14 C.F.R. § 415 (1998)

Falcon 9 Launch Services Agreement entered into between SpaceX and Orbcomm in 2012, Exhibit A 'Defined Terms' at p.24. (n.d.). Available at: http://www.sec.gov/Archives/edgar/data/1361983/000119312513112208/d468141dex102.htm

Freeland, S. (2010). Matching detail with practice: The essential elements of National Space Legislation. *Proceedings of the International Institute of Space Law*, *540*, 541.

French Space Operations Act, No. 2008-518 (2008), Article 1, translation from Philippe Clerc 'Consequences of the French Space Law on Space Operations (FSOA) on CNES' Mission' in Contracting for Space, supra note 11

IISL-ECSL Symposium Transfer of Ownership of Space Objects: Issues of Responsibility, Liability and Registration held on 19 March 2012 at UN COPUOS Legal Subcommittee (Vienna). (n.d.). Available at: http://www.unoosa.org/oosa/en/ourwork/copuos/lsc/2012/symposium.html

International Space Elevator Consortium (ISEC). (n.d.). online At: http://www.isec.org

Launch Vehicle Performance Website. (n.d.). Retrieved from http://elvperf.ksc.nasa.gov/Pages/Vehicles.aspx

NASA awards launch services contract to SpaceX. (2008, April 22). NASA News Release. Available at: http://www.nasa.gov/home/hqnews/2008/apr/HQ_C08023_KSC_launch_services.html

National Laws Governing Commercial Space Activities. Legislation, Regulation, & Enforcement, 36 Nw. J. Int'l L. & Bus. 1 (2016). (n.d.). Available at: http://scholarlycommons.law.northwestern.edu/njilb/vol36/iss1/1/

NLS II Contract, Section C, Clause 5.3; and SpaceX contract with Orbcomm, supra note 14 at p.3

NLS II contracts are different, supra note 7: Section C, Clause 1 states that The Government will only accept those launch services that successfully deliver a separated and undamaged payload to the proper orbit conditions Section C, Clause 24 explains how full success, partial success, and failure will be treated in any given LSTO

Park, W. (n.d.). The Tragic Tail of Saddam Hussein's Supergun. *BBC Future*. Available at: http://www.bbc.com/future/story/20160317-the-man-who-tried-to-make-a-supergun-for-saddam-hussein

Redacted, N. L. S. II contract between NASA and Space Exploration Technologies. (n.d.). Available at: http://www.nasa.gov/centers/kennedy/pdf/541918main_nls-contract.pdf

Soucek, A. (2015). Space Law Essentials (vol. 1). Linde Publishing.

Space News. (2016). Available at: http://spacenews.com/spacex-aborts-ses-9-launch/

Tennen, L. (2010). Towards a New Regime for Exploration of Outer Space Mineral Resources. *Nebraska Law Review, 88*(794), 802.

The Permanent Court of International Justice, P.C.I.J. (ser. A) No. 10 (1927), the S.S. Lotus case (France v Turkey).

The UKSA has published a list of questions applicable to launch only activities. (n.d.). Available at: https://www.gov.uk/government/uploads/system/uploads/attachment_data/file/464931/Guidance_for_applicants_-_October_2015.pdf

Thompson, M. (2015). *A Space Traveller's Guide to the Solar System*. Bantam Press.

Treaty on Principles Governing the Activities of States in the Exploration and Use of Outer Space, Including the Moon and Other Celestial Bodies (27 January 1967) 15 U.N.T.S. 206

Treaty on Principles Governing the Activities of States in the Exploration and Use of Outer Space, Including the Moon and Other Celestial Bodies (27 January 1967) 15 U.N.T.S. 206

UNGA resolution 68/74 of 11 December 2013. (n.d.). Available at http://www.unoosa.org/pdf/gares/A_RES_68_074E.pdf

UNGA resolution 68/74 of 11 December 2013, supra note 34

United Nations, Practice of States and international organisations in registering space objects, UN Doc. A/AC.105/C.2/L.255, 8, paras 46 & 49

United Nations Legal Subcommittee on the Peaceful Uses of Outer Space. Schematic overview of national regulatory frameworks for space activities, A/AC.105/C.2/2014/CRP.5 (Vienna, 2014). (n.d.). Available at: http://www.unoosa.org/pdf/limited/c2/AC105_C2_2014_CRP05E.pdf

US Code, 51 U.S.C. § 50904 (a)(3)-(4)

Wheeler, J., & White, S. (2012). *Satellite launch contract: Checklist, Checklist*. Practical Law Company Ltd. Available at: www.practicallaw.com

Yamagami, T. (n.d.). *Research on Balloons to Float over 50km Altitude*. ISAS. Available at: http://www.isas.jaxa.jp/e/special/2003/yamagami/03.shtml

ENDNOTES

1 Activities that are carried out by government bodies, such as national space agencies, are directly attributable to the State and therefore do not require independent authorisation to ensure conformity with international law.

2 NLS II Contract, Section C, Clause 5.3; and SpaceX contract with Orbcomm.

3 Some national space laws address inter-party liability in their licensing requirements. The U.S Commercial Space Launch Competitiveness Act 2015 s.107 and the French Space Operations Act, No. 2008-518 (2008), Article 20 both require reciprocal waivers of liability. *See further* R. Bender *Launching and Operating Satellites: Legal Issues* (Kluwer Law Int., 1998).

4 Activities that are carried out by government bodies, such as national space agencies, are directly attributable to the State and therefore do not require independent authorisation to ensure conformity with international law.

5 In the US, there are further controls by the Federal Communications Commission (FCC) in order to obtain spectrum for telemetry and communications. Since spectrum is needed to launch, this is a big issue and something to consider as you are working towards launch.

6 A good example of this are the agreements concerning the legal regime of the Ariane launcher from the Kourou Space Facility in the French Guyana, involving France, the European Space Agency (ESA) and the ESA Member States that have accepted the Ariane Declaration.

Chapter 4
Suborbital Spaceflight:
Technical Aspects

Gerold F. L. Fuchs
Austro Control, Austria

ABSTRACT

One of the major historical milestones of spaceflight, suborbital flights played a vital role in the development of technologies and procedures for the operation of spacecraft. Today suborbital flights still play an important part in scientific and commercial fields, and in this chapter, the basic terms and values will be defined and also explained in the historical context. Small launching system for scientific applications called sounding rockets have a significant importance due to an unbeatable flexibility and cost effectiveness and the new and emerging field of commercial human flights to the edge of space is growing as an economically sound business model for private participants. Either of them will be detailed in the appropriate sub-chapters, followed by an outlook on the future possibilities of suborbital flights.

INTRODUCTION

Along with the old dream of flying and fuelled by the urge to explore new worlds and surroundings, many visionaries thought about the possibility not just to fly, but leave our planet and go forth into space. To reach the goals of leaving our planet for longer periods and even land on other celestial objects, researchers and engineers had to master one step after the other on the ladder of technological evolution. One important step were suborbital flights.

The definitions of the parameters in suborbital flights are sometimes a bit blurry, so they will have to be explained in the following sub-chapter. Whilst suborbital flights had a vital impact on the history of space flight there are two aspects that keep them interesting for future applications: Scientific applications and commercial human flights. Each of the applications will be detailed in the appropriate sub-chapters, followed by an outlook on the future possibilities of suborbital flights.

DOI: 10.4018/978-1-5225-7256-5.ch004

Orbit

An orbit is, simply explained, the curved path of an object around a point in space, such as the path of Earth around our sun, or the path of a satellite around the Earth.

As an example, Isaac Newton's cannon ball depiction (Figure 1) can be used to illustrate the role of the force of gravity and speed in planetary motion. (Newton, 1728) The thought experiment depicts a cannon placed on top of a very high mountain with no atmosphere (and thus resulting drag).

If the velocity is very high, the cannon ball will leave Earth on a hyperbolic trajectory depicted by path (E). If the velocity is not enough to leave the gravitational sphere of influence of Earth, but equal to or greater than the so-called "orbital velocity", the cannon ball will enter into an orbit around the Earth (C & D) following an elliptic or circular path, depending on the speed. If the cannon ball does not achieve orbital velocity, it will return back to Earth such as trajectories A & B illustrate.

Given a distance from a planet, a body needs to have the correct speed, so that the gravitational pull is matched by the centripetal force. In other, more simplistic words: an orbiting spacecraft is constantly falling towards the planet it orbits, but keeps missing it. To reach a stable orbit around Earth, a spacecraft has to be accelerated to the correct orbital velocity. For the negligible small mass ratio between a planet and a spacecraft, this velocity depends basically only on the mass of the planet and the desired orbital altitude. For the theoretical model of an orbit height along the surface of the earth (at altitude zero) this value would calculate to roughly 7,9 km/s.

Alas, "cruising on altitude zero" is not an option for a real planet especially one with the presence of a dense atmosphere for various reasons (drag, excessive heat through friction, collision with terrain or buildings, noise abatement, etc.)

Figure 1. Newton's Cannon ball
Source: Author's Elaboration

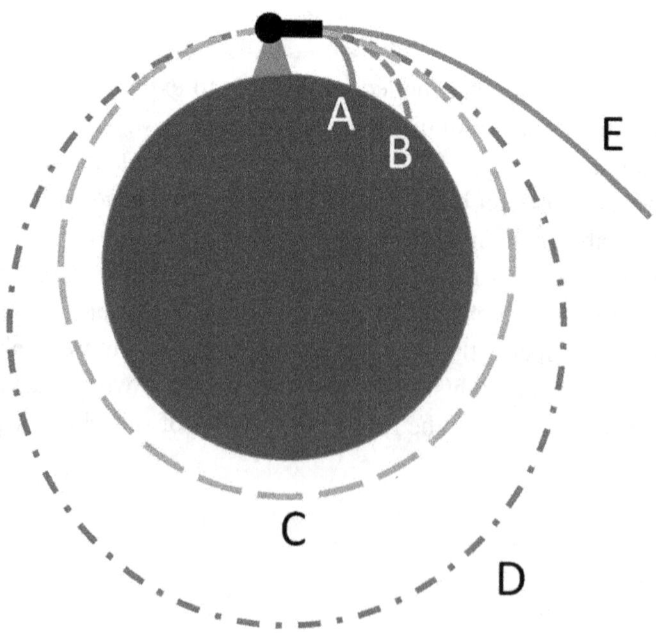

Practical orbit altitudes around the earth usually start at around 200 km above its surface. At this altitude, the drag exerted by the few remaining air molecules upon a spacecraft is small enough to plan for extensive flight operations. However, this small remaining drag force leads to a constant degradation from the selected orbit altitude, and warrants the necessity compensation by using additional rocket thrust inputs, known as “re-boosting”.

The more precise definition of “suborbital” would mean that a body (spacecraft) is launched away from another body (planet) with a speed lower than necessary to enter a stable orbit. For practical operations on earth, the term “suborbital” excludes the craft which has already reached the mechanically correct orbit altitude and speed, but entered a degrading orbit due to drag. In the most simplistic definition, a suborbital space flight is a flight which flies into space, but will not be able to finish one orbit around the planet.

Space

As the speed issue is already discussed, the second question is the definition of “space”. The Earth’s atmosphere does not feature a strict boundary, so the definition of space is more or less a man-made notion. Boundary definitions of the atmosphere take into account the change in atmospheric properties, a change in electrodynamic properties of particles, a noticeable effect upon re-entering vehicles among other factors. The United States Air Force for instance has defined an astronaut as someone flying higher than 50 miles (around 80 km) above mean sea level. (NASA, 2005)

One definition was also proposed for the boundary of space to be the lowest perigee attainable by an orbiting space vehicle (meaning it must at least complete one orbit). If aerodynamic drag is considered, this altitude would calculate to approximately 150 km for a circular orbit and roughly 130 km for an elliptical orbit with high eccentricity.

Finally, the most commonly used and accepted definition is the Kármán line, which lies at an altitude of 100 km. The physicist and engineer Theodore von Kármán (1881–1963) calculated the altitude in which a theoretical aircraft would have to fly so fast to create aerodynamic lift that it already reaches orbital velocity. To have a value, that is more easily to remember and to use, he rounded it to 100 km. The Kármán line was accepted by the Fédération Aéronautique Internationale, an international standard-setting and record-keeping body for aeronautics and astronautics and also by the United States National Aeronautics and Space Administration (NASA).

HISTORY OF SUBORBITAL FLIGHT

Ballistic Missiles

The principle of a rocket was already known to some ancient civilizations where it was used for fireworks as well as for warfare. When ignited, the fuel burns off and produces hot gas, which (according to Newton’s third law) produces a thrust force that propels the rocket forward as a reaction. After the fuel is burnt off, the thrust force is zero and the rocket follows along a ballistic curve. This is then the flight path any mass (like a thrown stone) would take. Thus, the term “ballistic missile” emerged.

Solid fuel and aerodynamic stabilization foils were the only technology available for centuries. At the end of the 19th century engineers and scientists started to develop flying prototypes featuring more modern materials, manufacturing and control mechanisms. Along with them some pioneers of science, like Konstantin Tsiolkovsky developed the mathematics for rocket flight and orbital mechanics.

The American engineer Robert Goddard published a ground-breaking patent for a modern rocket in 1914 and launched the first rocket with liquid fuel in 1926. (US Patent, 1914). At the same time scientists in many countries created major contributions to the field of rocket science that created the basis for modern space flight. Hermann Oberth, Sergey Korolyov and Eugen Sänger are some of those who's visions prepared the way for further technological progress.

In the 1930s and 1940s the political tensions that led to the Second World War contributed to the evolution of rocket technology. Rockets and missiles were used as small unguided devices to deliver explosives, as propulsion systems for various unusual aircraft designs of that time and even as guided missiles to shoot down aircraft.

One design of that time stands out from all others in a technological sense: The German A4, later named V2 (Figure 2). The German military identified the rocket technology as a promising new weapon and dedicated immense effort and resources into the development program. Under the technical lead of Wernher von Braun (a student of Hermann Oberth) a missile was developed that set the standard for years to come.

The A4 flew first on October 3rd, 1942, featured a rocket motor working with an ethanol/water mixture and liquid oxygen, high performance fuel pumps, an evaporation-cooled nozzle system, gyroscopic stabilization and analogue computer technology for the guidance system. The complexity of those systems marked a milestone on the way to space. (Hollingham, 2014)

On June 20th, 1944, a V2 (MW 18014) was launched vertically and crossed the Kármán line, becoming the first man-made object to travel into space. (Neufeld, 1995) At the time, the most common flight profile aimed for a maximum altitude of approximately 80 km, as the main goal was to bridge the distance between Germany and Great Britain and deliver a warhead of 1.000 kg.

After the collapse of the "Third Reich" in 1945, several scientists and engineers working in the V2-program, along with some of the plans and hardware were seized by the allied powers. The emerging rivalry between Soviet Russia and the United States fuelled the so-called "space-race" in the Cold War era. It was largely believed that there was a strategic advantage in mastering space flight as it could be used for spying on another country and putting explosives on an enemy position anywhere in the world.

In those times a new category of weapon was established in the military arsenals of the superpowers and later in other forces as well: the guided ballistic missile. These were the precursors to the Intercontinental Ballistic Missiles (ICBMs) and Intermediate Range Ballistic Missiles (IRBMs). The idea behind the systems was a missile that could deliver one or more warheads to any location worldwide with a precision of a few meters. There were also other plans for conquering space, such as setting up observatories of different nature, space stations and even moon-based habitation modules.

While von Braun helped developing ballistic missiles for the US Army in the 1950s, he had already published essays and books about space travel, space stations and more inspiring many of his time. Ironically the "Redstone" ballistic missiles of the US Army formed the basis of the first manned rockets. (Cassidy, Johnson, Leveve & Miller, 1964)

Figure 2. The V2
Source: US Air Force

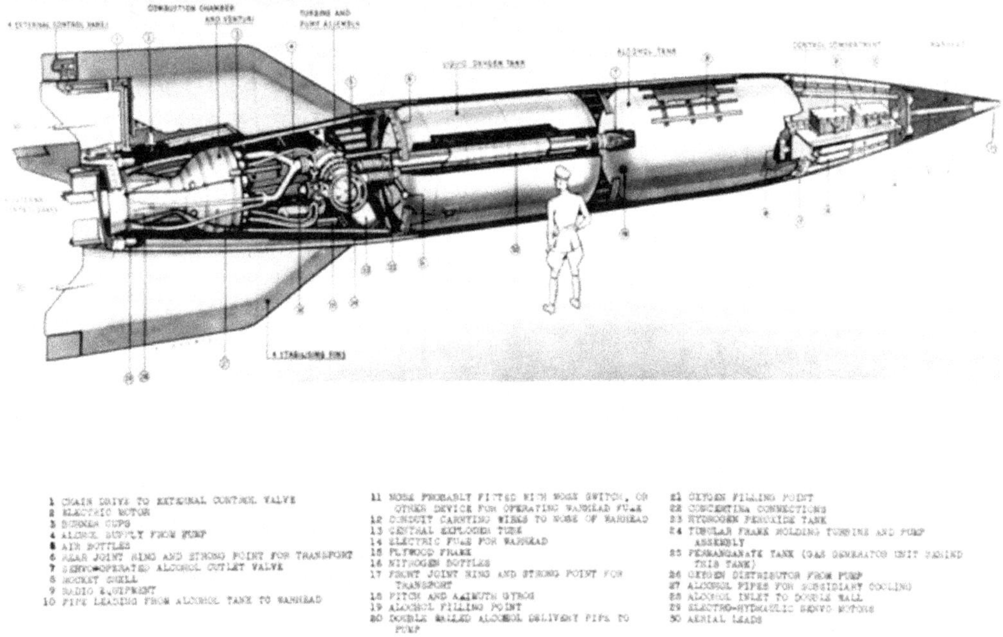

Manned Spaceflight: Plane Based

The most notable early manned suborbital flights took place within the United States' X-15 program running from 1959 to 1968 (Figure 3). The X-15 was a hypersonic aircraft powered by rockets, designed and manufactured by the aerospace company "North American". The three craft that were built were used primarily by the US Air Force and the National Aeronautics and Space Administration (NASA). Based on a concept study by Walter Dornberger, a German engineer out of the V2 program, the X-15 plane was designed to investigate high speed and high-altitude flight regimes. In a normal test-flight mission the plane was carried under the wing of a B-52 jet bomber to an altitude of around 13 km. After being released from the B-52 "mother-ship", the X-15 started its built-in rocket motors, propelling it to high speeds and high altitude (not necessarily at the same time of course). (Jenkins, 2007)

This experimental plane was designed to open the flight envelope to extreme values. As such, several new systems were invented to enable safe operation in this new environment. The most significant step forward was the "reaction control system" for low air density which used rocket thrusters along with a "stability augmentation system" which helped the pilot maintain attitude control. The development and testing of these systems was a necessary phase to master the later flights of the space program up to the era of the space shuttle.

Of the 199 flights performed, 13 flights took the plane above 80 km and two flights reached an altitude of 105,9 and 107,8 km respectively, taking the vehicle past the Kármán line. (NASA, 2014). It should be noted that the record for the fastest manned aircraft of Mach 6,7 was also established within this program and has not been broken to date.

Figure 3. The X-15 during launch from the carrier airplane
Source: US Air Force

Manned Spaceflight: Missile Based

In the 1950s the space race between the USA and the USSR posed the question of who can put a spacecraft, ideally a manned one into orbit first.

The Soviet space program was based upon the usage of the intercontinental ballistic missile R-7 Semyorka (Figure 4), designed by Sergey Korolyov in 1953. This missile was modified to carry a payload into orbit and after the successful launch of the Sputnik satellite in 1957 the first human, Yuri Gagarin was sent into orbit in 1961.

The United States stepped up their efforts into their space program and transferred it to the newly founded NASA. With "Project Mercury" running from 1958 to 1963, the United States established their first human spaceflight program whose primary goal it was to put a man into orbit around the Earth.

As the basis for the launch vehicle, the United States also chose to use a ballistic missile out of the inventory of the army – the Redstone rocket. (NASA, 2015) Unlike the soviets, the first Mercury capsules were sent on suborbital flight profiles. The first American Mercury astronaut, Alan Shepard, was sent to space in 1961 on a ballistic suborbital flight reaching an altitude of 187,5 km lasting only 15 minutes and 22 seconds. (Swenson, Loyd, Grimwood & Alexander, 1989) The following year, John Glenn flew on the first manned orbital mission of the United States on a Mercury-Atlas rocket.

Figure 4. The R-7
Source: Wikipedia

SUBORBITAL SPACEFLIGHT TODAY

Motivation

While suborbital flights were a necessary stepping stone, contributing vital knowledge and experience to the space programs, the question is: why should anyone perform such a flight today? The answer is research and tourism. The flight profile offers an opportunity to the suborbital experience including a descend in freefall, which will establish a short time of weightlessness or "microgravity" for a lower price than an orbital mission.

Microgravity

Microgravity is another term for weightlessness or "zero-g", indicating that gravitational forces are not completely zero but just very small. (Chandler, 1991) For a number of scientific research questions, conducting experiments in a reduced gravitational environment is highly valuable.

To be completely free from gravitational effects in a stationary way, a location is required that is very far from the Earth. This has only been achieved by interstellar probes but is of course highly impractical for any basic research projects, especially when there is a need to bring the experiments back to Earth.

According to Albert Einstein's Theory of General Relativity, in free fall, there is no observable distinction between inertial motion and motion under the influence of the gravitational force. (Einstein, 1916) As was described in Einstein's elevator experiment, for an observer in a small enclosed room, it is impossible to decide, by mapping the trajectory of bodies, whether the room is at rest in a gravitational field, or in free space aboard a rocket that is accelerating at a rate equal to that of the gravitational field. (Wheeler, 1990)

The practical application for this finding is that a free fall status near Earth is a valid replacement for zero-gravity, even though some gravitational effects still remain. The best condition for experimenting resembles the indefinite microgravity as it is present on-board a spacecraft orbiting the earth. However, conducting research in orbit is expensive. Even if the task is performed on an unmanned craft, there are significant cost factors to be taken into consideration such as cost of launch, trajectory control and surveillance, communication, re-entry systems etc. Temporary microgravity is an alternative for some applications where short time exposure is sufficient.

Within an atmosphere, free-fall is restricted by aerodynamic drag forces. As soon as such a force is acting upon a body, the state of weightlessness is corrupted. One technology to work around that problem is free fall in a so called "drop-tower" or "drop-tube". To enable a probe being exposed to several seconds of free fall, the experiment is dropped in a vertical tube, that has been evacuated thus minimizing aerodynamic drag.

A technique that produces longer experimenting time is called parabolic flying, or most commonly referred to as "zero-g flight". (Karmali & Shelhamer, 2008), depicted in Figure 5. For these kinds of flights an airplane pulls upward from level flight to a pitch angle of around 45 – 60 degrees and then steers along a parabolic trajectory until it is flying downward again in a similar angle as at the begin of the manoeuvre. During the parabolic phase of the manoeuvre zero-g is experienced in the aircraft.

Basically, this manoeuvre can be performed by almost every airplane, but the higher the possible airspeed the longer the microgravity phase can be established. For actual experimentation and training of astronauts, large passenger jets are used. An aircraft used for such flights will feature a modified cabin stripped of nearly all seats to give the maximum free space inside. With a smooth execution of the parabola zero-g can be maintained for around 25 seconds.

The third method of creating microgravity is the ballistic rocket, which also experiences weightlessness in the phase of thrustless coasting in space until aerodynamic drag kicks in at re-entering the atmosphere. These sorts of rockets will be discussed more detailed in the next sub-chapters.

Sounding Rockets for Research

Sounding in this context represents "taking a measurement". Sounding rockets are the most commonly used vehicle for suborbital research today and provide anywhere from 5 to 30 minutes of flight time and access to a high-quality microgravity environment depending on the target altitude or space environment depending on the goal of the mission. In addition, these vehicles also provide access to areas of the atmosphere that are too high for high altitude balloons and too low for satellites. Those rocket systems are tailored to fulfil a specific task or series of tasks and some can reach altitudes up to 1.500 km.

Typical mission tasks for sounding rockets are (Eberspeaker, 2013):

- Technology demonstrations for various systems
- Atmospheric research

Figure 5. Parabolic Zero-G flight
Source: Author

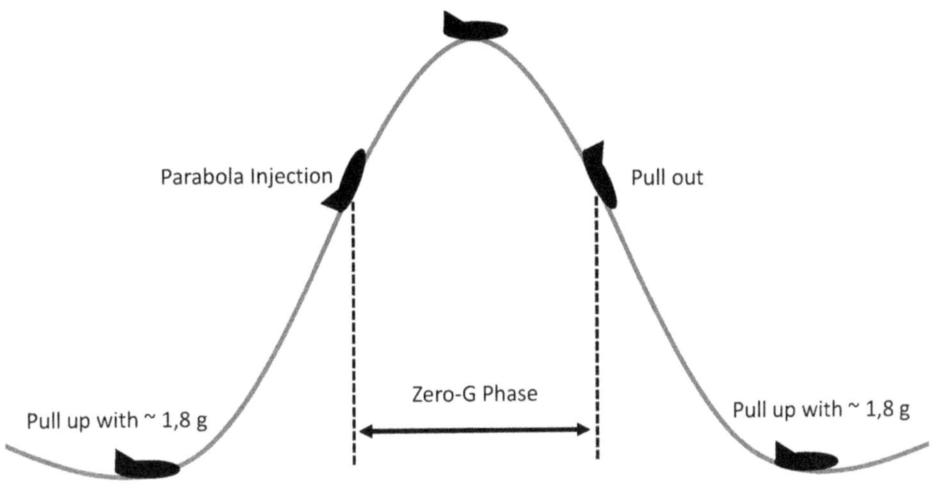

- Microgravity research
- High altitude radiation research
- Magnetic field research
- Astronomy
- And many more.

They are expendable rockets and usually only the payload section is recovered by parachutes. Depending on payload size and desired altitude they feature one to four stages driven mostly by solid fuel motors. These design specifications make sounding rockets comparably cheap with maximum price tags in the range of millions. (Jurist, 2008) Due to the smaller size and complexity they have lead times of just some months to reserve a flight, compared with years if a ride with an orbital mission is required.

One of the most powerful sounding rockets is the four staged “Black Brant” (Figure 6), manufactured by Bristol Aerospace, now Magellan Aerospace. (Magellan Aero, 2018). Over 800 rockets of various versions have been launched since 1961. The largest version is the “XII” with 15 meters length and a launch mass of 5.300 kg, reaching an apogee of 1.500 km. (NASA, 2015)

Rocket Operations

Most sounding rockets are launched from established facilities in many countries that normally offer complete launch infrastructure and telemetry. These facilities are mostly used as test ranges for a multitude of aerospace projects and are located in sparsely populated areas.

If necessary, researchers are also able to conduct their work at sites all over the world, since sounding rockets do not need a great deal of infrastructure to launch. Small to medium sounding rockets can even be launched from ships, which open the door for scientists to reach even remote parts of the world. In addition, sounding rockets also allow researchers to conduct experiments at a specific time and place, which is beneficial if data is needed from a specific event or phenomena that may only occur at certain times of the year.

Figure 6. The Black Brant XII
Source: NASA

A non-exhaustive list of some of the most renowned launch ranges for sounding rockets include:

- Andøya Space Center sites of Andøya and Svalbard, Norway
- Poker Flat Research Range of the University of Alaska Fairbanks, USA
- Esrange Range in Kiruna, Sweden
- Wallops Flight facility, Virginia, USA
- Woomera Instrumented Range, Australia

When sounding rockets are operated, a lot of safety measures have to be observed. A risk analysis must be performed to determine the hazards to third parties such as people on the ground, ships or air traffic. In cases that fly above the Kármán line, collision hazards with spacecraft or satellites must be considered.

There is also an operational risk that would not come to mind as easily: political tensions.

A scaring event became famous as the “Norwegian rocket incident” in 1995, when a Black Brant XII sounding rocket was launched from the Andøya Rocket Range in Norway. (Pry, 1999) It was equipped to study the aurora borealis over Svalbard on a high northbound trajectory. By coincidence it used initially the same corridor that ICBMs from the US would use to launch a nuclear attack on Moscow. The rocket even reached an altitude of 1.453 km like a submarine-launched Trident missile of the US Navy would do. Russian nuclear forces were of course on high alert, but on highest Russian level it was luckily determined that there was no nuclear attack and no retaliation strike is necessary.

Compared to the possibility of a nuclear war another category of unwanted side effects of rocket launches appears quite harmless: Strange light phenomena in the night sky. (Linder, 2009) It happened several times in history that rockets with malfunctioning thruster controls produced exhaust smoke in spiral form (due to an oscillatory movement). (Spell, 2011) When this smoke clouds are high enough, but still illuminated by the setting sun many people interpret it as a sign of UFO activity, mostly inspired by science fiction phantasies about extra-terrestrial visitors.

Anatomy of Sounding Rockets

There are multiple different types of sounding rockets available today (Figure 7). They come in a variety of sizes varying from the single stage series of rockets that span just over 2 meters that barely reach space to multi-stage rockets carrying payloads with hundreds of kilograms.

The sounding rocket is generally separated into two major parts as shown in Figure 8: the payload and the rocket motor.

The rocket motor is mostly of a solid fuel type. Solid propellants used in sounding rockets are often excess military motors which makes them relatively cheap available. This type of motor is also cheap in design and manufacturing and storable for long time. Also, the operation is very easy and is fairly straightforward with little to no moving parts. However, they might have issues in controllability. Once ignited, the propellant burns until used up and with a thrust value that cannot be adapted according to the situation. Rocket engines using fluid propellants in the contrary, are highly controllable, but also highly complex in design and manufacturing and not a prime selection for suborbital sounding missions.

Figure 7. Sixteen different sounding rocket vehicles, ranging from a single stage Orion to a four stage Black Brant XII, are available to fulfill science altitude requirements of 100 to 1.400 km. Payloads up to 1500 lbs can be accommodated.
Source: NASA Wallops

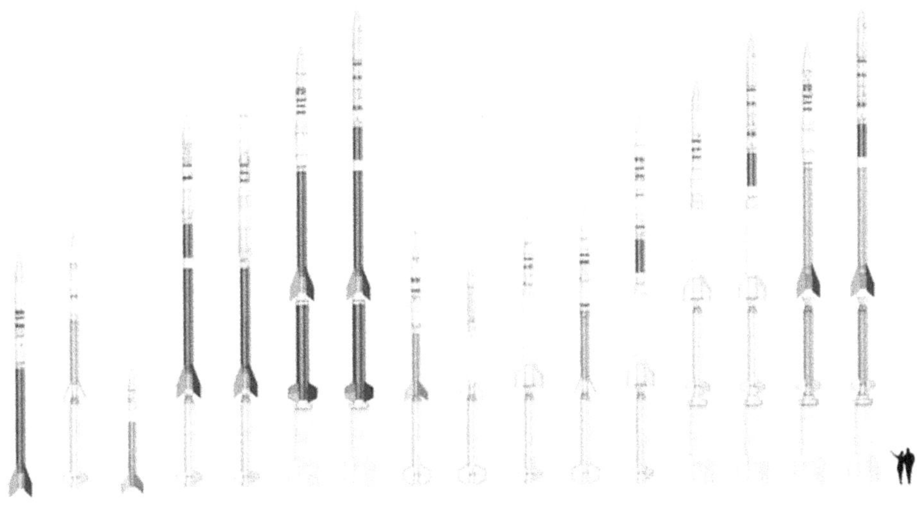

Figure 8. A typical configuration of a sounding rocket
Source: NASA

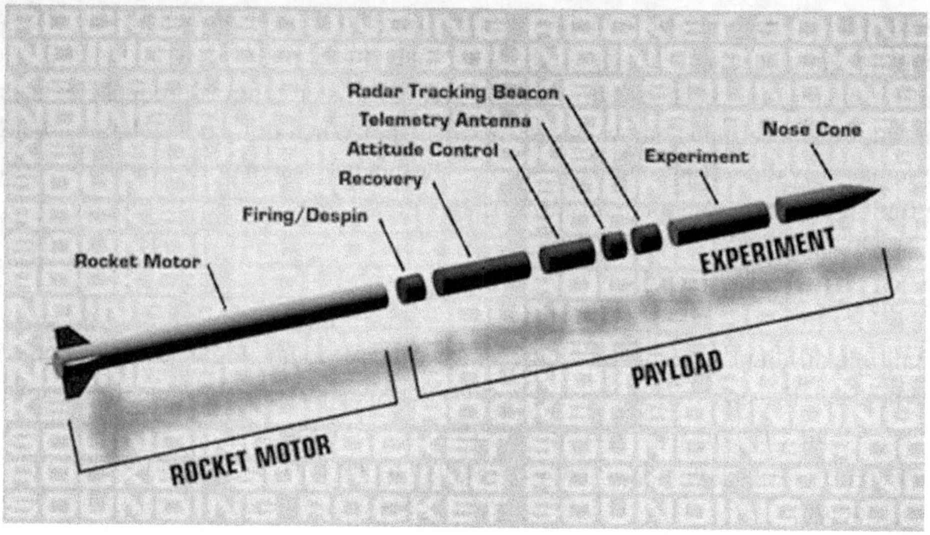

The payload section of the rocket is again divided into two components: the experiment and the support equipment needed for the flight itself. The experiment is primarily what scientists want to measure or test.

The support equipment consists of controls for the engine, trajectory, navigation, attitude as well as telemetry and communication devices to maintain a link with the ground control station. Due to the fast-paced evolution of electronics, the microcircuits and computers used nowadays deliver great performance in a minimal volume. An equipment part that cannot be miniaturized is the recovery system. Often, scientists want to bring back their experiments after the flight and use parachutes as the cheapest and easiest way for retrieval. Only in some cases, desired data is collected via telemetry and a recovery of the payload would be too expensive.

Flight Profiles and Launch Methods

A typical launch is displayed in Figure 12. The sounding rocket is usually launched from a fixed rail launcher taking the rocket to a parabolic flight path. The rail launcher supports the rocket in the beginning of the acceleration phase and provides guidance on the first 3 to 20 meters, as at low speeds the aerodynamic and/or thrust stabilization will not be effective enough.

After the rocket motor has burnt up its fuel, it separates from the payload and falls back down to Earth. In case of a staged rocket the next stage will ignite after the one before is burnt out and separated. By losing every piece of mass that is unnecessary for the rest of the mission, the remaining rocket can reach a higher acceleration and thus altitude.

A detailed calculation on how many stages are necessary and the speeds involved are to be found in literature revolving around the Tsiolkowsky equation. (Ціолковскій, 1903) (Moor, 1813)

After Brennschluss (a word taken from the German description for the time of cessation of fuel burning in a rocket) the payload continues on a trajectory formed by gravity only and will eventually reach the highest point of its journey. All data collected by the payload during this time can be recorded and stored on board, or is more often returned to Earth via telemetry links that allow researchers to examine the information in real-time.

When the payload falls towards Earth again it has to be slowed down and in some cases brought back with a parachute for retrieval. In case of very sensitive payload, the landing impact can be softened by air bags or similar system.

This is an example of the most typical mission profile (Figure 9) and, depending on the research goals my take on many deviations. To save fuel it is advantageous to use an aircraft as a launch platform, such as Pegasus. The Pegasus is an air-launched rocket developed by Orbital ATK capable of carrying payloads of up to 443 kilograms into low Earth orbit. The vehicle consists of three solid propellant stages and an optional monopropellant fourth stage. Pegasus is released from its carrier aircraft at approximately 12.000 m and its first stage has wing and a tail to provide lift and attitude control while in the atmosphere. (Orbital ATK, 2018)

Apart from the launch place and method also the trajectory reflects the manifold of scientific questions. To gain maximum altitudes a steep launch angle is chosen, with a shallow angle higher speed within the atmosphere and at re-entry will be the consequence.

Deviations in the last phase of the flight were already mentioned and rotate around the main decision if the payload needs to be recovered at all.

COMMERCIAL HUMAN FLIGHTS

The second big motivator for performing suborbital flights today is the opportunity to put people in space in a relatively easy way.

Figure 9. A Typical flight profile for a sounding rocket
Source: NASA

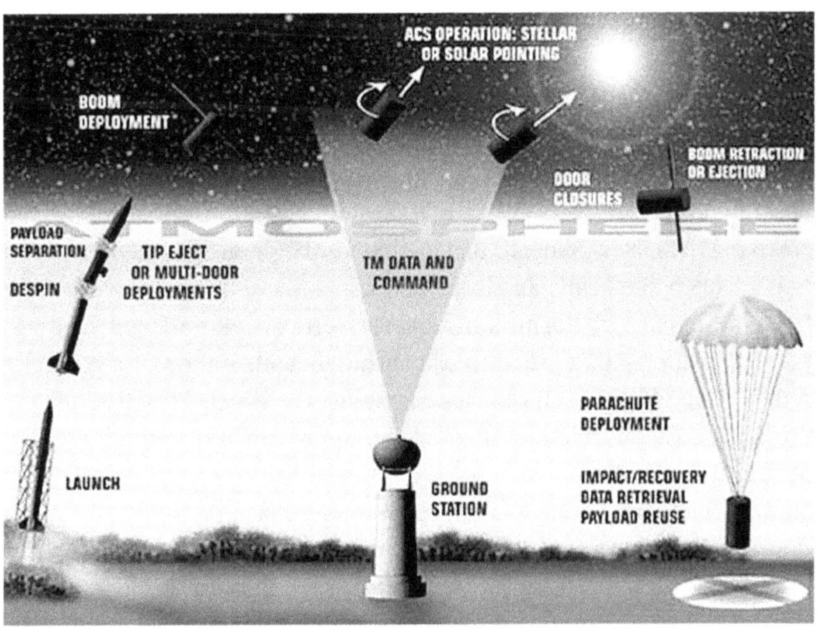

Due to the high costs of development and operation of spacecraft, combined with the immense military and political importance, space programs have traditionally been in the hands of national governments for the majority space flight's short history. Only in rare cases have private persons allowed to pay a significant amount of money and be part of a space mission. Since 2001 beginning with American millionaire Dennis Tito, a few self-funded space travellers were part of flights on-board Soyuz spacecraft of the Russian Space agency. Still they had to complete at least the cosmonaut basic training for every crew-member and had to show mental and physical fitness for the mission. As they are still not considered professional astronauts or cosmonauts, the term "Space flight participant" was forged.

For some years now space travel can be offered on the basis of a pure leisure flight to enjoy the sensation of space flight and relatively long weightlessness. (Wall, 2013) To keep such an offer affordable and accessible to more people, suborbital operations deliver a great solution.

Keeping these flights commercially viable it will be necessary to base all designs on reusable Suborbital Launch Vehicles (rSLVs). With new rSLVs, depending on the vehicle, human tended payloads can also be accommodated, lowering the cost of development and opening up the possibility of what kinds of research can be conducted. Suborbital research can provide human access to the microgravity environment for a fraction of the cost of going to orbit, but do not provide extended periods of time in space. Unlike sounding rockets, these vehicles have quick turnaround times and with the capability to fly every week if needed. The cost of these platforms are at a fraction of the cost of a sounding rocket, although they cannot deliver the same duration of microgravity time.

As usual, when a new chapter of technology and business is opened in history there are lot of ideas and companies striving for success. Several interesting designs have been tested, however only few ideas are still active today. In the following section, examples will be limited to only some of the advanced systems at the time this book was written.

Capsule Designs

A reusable capsule design would at least feature a one staged vertical take-off and landing rocket. The capsule with the passengers would launch similar to standard spacecraft and then separate from its booster stage and both parts, booster and capsule would use a parachute for descent. By definition both parts would be reusable, but still a parachute landing puts some damage on parts and causes some maintenance and repair work.

A good example of a successful capsule design is the "New Shepard" vehicle by Blue Origin. It features a capsule design that will take spaceflight participants and payloads to the edge of space and parachute back down to Earth. New Shepard had its first test flight in 2015 and multiple more successful launches from its US launch site near Van Horn, Texas.

Blue Origin's New Shepard platform is a reusable vertical take-off and landing spacecraft. The New Shepard system is made up of a pressurized capsule that sits on top of a booster. The single stage booster is powered by Blue Origin's self-developed engine, the BE-3. The BE-3 uses a liquid hydrogen and liquid oxygen. The booster accelerates upwards for approximately two minutes and 40 km altitude before the engine cuts off.

If nothing abnormal happens during the powered ascent phase of flight, the system has an escape system which allows participants to safely abort the mission. The escape system is powered by a solid rocket motor that provides 70,000 lbs. of thrust in a two second burn to move the capsule away from the danger zone. It was successfully tested at the point of highest dynamic pressure on the vehicle at supersonic speeds.

Once the engine cuts off, the capsule separates from the booster and coasts up to an altitude of more than 100 km. The pressurized capsule's interior has a volume of approximately 15 m^3; over 10 times the room that Alan Shepard (after whom this platform is named) had on his Mercury flight. The capsule has room for six spaceflight participants and allows them to move around the capsule during the few minutes of the weightless period of the flight. The capsule also has large windows to give the spaceflight participants a view of the curvature of Earth and space.

The crew capsule then re-enters the atmosphere and descends back to Earth under three independent parachutes that provide redundancy in case one fails. In addition, the capsule has a retro-thruster system that cushions the final descent to land.

After the capsule has separated, the booster falls freely for a few minutes before beginning an autonomously controlled rocket-powered vertical landing on the landing pad. As the booster re-enters the atmosphere, air flows through a ring located on the top of the booster, passively moving the centre of pressure to help control the rocket through descent. The booster also is designed with four wedge-shaped fins that further assist aerodynamic stability. These fins are controlled by hydraulic actuators that allow them to work through speeds up to Mach 4 to steer the booster back to a controlled landing. To help slow down the booster during descent, eight large drag brakes are deployed to slow the vehicle's speed by half. In addition, a key feature of the BE-3 engine that powers the rocket is its ability to restart, allowing it to slow the booster down to just about 8 km/h for landing.

The New Shepard's booster ability to deploy its landing gear and land after delivering the capsule to space allows for a truly reusable system. With the capsule landing under parachutes, both elements of the New Shepard suborbital flight platform can be ready to be used again. (Blue Origin, 2015). The company has stated the module only requires minor maintenance in between flight that will incur costs in the "low thousands of dollars." (Foust, 2016)

Spaceplane Designs

Spaceplane designs feature a spacecraft that can be flown like an airplane when operated within the atmosphere. The launch into space can be enabled by two ways. Vehicles launched by a rocket (Space Shuttle, Buran, X-37) are considered vertical take-off horizontal landing (VTHL). The X-15 as well as SpaceShipOne & SpaceShipTwo vehicles are rocket-powered aircraft, having been carried up to an altitude of several kilometres by an aircraft before being released. A "single stage" system featuring a horizontal take-off from a runway, flight to suborbital altitudes and landing on a runway again is highly desirable, but not yet built at the time of this publication.

Air launch provides some advantages and disadvantages. Disadvantageous aspects warrant a special modified airplane, which has to be re-worked by a licensed aviation company and certified by aviation authorities. All that is combined with some costs.

The main advantage is that the rocket does not need to fly through the low, dense atmosphere. Higher density at lower altitudes results in larger drag forces acting on the vehicle which then requires more thrust to continue the forward movement and which in turn means more mass from the required propellant used. Carrying the vehicle by an aircraft to a higher altitude and speed before launch enables the spacecraft efficient use of its engines allowing it to use more of its mass for payload or get the payload to higher altitudes. Reducing the mass of the spacecraft also reduces its associated cost. Air launch also offers the potential for aircraft-like operations such as launch on demand and operation from normal airfields. It also allows the aircraft to fly around weather conditions as well as fly to better launch points, and to launch a payload into any orbital inclination at any time.

SpaceShipOne

When the famous aircraft designer Burt Rutan with his company scaled composites, and started a project leading to the construction of SpaceShipOne, the foundation was laid for the most successful civilian spaceplane project to date.

The idea was to build experimental rocket propelled aircraft, launched from a mother ship aircraft (named "White Knight"). The spaceplane was designed for suborbital spaceflight capability. A special design feature is a "feathering" re-entry system into atmosphere that folds the rear half of the wing with the two tail booms upward to produce a stable aerodynamic configuration that also increases drag. Taking advantage from two configurations the system imitates both the performance of a winged vehicle and a capsule during its re-entry trajectory. Winged vehicles, such as the Space Shuttle, are able to provide gentle decelerations during re-entry, and allow for very accurate landings. However, winged vehicles require expensive, advanced thermal protection technology for the safety of the vehicle during the high temperatures incurred during re-entry. Capsule designs, on the other hand, require less advanced thermal protection systems and well suited for dynamic loading re-entries. However, astronauts in capsule platforms are subject to higher deceleration forces. In addition, capsules land via parachute which do not provide precision landings.

SpaceShipOne completed the first manned private spaceflight in 2004. That same year, it won the 10 million US-Dollar Ansari X Prize for the first non-government organization to launch a reusable manned spacecraft into space twice within two weeks and was immediately retired from active service, as it was a proof-of-concept prototype only. (Ansari, 2004) The first flight departed from the Mojave Air and Space Port which became the first commercial spaceport licensed in the United States. SpaceShipOne was flown by Michael Melvill to a height of 100 km which made him the first commercial astronaut. The prize-winning flight was flown by Brian Binnie who reached a height of 112 km and by this broke the record set by the X-15.

SpaceShipTwo

The entrepreneur Richard Branson founded the company Virgin Galactic and continued the project with the goal to offer regular suborbital spaceflights to paying passengers in the near future, aboard a newly design named SpaceShipTwo.

Virgin Galactic's SpaceShipTwo is an rSLV that is in development in Mojave, California and is projected to be flying with spaceflight participants and research as soon as the vehicle completes its flight test program. SpaceShipTwo is based on SpaceShipOne and will as well be launched from an aircraft too, and glide back for a landing on a runway.

Its carrier aircraft is a twin fuselage airplane named WhiteKnightTwo (Figure 10), that carries the Spaceplane in the space between the two fuselages. It is powered by four jet engines and is made primarily of carbon composite material which enables a lightweight but still strong structure. The Aircraft is operated like a normal jet airplane and carries its pilots in the right cabin. Currently, Virgin Galactic

Figure 10. WhiteKnight and SpaceShipOne flight 17P
Source: Don Ramey Logan

does not carry people in the left fuselage during flight or launch operations, though it would be possible to install passenger seats there as well.

A typical mission profile will start with WhiteKnightTwo taking off on a conventional runway with SpaceShipTwo mated in the middle position beneath its wing and climbing to an altitude of approximately 15.000 m. Once the crew has performed all necessary vehicle and safety checks, the pilots of WhiteKnight will release SpaceShipTwo from the aircraft. Within a few seconds after the drop, SpaceShipTwo's rocket motor will start up.

SpaceShipTwo will carry crew and spaceflight participants (or research crew with experiments) to high altitude powered by a hybrid rocket motor. Hybrid means it combines elements from both solid rocket motors and liquid rocket engines. The rocket motor uses a solid hydroxyl-terminated polybutadiene (HTPB) rubber fuel and liquid nitrous oxide oxidizer. Solid motors and liquid engines both provide their own advantages and disadvantages for spaceflight. A hybrid rocket motor provides benefits from both types for human spaceflight and does not require the complex plumbing and cryogenic tanks necessary in a pure liquid engine. Especially it allows the motor to shut off quickly and safely at any time during its boost, unlike solid motors.

With the rocket firing on full thrust level the spacecraft will then accelerate to three times the speed of sound, or Mach 3. At this time, the spaceflight participants will feel moderate G-forces as the vehicle propels them to space. After the rocket motor has fired for approximately 60 seconds, the pilot will shut off the motor and SpaceShipTwo will continue to coast upwards driven by its own momentum until it reaches its apogee. While coasting, SpaceShipTwo uses the same unique system as its predecessor and will 'feather' its wings in preparation for the re-entry. When SpaceShipTwo is in its feathered configuration, the entire tail structure of the spacecraft is rotated upwards about 60 degrees.

As the craft continues to coast upwards, spaceflight participants will be able to experience weightlessness and SpaceShipTwo will manoeuvre to give the participants a good view of the Earth from its twelve cabin windows.

After several minutes of weightlessness, spaceflight participants will have to return to their seats for the descent and landing phase. After re-entry, in an altitude of around 21,000 m, SpaceShipTwo resumes its original configuration with wings and tail in normal position and glides back to a runway, landing from where it took off.

The first of these spaceplanes, the VSS Enterprise, was intended to commence its first commercial flights in spring 2015. The company suffered a setback when the Enterprise broke up over the Mojave Desert during a test flight in 2014 killing one of the two pilots. The feathering mechanism was operated outside of its limits, which lead to a structural failure. In an investigation by the National Transportation and safety board (NTSB) a human factors problem at operating the feathering system was identified as the root cause. (NTSB, 2015)

A second spaceplane, the VSS Unity, was unveiled in February 2016 and had its first powered flight on April 5th, 2018.

Future Applications of Suborbital Flights

For sounding rockets there will continue to be a market in the future as they can provide relatively cheap flights, tailored to the researchers' needs. The technological evolution in materials and electronics will help those systems grow in efficiency and precision and eventually even lower the launch costs.

As soon as reusable suborbital launch vehicles come out from the development phase and enter into a scheduled commercial business model, they will establish themselves as a fixed variant of leisure flights. Even before Virgin Galactic was finished testing SpaceShipTwo, the company had taken reservations of approximately 700 people. Ticket prices for a flight to the edge of space and back range from USD 150,000 to USD 250,000 and with every flight up to six spaceflight participants plus two crew can be taken on board. (Wall, 2013)

If these kinds of reusable suborbital spaceflights become more established, and the technology matures, plans may evolve to develop them further to suborbital point-to-point flights. If the technology exists, it could be possible to offer suborbital flights from hub to hub around the world in a short amount of time. For example, a flight from Washington, D.C. to Tokyo in 2 hours instead of 14 hours could then be possible.

However, from the author's point of view, the technical problems may be solved with some likelihood within the next 10-20 years, but the question will be: who will want to pay for it? There was a small market for the first and only civilian jet airliner the Concorde, taking passengers across the Atlantic and to other destinations with a cruising speed of more than Mach 2. For a small number of people, it was no problem to finance the tickets, but the plane was never considered a "cash-cow" by the operating airlines. At the time this chapter is written there are new concepts for civilian supersonic aircraft, but mostly for the flight profile of a business jet. As there exists even a "business jet" version of the world's biggest and most expensive airliners, there may be a very minute flock of wealthy customers who can afford a private space ship.

Another application for future suborbital flight industry may be the conversion of one of those point-to point spacecraft to act as a carrier craft for an ascent stage into Low Earth Orbit. This concept was developed by the Austrian scientist, Eugen Sänger in 1933. (Sänger, 1933) A follow-on concept was developed in the 1980s in the form of a carrier plane releasing a spaceplane at 30 km and at a speed of Mach 7, but was discontinued for political and technological reasons.

CONCLUSION

In the tradition of functioning as a historical stepping stone of spaceflight, suborbital flights continue to inspire people. Small launching system for scientific applications will have a continued importance due to flexibility and cost effectiveness. The new and emerging field of commercial human flights will not only act as an economically sound business model but also act as a source of inspiration to people to think beyond their boundaries and look at our planet as a unique home and one world.

REFERENCES

ЦіолковскійК. (1903). Изслѣдованіе мировыхъ пространствъ реактивными приборами.

AnsariX. Prize (2004). Retrieved from https://ansari.xprize.org

Blue Origin. (2015). Retrieved from https://www.blueorigin.com/

Cassidy, J. L.; Johnson, R. I.; Leveye, J. C.; Miller, F. E. (1964). *The Mercury-Redstone Project*. NASA.

Chandler, D. (1991, May). Weightlessness and Microgravity. *The Physics Teacher, 312–13*. doi:10.1119/1.2343327

Eberspeaker, P., & Pierce, D. (2013, September 21). *An Overview of the NASA Sounding Rockets and Balloon Programs*. NASA.

Einstein, A. (1916). Die Grundlage der allgemeinen Relativitätstheorie. *Annalen der Physik, 49*, 769–822. doi:10.1002/andp.19163540702

Foust, J. (2016, June 16). *Space News - Blue origin flies New Shepard again*. Retrieved from http://spacenews.com/blue-origin-flies-new-shepard-again/

Global Solutions for the Aerospace Market. (n.d.). Retrieved from http://magellan.aero/product/rockets/

Hollingham, R. (2014, September 8). *V2: The Nazi Rocket that Launched the Space Age*. Retrieved from BBC: http://www.bbc.com/future/story/20140905-the-nazis-space-age-rocket

Jenkins, D. R. (2007). X-15: Extending the Frontiers of Flight. NASA.

Jurist, J. M. (2008, October 13). *The Commercial Suborbital Sounding Rocket Market: A Role for RLVs?* Retrieved from The Space Review: http://www.thespacereview.com/article/1228/1

Karmali, F., & Shelhamer, M. (2008). The dynamics of parabolic flight: Flight characteristics and passenger percepts. *Acta Astronautica, 63*(5-6), 594–602. doi:10.1016/j.actaastro.2008.04.009 PMID:19727328

Linder, A. (2009, December 9). Märkligt ljussken över Kiruna [Strange lightning over Kiruna]. *Norrländska Socialdemokraten*.

Loyd, S., Grimwood, J. M., & Alexander, C. C. (1989). 11-1 Suborbital Flights into Space. In This New Ocean: A History of Project Mercury. NASA Special Publication-4201 in the NASA History Series. NASA.

Moore, W. (1813). *A Treatise on the Motion of Rockets. To which is added, An Essay on Naval Gunnery.* Military Academy at Woolwich, London: G. and S. Robinson NASA (2005-10-21). A long-overdue tribute. NASA. Retrieved from https://www.nasa.gov/centers/dryden/news/X-Press/stories/2005/102105_Wings.html

NASA. (2014, February 28). *Armstrong Fact Sheet: X-15 Hypersonic Research Program*. Retrieved from NASA: https://www.nasa.gov/centers/armstrong/news/FactSheets/FS-052-DFRC.html

NASA (2015a). *Capability Catalog - Black Brant XII*. NASA. Retrieved 2015-08-18

NASA. (2015b, July 31). *Mercury-Redstone 3*. Retrieved from NASA: https://www.nasa.gov/mission_pages/mercury/missions/freedom7.html

Neufeld, M. J. (1995). *The Rocket and the Reich: Peenemünde and the Coming of the Ballistic Missile Era*. New York: The Free Press.

NTSB. (2015). *2015_spaceship2_BMG_abstract.pdf*. Retrieved from https://www.ntsb.gov/news/events/Documents/2015_spaceship2_BMG_abstract.pdf

Orbital, A. T. K. (2015). *Pegasus User's Guide*. Retrieved from orbitalatk.com

Pry, P. (1999). *Black Brant XII. War scare: Russia and America on the nuclear brink*. New York: Praeger.

Sänger, E. (1933). *Raketenflugtechnik*. München: Academic Press.

Spell, T. (2011). *Estimation of the Location, Trajectory, Size, and Altitude of the "Norway Spiral" Phenomenon*. Retrieved from http://www.spellconsulting.com/reality/Norway_Spiral.html

Wall, M. (2013). *Ticket Price for Private Spaceflights on Virgin Galactic's SpaceShipTwo Going Up*. Space.com.

Wheeler, J. A. (1990). *A Journey Into Gravity and Spacetime. Scientific American Library*. San Francisco, CA: W. H. Freeman.

KEY TERMS AND DEFINITIONS

Apogee: For any satellite of Earth, including the Moon, the apogee is the point of greatest distance to the host planet.

Ballistic: Behavior and effects regarding the motion of projectiles.

Hypersonic: Aerodynamic description of speed that is in a highly supersonic range, generally referred to speeds of Mach 5 and above.

Mach: Mach is a dimensionless quantity representing the ratio of flow velocity to the local speed of sound. An object traveling with Mach 1 has the same speed as the pressure (sound) waves that it is generating.

Microgravity: The term is a synonym for weightlessness and zero-g, but indicates that g-forces are not quite zero, just very small.

Sounding Rocket: A sounding rocket, sometimes called a research rocket, is an instrument-carrying rocket designed to take measurements and perform scientific experiments during its suborbital flight.

Spacecraft: A vehicle or machine designed to fly in outer space.

Spaceplane: An aerospace vehicle that operates as an aircraft in Earth's atmosphere, as well as a spacecraft when it is in space.

Chapter 5
Suborbital Spaceflight:
Legal Aspects

Yuri Takaya-Umehara
The University of Tokyo, Japan

ABSTRACT

The chapter outlines the legal aspects of suborbital spaceflight under international law with the aim of clarifying upcoming legal challenges to the existing space treaties. One of major problems lies in the applicability of air law to the spaceflight. In the case of using aircraft-type reusable launch vehicle (RLV), it fits into the definition of aircraft in air law. The other problems are also addressed, which include the different legal status of airspace and outer space, the lack of specific concepts and terms related the flights in the existing space treaties, and a need for international regulation to ensure safety in the flights. It concludes with recommendations for the present space traffic management.

INTRODUCTION

Suborbital flights highlight legal questions between space law and air law. When Reusable Launch Vehicle (RLV) such as the U.S. Space Shuttle started operational for scientific purposes, the questions mainly focused the definitions of outer space and spacecraft to clarify the applicability of space law to them. In the 1990s when private entities started entering into space activities, feasibly studies on the use of RLV for commercial purposes were explored and one of the representative business models was "space tourism". In order to provide commercial space flight to space tourists, further legal clarification was attempted to draw a line between astronauts and space tourists and the issues of state responsibility and liability were reviewed in the case of accidents caused by private entities.

On the other hand, a need for international regulatory framework for space transportation including the use of RLV for suborbital flights has been increasingly recognized in the name of Space Traffic Management (STM). It is a set of "rules of the road" to ensure safety in accessing to outer space by coordination of rules and to reduce risks posed by solving congestion of space objects and space debris in Earth orbit.

DOI: 10.4018/978-1-5225-7256-5.ch005

Taking them into consideration, the present Chapter consists of: types of suborbital flight from the legal perspective; the general international framework applicable to suborbital flight; and development of international regulatory framework - STM.

TYPES OF SUBORBITAL FLIGHTS FROM A LEGAL PERSPECTIVE

"Public" or "Private"

Legally speaking, there are two types of suborbital flights which are: public suborbital flights and private suborbital flights.[1] While the former is carried out by states with their authorized astronauts, the latter is divided into two types which are private suborbital flight and suborbital space tourism. While the term "tourism" is defined by the World Tourist Organization (WTO) and the U.S. Statistical Committee in 1994 as "[t]he activities of persons travelling to and staying in places outside their usual environment for not more than one consecutive year for leisure", (Launius & Jenkins, 2006) space tourism is defined by scholars as "any commercial activity offering customers direct or indirect experience with space travel" (Hobe & Cloppenburg, 2004) which does not exclude parabolic flights. (Von der Dunk, 2007) Recognizing the vague concept of space tourism, the term "private spaceflight" is to be used in the present legal study (Von der Dunk, 2015).

"Plane-Type" vs. "Rocket- Type"

With the advances in space technology, the suborbital flights have become certain type of transportation activities, different from satellite missions, and legal issues are often identified in legal gaps between the existing rules for aviation and space transportation. In short, there is the possibility of applying air law to the use of RLV for orbital flights.

Such a problem is derived from using certain types of launch vehicle for the flights. In general, there are two types of launch vehicles originated in the beginning of space era: a rocket with a capsule on a top which was used for the Mercury manned exploration in 1961; and an aircraft-type vehicle which are equipped with a reaction control system, used to explore the lower end of the hypersonic flight regime in 1963. (Jakhu, Sgobba & Dempsey, 2011) Those two models still remain appropriate for the suborbital flights and most of legal issues debated in the 1980s were related to the aircraft-type vehicle such as Space Shuttle developed by the U.S. In the 1990s, similar legal studies attracted attention again when private entities started entering into space business by planning the use of the aircraft-type vehicle for the flight, namely, space tourism. Afterwards, with the increasing number of actors involved in suborbital flights, a need for international regulatory framework "Space Traffic Management (STM)" has been recognized to ensure safety in space transportation.

In the context of a distinction between manned and unmanned suborbital flight, it should be noted that neither Inter-Continental Ballistic Missiles (ICBMs) nor Fractional Orbital Bombardment System (FOBS) are RLV to which space law applies, even though they reach into Low Earth Orbit (LEO) and de-orbit before completing their first full orbit.

THE GENERAL INTERNATIONAL FRAMEWORK APPLICABLE TO SUBORBITAL FLIGHTS

UN Space Treaties

The legal basis for the use of RLV for suborbital flights is found in Article I of the Outer Space Treaty of 1967 (OST) which grants the right for free exploration and use of outer space to any states, even if a state does not ratify the treaty. It stipulates that "[t]he exploration and use of outer space, including the Moon and other celestial bodies, shall be carried out for the benefit and in the interests of all countries, irrespective of their degree of economic or scientific development, and shall be the province of all mankind."[2] In order to implement fundamental principles in OST, the following treaties were also elaborated: the Rescue Agreement of 1968 (RA), the Liability Convention (LC) of 1972, the Registration Convention (RC) of 1975 and the Moon Agreement (MA) of 1979. Those treaties were unanimously adopted in the UN Committee on the Peaceful Uses of Outer Space (UN COPUOS) which is mandated to consider the issues derived from the peaceful uses of outer space. (COPUOS History, 2018)

In order to apply those treaties to the use of RLV for suborbital flights, legal challenges are to understand the definitional problems and procedural aspects of private suborbital flights from the perspectives of state responsibility and international liability.

The Definitional Problems

Suborbital Flights

While the suborbital flights have become feasible and recognized as new model for competitive space transportation, it is not defined in any space treaties. According to unofficial definition of the International Civil Aviation Organization (ICAO), it is "a flight up to a very high altitude which does not involve sending the vehicle into orbit". (Fenema, 2005) At the national law level, South African Space Affairs Act has the definition in the term "launching" which means "the placing or attempted placing of any spacecraft into a sub-orbital trajectory or into outer space". (Von der Dunk, 2011) In addition, some scholars define it as "a flight up to a very high altitude beyond 100 km above sea level but in which the vehicle does not go into orbit" (Jakhu, Sgobba & Dempsey, 2011) or "the launch of an object or objects into outer space without that object or such objects completing one or more orbits around the Earth. (Fenema, 2005) Thus, it is still vague whether or not the suborbital flights reach into outer space. Further clarification needs to be made when any specific regulatory framework for the suborbital flight is to be established.

Outer Space

Outer space is not defined in any space treaties and its definition is still considered in the Legal Subcommittee of UN COPUOS. (UN Doc. A/AC.105/C.2/L.304/Add.2, 2018) Such a lack has not caused any problem before international courts as no case was brought to question the application of space law; (Contant-Jorgenson, Lala & Schrogl, 2006) however, considering the increasing number of spacefaring states as well as launch vehicles for the suborbital flights, the answer needs to be explored.

The definition and delimitation of outer space has been an agenda item of UN COPUOS since 1959 (UN Doc. A/4141, 1959) and the international community has repeatedly tried to clarify the boundary between air space and outer space. In 2018, pursuant to the General Assembly resolution 72/77, the Legal Subcommittee of the UNCOUPOS on 9 April 2018 reconvened its Working Group on the Definition and Delimitation of Outer Space (UN Doc. A/RES/72/77, 2017) with accumulated outcomes of the past studies for the same theme.[3] However, the 2018 report indicates low possibility to reach an agreement over the boundary.

This boundary question is one of the oldest issues in international space law. In the Cold War era, the U.S. and the former Soviet Union were the only spacefaring states and proposed different approaches to define outer space which were: (1) functionalist approach (proposed by the U.S.) and (2) spatialist approach (proposed by the Soviet Union). The former focuses on the goal of activity or function of the vehicle itself, while the latter prefers to establish "a demarcation line between air and outer space". (Gorove, 2000) Gradually the spatialist approach has been explicitly taken in national space legislations (for example, Australia, Denmark and Kazakhstan) that demarcates 100-110 km at altitude from the sea level where an object starts orbiting around the earth.

In the 1990s, scholars explored the third approach "not to define" outer space or the fourth "effective" or "hybrid" approach. (Cheng, 1997) (Christol, 1992) The fourth one focuses on (a) the intended purposes, or (b) the effects of hybrid vehicular activity, or both (a) and (b). However, the fourth approach might cause a further confusion. For example, when the Spaceship One[4] won the first X-Prize, it was operated under a flight permission as a glider from the U.S. Federal Aviation Administration (FAA) as well as a launch license from the FAA Office of Commercial Space Transportation (AST) as a RLV, in another word, spaceplane. (Jakhu, Sgobba & Depsey, 2011)

Furthermore, another approach attempted by UN COPUOS was to fix the line at 100-110 km altitude according "the right of innocent passage" which allows other state to fly through the underlying airspace above foreign territories. (Gorove, 2000) Although it seems appropriate that the right of innocent passage also serves to ensure the right of free access to outer space, it needs to be reviewed in the context of manned spaceflight. (Von der Dunk, 2015)

The reason for such a long debate over the boundary between air space and outer space is found in the limitation of sovereignty in the national airspace. Every state has complete and exclusive sovereignty (the supreme power to govern its own territory) over the airspace above its territory under the Chicago Convention of 1944[5] reinforcing Article 1 of the Paris Convention of 1919. In contrast, no sovereignty is allowed to be claimed by any state in outer space according to Article II of the OST that stipulates "[o]uter space, including the moon and other celestial bodies, is not subject to national appropriation by claim of sovereignty, by means of use or occupation, or by any other means". Therefore, drawing a line between air space and outer space is not necessarily prioritized by all states. The U.S. does not find a need for the demarcation in order to extend its air space activity into outer space activity, while Russia and China emphasize a need to delimitate outer space to ensure safety in space activities under international regulation. Thus, the boundary issue is rather complex as it involves with the matter of state sovereignty.

To avoid any confusion, it needs to be added that such disagreement lies in the need for a line between air space and outer space, not in the existence of outer space itself. For example, if the height is around 120 km above the sea level, all states agree that airspace already ends, and outer space starts in the context of international law.

Space Object

What is the legal status of the aircraft-type RLV used for the suborbital flights? In the space treaties, the term "space object(s)" is defined in Article I(d) of the Liability Convention of 1972 stipulating that "[space object(s)] includes component parts of a space object as well as its launch vehicle and parts thereof." While this expression is quite broad, there is no doubt that any type of space vehicle used for the suborbital flights falls into the definition. However, according to the definition of "aircraft" in the Chicago Convention of 1944, the aircraft-type launch vehicle leaves the possibility of falling into the definition of aircraft that is "[a]ny machine that can derive support in the atmosphere from the reactions of the air other than the reactions of the air against the earth's surface." (ICAO, 2016) As the aircraft-type RLV has wings to depart from and land on airport/spaceport, if a state takes the spatial approach explained, air law might be applied; while the functional approach or the effective approach, space law might be applied. Thus, as long as the legal status of the vehicle is vague, its legal status is at the discretion of operating states.

Then, if the aircraft-type RLV is categorized as aircraft, how the operating state could excise jurisdiction and control (e.g. claim the ownership) over the RLV in outer space? In principle, if the suborbital flights by the RLV are carried out for space mission, the RLV need to be registered in national registry as well as in the UN Register of Space Object under Article VIII of the OST and Article II (3) of the RC of 1975. [6] However, in practice, due that suborbital flights do not orbit around the Earth, their vehicles are not necessarily registered. For example, the U.S. FAA registers its licensed vehicles only in its national registry but not in the UN Register.[7] The main reason is the short duration of suborbital flights departing and landing on the same state. As it is not explicitly required to register space objects in both national registry and the UN Register, the latter contain information submitted by the State of registry, not covering all objects launched into outer space. (Hobe, 2009).

Space Tourists

Different from astronauts, passengers for space tourism, namely, "space tourists", are not defined in space treaties. In Article V of the OST, astronauts are defined as "envoys of humankind"[8] and state parties to the OST are obliged to render all possible assistance in the event of accident, distress, or emergency landing on the territory of another state party or on the high seas.[9] While the same assistance are required to give astronauts' activities in outer space and on celestial bodies,[10] the RA of 1968 uses the terms "the personnel of a spacecraft". In this respect, it is questionable whether space tourists are included in those "astronauts," "personnel of a spacecraft" and "envoy of mankind". (Hobe, Goh & Neumann, 2007).

Furthermore, the state of registry of a space object maintain jurisdiction and control over it as well as its "personnel thereof."[11] Although space treaties lack clear distinction between astronauts and passengers, some distinction is found in the following legal instrument (Von der Dunk, 2011), the 2001 Principles for processes and criteria for selection, assignment, training and certification of International Space Station (ISS) crew under the Intergovernmental Agreement of 1998 ruling the ISS. (IGA, 2001). It has a definition of "ISS crewmembers" which include ISS expedition crew and "visiting crew". Further clarification may be required in near future when more spacefaring states get involved in suborbital flights sending non-astronauts into space.

Thus, the general framework applicable to suborbital flights has the definitional problems in applying of UN space treaties to private suborbital flights, in particular, to the use of aircraft-type RLV for space tourism. Taking it into consideration, the following section explains procedural aspects in the context of state responsibility and international liability related to the suborbital flights.

Procedural Aspects of Private Suborbital Flights

State Responsibility

According to Article VI of the OST, state parties bear international responsibility for national pace activities including both governmental and private missions, and for assuring that national activities are carried out in conformity with obligations under the OST. In order to implement the state responsibility, the activities of private entities need authorization and continuing supervision by the "appropriate" state party. In the case that the activities are to be carried out by an international organization, responsibility for compliance with the OST is to be borne both by the international organization and by the states parties to the treaty participating in such organization.[12]

In terms of private suborbital flights, private entities to obtain authorization from and continuous supervision by the "appropriate" states before starting their business. In general, it is the state where the private entities are registered under national legislation and the appropriate state has the discretion to decide the way how to authorize them. As to the continuing supervision, no specific means for implementation is explicitly required in the OST as it is subject to the discretion of states. The only conditions are that the duration for supervision needs to cover the entire duration of the activity by non-governmental entities; and states have to provide adequate means for receiving information, intervention or deterrence and punishment to ensure the conformity with the provision of the OST. (Hobe, 2009).

For states, one of the best measures for compliance with this obligation is to enact national space legislation with a licensing system for commercial space activities. (Hobe, Goh & Neumann, 2007). For example, the U.S. has the most well-established license system pursued by the FAA that issues license and permit for commercial space transportation. As the enactment of national space legislation for authorization and continuous supervision is directly related to the competitiveness of private entities in international market, it is important to take suborbital flights into consideration when such legislation is to be drafted. As of 2018, 22 states have national space legislation, while not all of them include private space legislation. (UNOOSA, 2016).

Recognizing a need for further development of national space legislation, the following section explains international liability issues related to the use of aircraft-type RLV for private suborbital flights.

International Liability

When any suborbital flight is resulted into a crash on the Earth, or in outer space, which state is liable for that damage? According to Article VII of the OST and Article I (c) of the LC of 1972, states would be liable "if" they are "launching state" that are: the state that launches or procures the launching; or on whose territory or facility an object is launched. As to the damage caused to the surface of the Earth and to aircraft in flight, launching states are absolutely liable for the damage (even no fault is found on their side)[13]. On the other hand, if the damage is caused in outer space, launching states would be liable if the damage is due to its fault.[14]

In the case of damage caused by the aircraft-type RLV for suborbital flights, the definitional question is to be asked again whether or not it is a space object and the next question is whether or not their departure from airport/spaceport are launching activities. If a state takes the functionalist approach, like the U.S., it is clear that the aircraft-type RLV for suborbital flight is a space object because the vehicle is designed to function in outer space. Even if a state takes the spatialist approach in defining the outer space, while it is the discretion of the state that authorizes and continuously supervises the suborbital flights, the state cannot escape from its international liability by insisting that the aircraft-type vehicle is an aircraft. Why? One of the reasons lies in Article 27 of the Vienna Convention on the Law of Treaties of 1969 stipulating that "[a] party may not invoke the provisions of its internal law as justification for its failure to perform a treaty." (VCLT, 1969). Furthermore, considering that the scale of damage caused by the suborbital flight vehicle might be huge, the third-party victim needs to be compensated immediately[15] according to the Preamble of the LC of 1972[16] which aims to ensure "the prompt payment of a full and equitable measure of compensation to victims of such damage".

Thus, recognizing that launching states be liable for the damage under Article VII of the OST and the LC of 1972, transparency in suborbital flight activities should to be enhanced. For example, possible transparency measures could be the registration of the RLV for suborbital flights in the UN Register of Space Objects as well as the notification of the UN Secretary General about suborbital flight operation under Article XI of the OST.

In conclusion, suborbital flights, particularly the use of aircraft-type RLV for private suborbital flights, highlight legal challenges to the existing general framework from the definitional aspects to the procedural aspects. Recognizing the lack of explicit definition on key terms related to the activities, the following section focuses international regulatory framework being explored to ensure safety in space transportation.

DEVELOPMENT OF INTERNATIONAL REGULATORY FRAMEWORK: STM

The present chapter starts with explaining two types of suborbital flights: departing from and landing on the same point; or flying between different points like civil aviation. While most of present space missions fit in the former, the latter is potential for future space business, enabling to transport carriage and passengers between New York and Tokyo in 2-3 hours. Such a business model by the aircraft-type RLV has attracted attention in the civil aviation industry since the 1990s. (Collins, Iwasaki, Kanayama & Ohnuki, 1994). For this reason, in 1995, the U.S. transferred the Office of Commercial Space Transportation (AST) from the Department of Transportation (DOT) to FAA for licensing private entities to start space launch activities.

On the other hand, a need for international regulatory framework for space transportation was recognized when suborbital flights became feasible. Space transportation consists of the launch and return phases that need safety in the transit and re-entry of airspace by avoiding possible interference with aircraft in flight. As to the administrative aspects, civil aviation are regulated by national law under national air traffic management agencies, as well as under international specialized agency, namely, International Civil Aviation Organization (ICAO) or regional organizations such as European Aviation Safety Agency (EASA) for aviation safety and Eurocontrol for air navigation. (Chicago Convention, 1944) (Fenema, 2015). Thus, the rise of space transportation requires further coordination between the existing framework for civil aviation, Air Traffic Management (ATM)[17], and the entry of suborbital flights into space transportation industry needs STM.

Recognizing that suborbital flights need to be pursued under international regulatory framework, the present chapter focuses international efforts to establish the framework for space transportation and explains: an overview of STM and challenges to suborbital flights in the Low Earth Orbit (LEO) environment.

Overview of STM

STM is a concept to provide a framework for the safety, stability and sustainability of space activities in the future, including technical as well as regulatory elements. (Schrogl, Jorgenson, Robinson & Soucek, 2018). It aims to provide appropriate means for conducting space activities without harmful interference, supporting the universal freedom to use outer space under the OST. (Contant-Jorgneson, Lala, & Schrogl, 2006).

As to the academic research aspects, the term appeared in the 1980s and the first concept was formed through two workshops organized by the American Institute of Aeronautics and Astronautics (AIAA) in 1999 and 2001. (Lubos, 1982). Further study was taken over by the International Academy of Astronautics (IAA)from 2001 to 2006 which released a report "the IAA Cosmic Study on STM". (Aoki, 2009) (Schrogl, 2008). According to the report, the definition of STM is "the set of technical and regulatory provisions for promoting safe access into outer space, operations in outer space and return from outer space to Earth free from physical or radio-frequency interference". (Contant-Jorgneson, Lala, & Schrogl, 2006). This definition is taken over by the latest IAA publication of 2018, which provides comprehensive study on the STM. (Schrogl, Jorgenson, Robinson & Soucek, 2018).

In order to achieve the goal of the STM, a need for various measures from the technical perspective have been recognized including space debris mitigation measures, frequencies and slot management, the pre-launch notification system, and concrete space traffic rules[18]. Furthermore, the followings were also identified as important STM components which are: knowledge of the orbital environment; standard formats for data exchange; best practices; ownership and operating characteristics of spacecraft; and space traffic control service. The list below provides further explanation. (Ailor, 2015).

- Knowledge of the orbital environment: information including weather (on Earth and in outer space) to alert satellite operators of natural events threatening normal operations (e.g. the U.S. Space Surveillance Network (SSN), Russia's International Scientific Optical Network (ISON));
- Standard formats for data exchange: international standards for exchange of data (e.g. ISO 26900:2012 (CCSDS 502.0-B-2) CCSDS Orbit Data Messages);
- Best practices: rules-of-the-road practice;
- Ownership and operating characteristic of space flight: the catalogues of tracked objects serve to identify who is responsible for the tracked space objects;
- Space traffic control service: a service that combines data from multiple sources to develop accurate estimates of where are all tracked objects.

Those STM measures and components are still increasing due to an upcoming trend with the issue of space debris. After the space debris issues entered into the scope of STM in the context of ensuring sustainable uses of outer space, public awareness of a need for international regulatory framework, STM, has been enhanced rapidly and a particular emphasis is put on the integral framework with Space Situational Awareness (SSA) which catalogue space debris and objects in outer space. An example is the U.S. Space Policy Directive 3 (SPD-3) which tasked the Department of Commerce with establishing a civil system for SSA and STM in order to ensure safety, stability and operational sustainability in space activities. (Space Policy Directive-3, 2018) (US Office for Space Commerce, 2018).

Taking it into consideration, the following section introduces challenges in terms of the use of LEO.

Challenges to Suborbital Flights in the LEO Environment

It is doubtless that LEO has increasingly become congested and contested more than other Earth orbits. As suborbital flights are pursued in air space and outer space, the present section outlines the existing challenges to suborbital flights in LEO which are: coordination between Air Traffic Management (ATM) and STM; and space debris management in LEO.

Coordination Between ATM and STM

When aircraft are operational in national airspace, they need to follow the rules under Standards and Recommended Practices (SARPs) of ICAO.[19] In other words, SARPs apply to all flights in national airspace through national implementing legislation.[20] Though military aircrafts are not subject to ICAO regulations[21], coordination between military and civil air traffic control agencies is taken by the same "safety first" standard[22].

In order to create a single regulatory regime for STM where suborbital flights enter into its scope, ICAO needs to coordinate the existing regulatory rules. In particular, as the Chicago Convention of 1944 applies to the international air navigation departing from and landing at the difference places, namely transportation, it needs further clarification whether or not the space transportation by suborbital flights would be subject to international air law. (Fenema, 2005). In this respect, ICAO's involvement is indispensable to explore the possibility of integrating and coordinating the existing regulatory regime related to ATM and STM for future suborbital flights.

Space Debris Management in LEO

The use of LEO by private entities accelerates space debris issues which has become a real threat to the LEO environment. According to the Inter-Agency Space Debris Coordination Committee (IADC), presenting at the Scientific and Technical Subcommittee of the UNCOPUOS in 2018, (IADC, 2018), the implementation level of the IADC Space Debris Mitigation Guidelines in LEO is considered as insufficient and no apparent trend towards a better implementation is observed", compared with Geostationary Orbit (GSO).

According to the UCS Satellite database, the total number of identifiable satellites in LEO is 1,186. Within this context of an already crowded environment, recent plans for Large Satellite Constellations (LSC) highlight the risks of space debris collision in LEO. (UCS Satellite Database, 2018) The U.S. Federal Communication Commission (FCC) accepted applications for the approval of LSC-based frequencies from private entities such as SpaceX, OneWeb, Telesat, O3b Networks, Theia Holdings in March 2017. As of April 2018, 12 entities have applied and 4 of them obtained the approval for the use of LEO frequencies by LSC with the massive number of LSC satellites which are: (1) OneWeb - June 2017 - 720 satellites, (2) Telesat - November 2017 - 117 satellites, (3) SpaceX - March 2018 - 4435 satellites, (4) O3b - May 2018 - additional 26 satellites, and (5) Karousel LLC - August 2018 - 12 satellites.[23] In total FAA approved the use of 5462 satellites for LSC to the 5 private entities; however, the feasibility of such an intensive use of LEO is still in question.

Facing the high possibility of continuous debris collision in LEO, Active Debris Removal (ADR) has been explored as effective measures in the situation that "the debris environment in LEO has reached a point where the debris populations will continue to increase even if all future launches are suspended". (Ailor, 2015). Targets of ADR are dysfunctional satellites and orbital launch stages in LEO that would be the sources of further debris objects in the event of a collision. (Ailor, 2015). From the legal perspective, the use of ADR needs a consent of states who holds the ownership of space objects even they are dysfunctional already in order not impede the right of any state for free exploration and use of outer space in Article I of the OST of 1967.

Thus, the concept of STM is not only for a safety in space transportation but for the sustainable use of the LEO environment. In terms of building international regulatory framework, complexity lies in: the limited number of spacefaring states who have the capabilities to identify, monitor and catalogue space debris as well as space objects deployed in outer space; the entry of private entities into new business with LSC; and the technical difficulty to mitigate space debris by ADR.

In conclusion of the present chapter, STM needs notification system and data exchange standards; framework for handling and protecting registered space object data; addressing liability related to ADR; and space traffic control service provider with international agreement. (Ailor, 2015). In terms of legal instruments effective for STM are not restricted to the existing space treaties which are flexible but abstract, the introduction of non-binding instruments such as UN Space Debris Mitigation Guidelines and Draft International Code of Conduct for Outer Space Activities of 2014 serves to be part of STM. Considering that the establishment of STM system needs further coordination among the existing regulatory framework under ICAO, International Telecommunication Union (ITU), and other regional or national administrations, one of practical proposal is to establish an international regulatory organization specialized in STM, which is similar to ICAO but more concentrated on Space Situational Awareness (SSA) gathering and sharing the catalogued data of space debris.

CONCLUSION

The more suborbital flights enter into international market for transportation services, the more legal aspects need to be studied to avoid any legal confusion or conflict in pursuing the flights. In particular, the use of aircraft-type RLV for the flights highlights challenges to the existing legal instruments particularly in air law and space law.

Taking it into consideration, Chapter I first categorized suborbital flights from the legal perspective into two types, "public or private" and "plane-type or rocket type" in order to clarify that the use of aircraft-type RLV for private suborbital flights raises more legal questions than the use of rocket-type RLV for public suborbital flights. After the clarification, Chapter II introduced the general international framework applicable to suborbital flights. Starting with the UN space treaties, the chapter mainly focused the definitional problems particularly related to the use of aircraft-type RLV. The last section outlined the procedural aspect private suborbital flights in the context of state responsibility and liability. Lastly, Chapter III linked the development of international regulatory framework, namely, STM and suborbital flights. Tracing the accumulation of academic as well as institutional efforts for building STM, the chapter provided an overview of STM and identified the existing challenges to the suborbital flights in the LEO environment which are: a need for coordination between ATM and STM; a need for space debris management in LEO. In particular, an emphasis was put on the pressing situation that LSC use a massive number of satellites and increase risks for continuous collision in LEO after their deployment.

While legal studies of suborbital flights have mainly focused the definitional problems between air law and space law, attention needs to be paid to the expanding scope of STM which accommodate space debris issues including the deployment of LSC in LEO. Recognizing a need for international regulatory framework for STM, further coordination is necessary in the existing regulatory instruments, binding or non-binding, national or international, in order to ensure a safety in the suborbital flights as well as sustainability in space activities. A key approach is to involve private entities in the STM framework in order to ensure their implementation of STM rules.

REFERENCES

Agreement among the Government of Canada, Governments of Member States of the European Space Agency, the Government of Japan, the Government of the Russian Federation, and the Government of the United States of America concerning Cooperation on the Civil International Space Station (hereafter Intergovernmental Agreement), Washington, done 29 January 1998, entered into force 27 March 2001; Space Law e Basic Legal Documents, D.II.4.

Agreement Governing the Activities of States on the Moon and Other Celestial Bodies, 5 December 1979, 1363 *UNTS* 3 / 18 *ILM* 1434.

Agreement on the Rescue of Astronauts, the Return of Astronauts and the Return of Objects Launched into Outer Space, 22 April 1968, 19 UST 7570, 672 UNTS 119.

Ailor, W. (2015). Space Traffic Management. In Handbook of Space Security. New York: Springer Science+Business Media. doi:10.1007/978-1-4614-2029-3_58

Aoki, S. (2009). Space Traffic Management for the Prevention of Weaponization of Outer Space. In *Proceedings of the International Institute of Space Law*. American Institute of Aeronautics and Astronautics.

Cheng, B. (1997). *Studies in International Space Law*. Oxford, UK: Clarendon Press. doi:10.1093/acprof:oso/9780198257301.001.0001

Christol, C. Q. (1992). Legal Aspects of Aerospace Planes. In *International Conference on the Law, Policy, and Commerce of International Air Transport and Space Activities. The Highways of air and outer space over Asia*. Dordrecht: Martinus Nijhoff Publishers.

Collins, P., Iwasaki, Y., Kanayama, H., & Ohnuki, M. (1994). Commercial Implications of Market Research on Space Tourism. *The Journal of Space Technology and Science.*, *10*(2), 3–11.

Contant-Jorgenson, C., Lala, P., & Schrogl, K. U. (2006). *Cosmic Study on Space Traffic Management*. Paris: International Academy of Astronautics. doi:10.1016/j.spacepol.2006.08.004

Contant-Jorgneson, C., Lala, P., & Schrogl, K.-U. (2006). The IAA Cosmic Study on Space Traffic Management. *Space Policy*, *22*(4), 283–288. doi:10.1016/j.spacepol.2006.08.004

Convention on International Civil Aviation (Chicago Convention), Chicago, 7 December 1944; 15 UNTS 295; TIAS 1591; 61 Stat. 1180; Cmd. 6614; UKTS 1953 No. 8; ATS 1957 No. 5; ICAO Doc. 7300; Annex 2, Rules of Air, Chapter I Definitions.

Convention on International Liability for Damage Caused by Space Objects, 29 March 1972, 961 *UNTS* 187.

COPUOS History. (n.d.). *The website of United Nations Office for Outer Space Affairs*. Available at: http://www.unoosa.org/oosa/en/ourwork/copuos/history.html

UN Doc. A/4141 of 14 July 1959, Part III, III.A, "Report of the Ad Hoc Committee on the Peaceful Uses of Outer Space".

UN Doc. A/RES/72/77, "International cooperation in the peaceful uses of outer space", 14 December 2017.

UN Doc. A/AC.105/C.2/2018/DEF/L.1, "Draft report, Annex II, Report of the Chair of the Working Group on the Definition and Delimitation of Outer Space", 13 April 2018.

UN Doc. A/AC.105/C.2/L.304/Add.2, "Draft report: V. Matters relating to the definition and delimitation of outer space and the character and utilization of the geostationary orbit, including consideration of ways and means to ensure the rational and equitable use of the geostationary orbit without prejudice to the role of the International Telecommunication Union", 13 April 2018.

Doc, U. S. Space Policy Directive-3, National Space Traffic Management Policy, issued on 10 June 2018. (n.d.). Available at: https://www.whitehouse.gov/presidential-actions/space-policy-directive-3-national-space-traffic-management-policy/

EUROCONTROL. (n.d.). *What is air traffic management?* Available from: http://www.eurocontrol.int/articles/what-air-traffic-management

Freeland, S. (2010). Fly Me to the Moon: How Will International Law Cope with Commercial Space Tourism? *Melbourne Journal of International Law*, 11, 90.

Gorove, K.M. (2000). Delimitation of Outer Space and the Aerospace Object – Where is the Law? *Journal of Space Law*, *28*, 11-16.

Hobe, S. (Ed.). (2009). *Cologne Commentary on Space Law: Volume 1. Outer Space Treaty*. Koln, Germany: Carl Heymanns Verlag.

Hobe, S., & Cloppenburg, J. (2004). Towards a New Aerospace Convention? Selected Legal Issues of "Space Tourism". *Proceedings of the Forty-Seventh Colloquium on the Law of Outer Space*, 377.

Hobe, S., Goh, G., & Neumann, J. (Eds.). (2007). Space Tourism Activities-Emerging Challenges to Air and Space Law. *J. Space L.*, *33*, 366.

International Civil Aviation Organization. Convention on Civil Aviation, 7 December 1944, 15 *UNTS*. 295. (n.d.). Available at: http://www.refworld.org/docid/3ddca0dd4.html

Jakhu, R. S., Sgobba, T., & Dempsey, P. S. (Eds.). (2011). The Need for an Integrated Regulatory Regime for Aviation and Space: ICAO for Space? Studies in Space Policy Vol. 7. Wien: Springer. doi:10.1007/978-3-7091-0718-8

Launius, R. D., & Jenkins, D. R. (2006). Is it finally Time for Space Tourism? *Astropolitics*, *4*(3), 253–280. doi:10.1080/14777620601039693

Lubos, P. (1982). Traffic Rules for Outer Space. In *Proceedings of the International Institute of Space Law*. American Institute of Aeronautics and Astronautics.

Schrogl, K.-U. (2008). Space traffic management: The new comprehensive approach for regulating the use of outer space—Results from the 2006 IAA cosmic study. *Acta Astronautica*, *62*(2-3), 272–276. doi:10.1016/j.actaastro.2007.09.001

Schrogl, K.-U., Hays, P. L., Robinson, J., Moura, D., & Giannopapa, C. (Eds.). (2015). Handbook of Space Security. New York: Springer Science+Business Media. doi:10.1007/978-1-4614-2029-3

Schrogl, K.-U., Jorgenson, C., Robinson, J., & Soucek, A. (Eds.). (2018). *Space Traffic Management – Towards a Roadmap for Implementation*. International Academy of Astronautics.

Sundahl, M. (2009). The Duty to Rescue Space Tourists and Return Private Spacecraft. *Journal of Space Law*, *35*, 163-200.

Takaya-Umehara, Y. (2014). State Responsibility and Liability for Air-Launch over the High Seas. In *Proceedings of the fifty-sixth Colloquium on the Law of Outer Space*. Eleven International Publishing.

Treaty on Principles Governing the Activities of States in the Exploration and Use of Outer Space, Including the Moon and Other Celestial Bodies, 27 January 1967, 610 *UNTS* 205.

Tronchetti, F. (2011). Regulating Sub-Orbital Flights Traffic: Using Air Traffic Control as a Model. *Proc. Int'l Inst. Space L.*, *54*, 176.

UN Doc. Technical Presentations made at the Scientific and Technical Subcommittee, "The Inter-Agency Space Debris Coordination Committee (IADC) -An overview of the IADC's annual activities". (n.d.). Available at: http://www.unoosa.org/documents/pdf/copuos/stsc/2018/tech-04E.pdf

UN Office for Outer Space Affairs. (n.d.). *National Space Law Collection*. Available from: http://www.unoosa.org/oosa/en/ourwork/spacelaw/nationalspacelaw/index.html.

Union of Concerned Scientists. (2018). *UCS Satellite Database*. Author.

U.S. Office for Space Commerce. (2018). *STM Remarks from EUSpace for Business Conference.* Available at: https://www.space.commerce.gov/stm-remarks-from-euspace-for-business-conference/

Van Fenema, H. P. (2005). Suborbital Flights and ICAO. *Air and Space Law*, *30*, 405.

Van Fenema, H. P. (2015). Legal aspects of launch services and space transportation. In *Handbook of Space Law* (p. 408). Edward Elgar. doi:10.4337/9781781000366.00016

Vienna Convention on the Law of Treaties, opened for signature on 23 May 1969, 1155 *UNTS* 331.

Von der Dunk, F. G. (2007). Passing the Buck to Rogers: International Liability Issues in Private Spaceflight. *Nebraska Law Review*, *86*(2).

Von der Dunk, F. G. (2011). Space tourism, private spaceflight and the law: Key aspects. *Space Policy*, *27*(3), 146–152. doi:10.1016/j.spacepol.2011.04.015

Von der Dunk, F. G. (2015). *Legal aspects of private manned spaceflight. In Handbook of Space Law* (p. 669). Cheltenham, UK: Edgar Elgar Publishing.

Von Der Dunk, F. G., & Tronchetti, F. (Eds.). (2015). *Handbook of Space Law*. Cheltenham, UK: Edward Elgar Publisher. doi:10.4337/9781781000366

ENDNOTES

1 Including orbital spaceflight, legal analysis by Prof. Frans G. von der Dunk categorized five types of private spaceflight which are: orbital space tourism, sub-orbital space tourism, sub-orbital private spaceflight, hotels in orbit, and private flights to the Moon.

2 *See*, Art. I, Outer Space Treaty.

3 *See*, UN Doc. A/AC.105/C.2/2018/DEF/L.1, "Draft report, Annex II, Report of the Chair of the Working Group on the Definition and Delimitation of Outer Space", 13 April 2018. The working group considered the issue through the accumulated outcomes in the past as follows: (a) national legislation and practice relating to the definition and delimitation of outer space (A/AC.105/865/Add.20 and A/AC.105/865/Add.21); (b) questions on suborbital flights for scientific missions and/or for human transportation (A/AC.105/1039/Add.10 and A/AC.105/1039/Add.11); (c) note entitled "Definition and delimitation of outer space: views of States members and permanent observers of the Committee (A/AC.105/1112/Add.4 and A/AC.105/1112/Add.5); (d) working paper prepared by the Chair of the Working Group on the Definition and Delimitation of Outer Space entitled "Promoting the discussion of the matters relating to the definition and delimitation of outer space with a view to elaborating a common position of States members of the Committee on the Peaceful Uses of Outer Space" (A/AC.105/C.2/L.302); (e) working paper submitted by the Russian Federation entitled "The challenging context of considering all aspects of the delimitation of airspace and outer space: arguments for adding dialectical elements to, and setting new analytical trends in, discussion of the issue" (A/AC.105/C.2/L.306); (f) conference room paper entitled "Suborbital flights and the delimitation of airspace vis-à-vis outer space: functionalism, spatialism and State sovereignty", submitted by the Space Safety Law and Regulation Committee of the International Association for the Advancement of Space Safety (IAASS) (A/AC.105/C.2/2018/CRP.9).

4 Spaceship One is an experimental air-launched suborbital spaceplane using a hybrid rocket propulsion

5 Airspace is defined as the airspace over "the land areas and territorial waters adjacent thereto under the sovereignty, suzerainty, protection or mandate of such State". Article 1 and 2, Convention on International Civil Aviation (hereafter Chicago Convention), 7 December 1944, 15 UNTS 295; TIAS 1591; 41 Stat. 1180; Cmd. 6614; UNTS 1953 No. 8; ATS 1957 No. 5; ICAO Doc. 7300.

6 To understand the importance of registering the aircraft-type launch vehicle in the UN Registry, *see*, Takaya-Umehara, Y. (2014). State Responsibility and Liability for Air-Launch over the High Seas. *Proceedings of the fifty-sixth Colloquium on the Law of Outer Space*. Eleven International Publishing. p. 669-679.

7 Based on the interview with Dr. George C. Nield, Associate Administrator, FAA, at International Symposium on Ensuring Stable Use of Outer Space, Tokyo, Japan, 3-4 March 2016.

8 Article V (1) of the Outer Space Treaty of 1967.

9 Article V (2) (3) of the Outer Space Treaty of 1967.

10 Article V (3) of the Outer Space Treaty of 1967.

11 Article VIII of the Outer Space Treaty of 1967.

12 Article VI of the OST.

13 Article II the Liability Convention of 1972.

14 Article III the Liability Convention of 1972.

15 Article IV of the Liability Convention of 1972.

16 The 4th paragraph in the Preamble of the Liability Convention of 1972

17 Air Traffic Management primarily consists of three distinct activities: Air Traffic Control; Air Traffic Flow Management, and Aeronautical Information Services. For further information.

18 In 2007, a set of rules for STM was elaborated through a project during the summer session of International Space University.

19 SARPs are stipulated in Annexes to the Chicago Convention. Convention on International Civil Aviation, 7 Dec. 1944, 15 UNTS 295 ; TIAS 1591.

20 Article 12 of the Chicago Convention.

21 Article 3(a) (b), Chicago Convention.

22 For example, Article 3 distinct state aircraft (military) and civil aircraft; however, it obliges state aircraft to operate with due regard for civil and commercial air traffic for the safety of civil and commercial air traffic. *See*, Van Fenema, H. P. (2015). Legal aspects of launch services and space transportation.

23 The other companies that have applied for the FCC approval include: (6) Kepler - 2-140 satellites, (7) LeoSat - 78 satellites, (8) Boeing - 1396-2956 satellites, and (9) Theia - 112 satellites.

Chapter 6
The International Space Station: Technical Aspects

Chris Nie
Lockheed Martin, USA

ABSTRACT

A new era of spaceflight dawned following the conclusion of the United States and Russian space race. This new era has been marked by the design, assembly, and operation of one of the greatest engineering feats mankind has accomplished, the International Space Station (ISS). The ISS is comprised of hundreds of thousands of kilograms of material built on the ground and transported to space for assembly. It houses an artificial atmosphere to sustain life in outer space and has been continually inhabited for over 15 years. This chapter describes the technical complexity of the ISS, the background of how it was assembled, its major systems, details of crew life onboard, commercial usage of the resource, and examples of mishaps that have occurred during the ISS's operation. The technical details of the ISS provide a glimpse into what future space stations that might orbit the Moon and Mars will resemble.

INTRODUCTION

The year is 1984, humans have walked on the moon, the space shuttle has been flying for three years, and the Russians are beginning to design a space station to house humans in low earth orbit, and the United States President Ronald Reagan announces an attempt for the National Aeronautics and Space Administration (NASA) to build Space Station Freedom, a U.S. space station in low earth orbit. Nearly a decade later a vote in the U.S. House of Representatives to end the still U.S.-based program fails to pass by just a single vote, 216-215 in favour of continuing the development of the space station. In the two decades it took to design and develop the space station, international partners were brought on to help support the technical, political, and financial growth of the program. In January of 1998 a series of agreements were signed to form the framework of what is now known as the International Space Station, or ISS. The first agreements included representatives from NASA, the European Space Agency (ESA), the Canadian Space Agency (CSA), and the Russian Space Agency. Considering that the collapse of the Soviet Union occurred earlier in the same decade, the political significance of these countries working

DOI: 10.4018/978-1-5225-7256-5.ch006

together was celebrated at the time and continues to be the framework to which the ISS is governed over 20 years later.

The ISS is humankind's first long-term foothold in space. It is not since November 1st, 2000, that every human had a foot on the planet Earth. The ISS has been continuously inhabited since November 2, 2000 with varying crews made up of astronauts from around the world totalling over 200 people from 18 different countries as of May 2016 (Garcia, 2016). As a coordinated effort of over multiple decades by NASA (U.S. space agency), Roscosmos (Russian space agency), ESA (European Space Agency, representing 22 European member states), CSA (Canadian Space Agency), and JAXA (Japanese Space Agency), the ISS has become a model for future space exploration to Mars and beyond (Center for the Advancement of Science in Space, 2015; *NASA's Journey to Mars: Pioneering Next Steps in Space Exploration*, 2015; "New Member States," n.d.). Its mission is to conduct scientific research in the unique environment of low-Earth orbit and act as a stepping stone for future human spaceflight exploration (Garcia, 2016).

This chapter will offer an overview on the essentials of the ISS design and operation, and of the legal framework governing the ISS activities.

BACKGROUND

The ISS currently orbits Earth at approximately 400 kilometres altitude, traveling at more than 7.5 kilometres a second, completing a full orbit around the globe every 90 minutes. It has been continuously inhabited by humans since November of 2000 and has been visited by more than 230 people from 18 countries as of November 2018. Taking up an area of 51 by 109 meters and having a mass of 420,000 kilograms, it can nearly cover a football pitch and is more massive than 320 automobiles. It contains 388 cubic meters of liveable volume and produces an average of 84 kilowatts for crews that vary between 3 and 6 normally, with a record of 13 for a short duration in July of 2009 (Moskowitz, 2009).

The orbiting station has been built to be a stepping stone to humanity's exploration of our solar system. Its primary purpose is to serve as a laboratory to study what happens to humans, plants, and technology that undergo long duration spaceflight. To accomplish this mission, the ISS was built over the course of 15 years with subsystems designed to keep humans and experiments happy and healthy while on-board.

The United States Congress designated the ISS a U.S. National Laboratory in 2005 in order to enable space research and development and increase access to conduct experiments on the ISS for commercial, academic, and government users. This designation led to the development of the Center for the Advancement of Science in Space, or CASIS, that manages the U.S. National Laboratory on the ISS. This effort, in addition to the experiments flown prior to CASIS and by other countries, has enabled over 2,400 experiments or investigations as of September 2017 to be conducted on-board, covering disciplines including: biology and biotechnology, earth and space science, educational and cultural activities, human research, physical science, and technology development and demonstration (National Aeronautics and Space Administration, 2018).

The subsystems critical to the ISS's operation are power and thermal control, communications, and environmental control and life support. The power and thermal control subsystem uses a football field of over 30,000 solar cells to create up to 31 kW of power and after taking in all this energy from our sun, keep the inside habitat a nice 20°C. The communications subsystem uses several parts of the electromagnetic spectrum to transfer voice, video, and scientific and engineering data between the ground and space. The environmental control and life support system uses several machines to create oxygen, reduce carbon dioxide, maintain humidity, and remove any contaminants in the air to keep all living things (not just humans) happy and healthy during their stay in space. All these systems working together allow the astronauts to perform research that creates the understanding humanity needs to reach deeper into space.

THE INTERNATIONAL SPACE STATION

Assembly of the International Space Station

The in-space assembly of the space station began with the launch of the Russian Zarya (Russian for "Sunrise") module in November of 1998 and continued for more than 10 years, 190 spacewalks, and over 30 missions by NASA, Roscosmos, ESA, CSA, and JAXA (Center for the Advancement of Science in Space, 2015). All major modules of the design were assembled in orbit by 2011, although module additions and replacements are still ongoing today. One of the greatest engineering feats of the ISS assembly is the integration of the modules in-space, most of which touched for the first time in-orbit, without being fit-checked on the ground. The assembly consisted of launching each segment individually on either a Russian Proton rocket, or the Space Transportation System (STS) orbiter, commonly known as the Space Shuttle. Once in orbit, a combination of remotely controlled robotic arms and Extra-vehicular Activity (EVAs, or spacewalks) were used to put the module in its designed position (W. J. Larson & Pranke, 2000).

Figure 1 displays the International Space Station's completed design, highlighting the contributions by each partner state, and Table 1 gives a brief summary of the modules used for living and research. Section 18 describes the utilization of each module.

NASA was largely involved in system engineering and integration, the establishment of safety requirements and plans and the planning for and coordination of the execution of the overall integrated operation of the Space Station. The segments NASA built include the truss, radiator, and solar arrays, as well as the pressurized modules Unity, Harmony, the Quest Airlock, and the mating adapter between the U.S. and Russian segments. The Russian segment consists of Zarya, the first module of the ISS, Zvezda, the Mini-Research Module, and the Multi-Purpose Laboratory Module. The ESA provided the Destiny, Tranquillity, and Columbus modules, and JAXA contributed Kibo (the Japanese Experiment Module, or JEM), the JEM Experiment Logistics Module, and the JEM Exposed Facility.

Several modules of the ISS provide a unique access to space. These modules are the Cupola, airlocks (Quest and JEM), and the robotic arm (the Mobile Servicing System, or MSS).

The Cupola is a panoramic window bay used as a control station during robotic operations (Wright, 2013a). It is the largest window ever built for space, constructed of four layers of fused silica and borosilicate glass that help maintain the pressure vessel of the module and protect against micrometeorite damage (European Space Agency, 2013). The location has become very popular by crew to take breathtaking photographs of Earth that help study Earth's surface.

The Quest airlock is the main exit and entry for crew at the beginning and end of spacewalks. It houses umbilical gear that provides power and life support system connections to the spacesuits when the crew is preparing for spacewalks. To prepare for a spacewalk, the crew reduces the amount of nitrogen in their body to reduce the risk of the formation of nitrogen bubbles in body tissues and is required due to the change in pressure the crew's body experiences between the 14.7 psia of the pressurized modules and the 4.3 psia of the U.S. spacesuit or the 5.8 psia of the Russian Orlan spacesuit (Eckart, 1996; National

Figure 1. ISS complete configuration showing each module and the partner responsible for its design and operation
Source: National Aeronautics and Space Administration, 2010

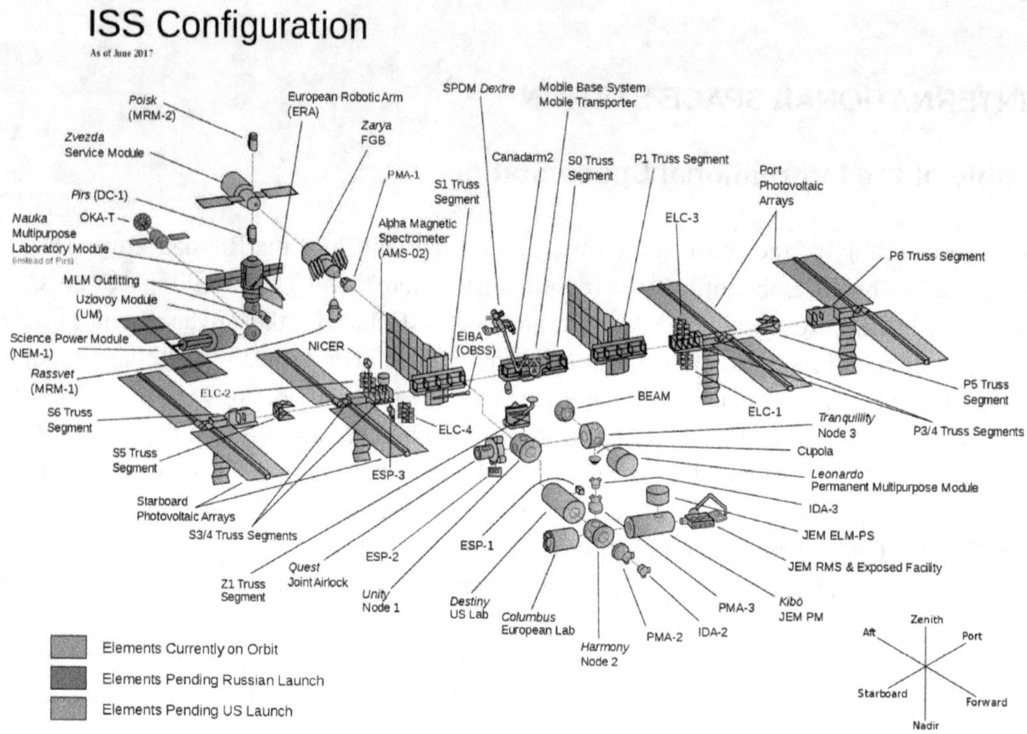

Table 1. Habitable modules of the ISS listed in order of launch date

Module Name	Agency	Launch	Description
Zarya (FCB)	FKA	Nov., 1998	First element, houses propulsion and power
Unity (Node 1)	NASA	Dec., 1998	Connects the U.S. and Russian segments
Zvezda	FKA	Jul., 2000	Living quarters, life support, and power
Destiny (U.S. Lab)	NASA	Feb., 2001	Primary research lab for U.S. payloads.
Quest (Airlock)	NASA	Jul., 2001	Primary spacewalk entry and departure
Harmony (Node 2)	NASA	Oct., 2007	Connects Columbus, Kibo, and docking port
Columbus	ESA	Feb., 2008	The primary research lab for ESA payloads
Kibo (JEM)	JAXA	May., 2008	The primary research lab for JAXA payloads
Poisk (MRM2)	FKA	Nov., 2009	Mini-Research Module 2, airlock and storage
Cupola	ESA	Feb., 2010	Seven-window observation module
Tranquillity (Node 3)	NASA	Feb., 2010	Hosts the Cupola and a docking port
Rassvet (MRM1)	FKA	May., 2010	Mini-Research Module 1, cargo and research
Leonardo (PMM)	ESA	Feb., 2011	Multipurpose Module to house payloads
BEAM	NASA	Apr., 2016	An experimental expandable habitat module

(NASA, 2010)

Aeronautics and Space Administration, 2010). They do this through an activity called "prebreathing" in the airlock, which consists of inhaling pure oxygen. Once the crew has their spacesuits on, the airlock depressurizes. Once the EVA is complete, the crew returns into the airlock and it re-pressurizes by releasing air and nitrogen into the airlock from oxygen and nitrogen canisters stored on-board. The ability to vary the amount of pressure held within the module is what allows astronauts to enter and exit the ISS without losing the standard atmosphere environment inside (Wright, 2013c).

The Japanese Experiment Module (JEM), nicknamed Kibo (Japanese for "Hope"), includes a scientific airlock that has a Remote Manipulator System (RMS) robotic arm on the outside to move experiments between the internal and external environments of the ISS (Wright, 2013b). This smaller airlock has been used to install deep space and Earth observation payloads onto the External Facility (EF) such as SEDA-AP (Space Environment Data Acquisition equipment-attached Payload) to measure Z-ray sources in space, and SMILES (Superconducting Submillimeter-wave Limb-emission Sounder) to observe trace gases in Earth's stratosphere. The JEM-EF is also used to launch small satellites into low-Earth orbit as shown in Figure 4.

Arguably the workhorse of the ISS, the MSS is the robotic manipulator that enabled the complex construction of the modules and the berthing of visiting vehicles (Kauderer, 2013). The MSS is a 7-degree-of-freedom (can translate and/or rotate in 7 different ways) manipulation system with latches on both ends that is 17.6 meters long when fully extended, and can move up to 116,000 kilograms of mass. It is broken up into three parts: the Space Station Remote Manipulator System (SSRMS, also known as the Canadarm2), the Mobile Base System (MBS), and the Special Purpose Dexterous Manipulator (SPDM, or Dextre). The SSRMS is the physical arm that is used to move supplies, equipment, crew on EVA, and free-flying spacecraft that are to dock with the ISS. The MBS is a platform that can be used during EVA to move supplies on a tracking system mounted to the truss. The SPDM is fully equipped with lights, video, and tool holders and is used for precise movement of equipment and an aide to crew during spacewalks as a robotic helper (National Aeronautics and Space Administration, 2010).

International Space Station Mission Architecture

This chapter will focus heavily on the space portion of the International Space Station, but this orbiting laboratory in the sky would have not been able to exist if not for the infrastructure created to support it from the ground and transportation to and from the ISS. The following section give a brief background of the ground architecture and transportation that has and will be used to supply and service the ISS.

Ground Architecture

An extensive network of ground antennas and operators is needed in order to communicate and operate the ISS. The Communications and Tracking (C&T) system provides this network and provides the ability for the hundreds of engineers and operators in mission control rooms around the world to communicate with the ISS and its crew on-board. In addition to directly communicating with the crew, this architecture allows payload science data to be downlinked to the ground and transmission with vehicles the need to rendezvous and dock with the station (National Aeronautics and Space Administration, 2015). The fastest connection currently set up to communicate from the ground to the station is 300 megabits per second (Mbps), faster than most home high-speed internet connections today (Jacobson & Hume, 2016). This data is communicated through space to a ground „node" and then relayed to the proper control center,

NASA's Mission Control Center (MCC) in Houston, Texas for example. Although the ISS is always in view of Earth in Low Earth Orbit (LEO), there are Loss of Signal (LOS) scenarios that happy on a day-to-day basis due to bandwidth constraints on the relay network. To reduce these bandwidth constraints space agencies are working to develop laser communication technology that would allow for point-to-point communications at higher speeds (Lafrance, 2015).

Transportation

Space agencies around the world have contributed to building the ISS. In addition to supplying the design and construction of the physical modules that comprise the station, many nations and companies have contributed to the transportation that is necessary to get and from the orbiting laboratory. Table 2, below, highlights the vehicles that have been used in the construction and operation of the ISS.

Subsystems

To support the crew and equipment on board, the ISS collects power from the sun, uses a system of satellites to communicate with Earth, and produces and reuses the environment necessary for the crew to stay alive.

Power and Thermal Subsystem

The ISS collects sunlight with its eight photovoltaic solar arrays that are 12 meters wide and extend 35 meters in length, producing the 84 kilowatts the station needs to power life support systems and experiments (Wright, 2013d). The solar arrays are built of a blanket-like material that was extended by a deployable truss structure once installed on orbit. Through the use of two gimbal mechanisms that rotate the solar arrays around the truss structure, like spokes on a bicycle wheel, the solar arrays are capable of constantly facing the sun. This enables the ISS to generate more power when it is not in the

Table 2. Vehicles used to construct and operate the ISS

Vehicle	Agency/Org.	Launch Vehicle	Crewed/Uncrewed	Dates Used
Space Shuttle	NASA	STS[1]	Crewed	Dec. 2000 – July 2011
Soyuz	FKA	Soyuz	Crewed	Oct. 2000 – ongoing
Progress	FKA	Soyuz	Uncrewed	Nov. 2000 – ongoing
Dragon	SpaceX	Falcon 9	Uncrewed	May 2012 – ongoing
Cygnus	Orbital ATK[2]	Antares and Atlas-V	Uncrewed	Sept. 2013 – ongoing
HTV[3]	JAXA	H-IIB	Uncrewed	Sept. 2009 – Jan. 2017
ATV[4]	ESA	Ariane 5	Uncrewed	Mar. 2008 – Feb. 2015

[1]Space Transportation System
[2]Acquired by Northrop Grumman in June 2018
[3]H-II Transfer Vehicle
[4]Automated Transfer Vehicle
(Garcia, 2018)

shadow of Earth (which occurs for 45 minutes during every 1.5-hour orbit). The ISS runs off of nickel-hydrogen batteries during the times when the Sun is blocked by Earth. The heat that is created due to the Sun's radiation, the reflection of sunlight of Earth's surface, and the power generated on board must be dissipated to maintain temperatures bearable by the crew and components on board. The thermal subsystem uses passive thermal control, insulation, surface coatings, heaters, and heat pipes to manage this heat. The insulation and surface coatings reduce the amount of heat coming in from the external environment, the heaters keep the crew and electrical components at an operational temperature, and the heat pipes and passive thermal control transfer the heat from areas that have too much to areas that have to little. The heat pipes that are internal to ISS are filled with water, and the external heat pipes are filled with ammonia. Both ammonia and water have high specific heat capacities. While the capacity of ammonia is higher, it is a toxic substance and having it on the external coolant loop creates a safer environment for the crew. The pipes transfer heat to the radiators, which through their large surface area radiate heat to space.

Communications Subsystem

The antennas are the main communication link between the ISS and ground stations. They transfer voice, telemetry, and command and control data to the station (Kauderer, 2013). The ISS transfers data (like voice, images, and science data) to the ground through a system called the Tracking and Data Relay Satellite System (TDRSS). TDRSS is a platform of satellites that are used by NASA and the U.S. government to provide communications for satellites and aircraft and operates in geosynchronous orbit (Wertz & Larson, 1999). The communications with ISS are not constant, because TDRSS is a shared asset with other satellites. It is common to lose the capability to communicate with the ISS for brief (approximately 30 minutes) periods throughout the day during normal operations. Although there is a time delay of about

Figure 2. ISS Solar panels deployed, capturing the Sun's energy and converting it to electrical power
Source: National Aeronautics and Space Administration, 2010

three seconds between ground and the station, this has not been found to significantly alter the way in which the crew and ground converse. Ultra-High Frequency (UHF) communications is used for close proximity operations with visiting vehicles and Extra-vehicular Activities (EVAs).

Environmental Control and Life Support Subsystem

A network of systems to provide a breathable atmosphere, water, food, waste, and hygiene to the crew is used to keep them alive and healthy during their mission. This system is called the Environmental Life Control and Support System (ECLSS) and is made up of two main components: the water recovery systems and the atmospheric generation and revitalization system.

The Water Recovery System (WRS) reclaims approximately 93 percent of wastewater using vacuum distillation and is broken down into the Urine Processor Assembly (UPA) and Water Processor Assembly (WPA), which recycle urine and used water into purified water and water to be used in the other ECLSS components ("International Space Station: Environmental Control and Life Support System," 2008). All human waste that cannot be recycled through these processes is packaged as trash and put into cargo vehicles that burn up upon re-entering Earth's atmosphere.

Figure 3. ISS Communication system architecture, showing how the ISS communicates to the ground using TDRSS, and how it communicates with visiting vehicles
Source: National Aeronautics and Space Administration, 2010

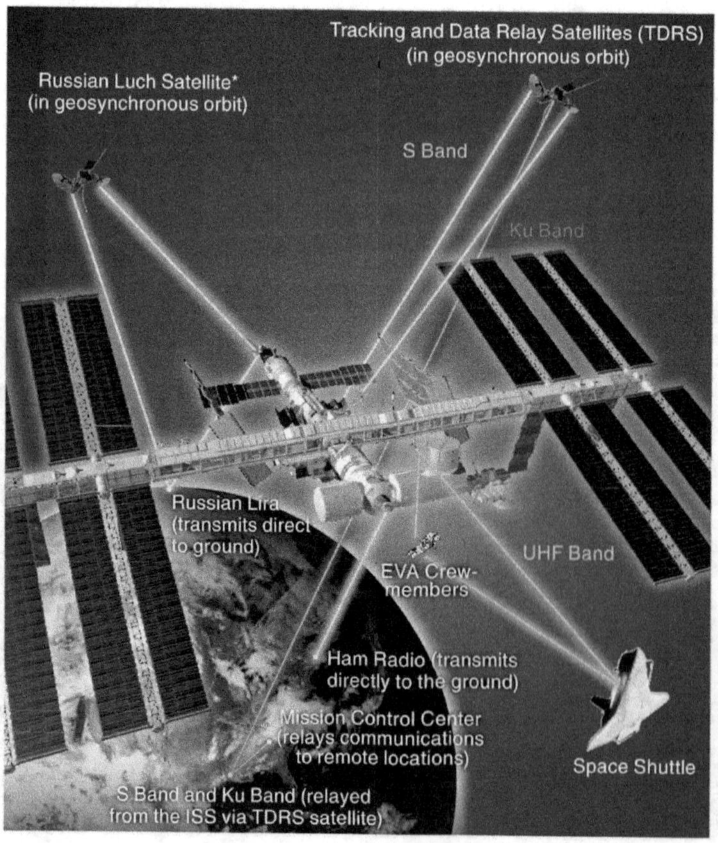

There are two oxygen-producing systems on the ISS, the United States version is called the Oxygen Generation System (OGS) and the Russian version is named the Elektron. Both produce oxygen through electrolysis, the process of splitting water molecules into hydrogen and oxygen molecules by applying a direct current to the water. The Russian system includes a backup that is a solid-fuel oxygen generator (SFOG), where lithium perchlorate can be ignited to rapidly produce oxygen in an high-energy chemical reaction (European Space Agency, 2010).

Both the U.S. and Russian segments have systems to remove carbon dioxide from the air for all of ISS. The U.S. system is the Carbon Dioxide Removal Assembly (CDRA) and the Russian system is the Vozduk. Both systems remove humidity from the processed air first, then use an absorbent bed (typically zeolite, an aluminosilicate) to remove the carbon dioxide. A backup carbon dioxide removal system uses lithium-hydroxide canisters; Lithium hydroxide reacts with carbon dioxide and scrubs the carbon dioxide out of the air, leaving lithium carbonate and water. Although used in the Space Shuttle program, these canisters are typically very heavy and require replacement. To scrub other dangerous particles from the air, activated charcoal beds are used to remove odours and trace contaminants.

Crew Activities

During nominal operations, the International Space Station operates with a crew of six that usually includes at least one NASA astronaut and one Russian cosmonaut, as well as representatives from CSA, ESA, and JAXA. The number of representatives from each nation is chosen on a mission to mission basis based on the Memorandum of Agreement between NASA and CSA, NASA and ESA, NASA and the Government of Japan and NASA and Roscosmos and through the management bodies created under the IGC and MOUs (*Intergovernmental Agreement (IGA) on Space Station Cooperation*, 1998). Astronauts from other countries fly based upon agreements with one of the space agencies. These crews rotate in groups of three about every six months based on available launches. The longest stay on ISS was completed in March of 2016 by astronaut Scott Kelly and cosmonaut Mikhail Korniyenko, who both lived and worked on the ISS for 342 days before coming back to Earth (National Aeronautics and Space Administration, 2016). The daily tasks are split between the crew, and include maintaining the subsystems of the ISS, conducting scientific research, exercising, educational outreach, eating, sleeping, and personal time (May, 2015). The crew schedule is based on Greenwich Mean Time (GMT) and although the ISS passes into Earth's shadow every 45 minutes, lights on board are left on during GMT day hours, and crew sleeping quarters are unlit during GMT night (Harrison, 2001).

With the completion of the major modules of the ISS, the crew can spend more of their time on conducting scientific research. Experiments in biology, Earth and space sciences, human research, and physical sciences are conducted in addition to technology demonstrations and educational activities (Rainey, 2016). There are hundreds of experiments that have been completed. Experiments have ranged from growing protein crystals, vaccine development, studying cosmic radiation, studying eyesight degradation in space, launching small satellites, and how to make a coffee cup work in microgravity. Many of the experiments occur in the lab modules where equipment such as cryogenic freezers have been modified to work in space in order to store samples and replicate labs here on Earth.

Commercial Use

As the ISS reached completion, the desire to fully utilize it as a resource for humanity grew and continues to grow today. The member states of the ISS have increased opportunities for researchers around the world to use the ISS as a lab in an effort to increase the accessibility of the resources the ISS provides. Many private ventures face obstacles and difficulties to utilize the station for science due to complex technical interface and strict safety requirements. Once an experiment reaches the ISS practical problems, like equipment failures on board of the ISS can negatively impact experimental results (Walter, 2012). The state member nations are working with the commercial industry to utilize the ISS, but many users are non-space companies and researchers that are not familiar with the development of space hardware (Tkatchova, 2011). Furthermore, there are high market risks for ISS on-board products and services.

The commercial projects that have successfully flown on ISS range from the testing of osteoporosis medical devices to protein crystal growth. The crystal growth project might lead to new medicine and treatments that could be worth billions in the future (Dinerman, 2008). The space agencies have made progress towards maximising the private utilization of the ISS in recent years (National Aeronautics and Space Administration, 2013). This growth has been helped by both universities and corporations that provide experiments and projects and is expected to continue in the future, for example, for NASA with SPACEHAB, Zero G Corporation, and the University of Colorado (Dinerman, 2008).

Experiments conducted onboard the space station require an immense amount of creativity and engineering skill. Scientists from the Stanford University School of Medicine are currently conducting an experiment that will study the effects of microgravity on cardiac stem cells or cardiomyocytes. These cells have been seeded in three six-well plates that must be carefully temperature controlled and immersed in a 5% carbon dioxide environment at all times. These environmental requirements are necessary in order to mimic the conditions of the human body and provide a unique challenge to those who are responsible for designing the experiment and ensuring its successful completion.

The crew member on station will remove all three of the six-well plates from an on-orbit incubator and will perform four to five media exchanges every eight days over a thirty-day period. The growth media that the cells are immersed in will provide essential nutrients needed for them to continue to develop and mature over the duration of the experiment.

In addition to media changes, the crew member will also perform microscopy of one of the six-well plates every eight days using a standard laboratory microscope that has been modified for use in space. The crew member will take several video images of each well and will relay these video images back to the science team on the ground. Imaging of the six-well plate will ensure that all cellular development is progressing as expected and will allow the scientists to make real-time observations. Ultimately, they will be looking for alterations in cellular structure, function and gene expression.

A few days prior to experiment completion, the crew member will preserve one of the six-well plates and will perform one last media change on the other two live plates prior to them being sent back to Earth. Ultimately, this experiment will provide further insight into the cardiovascular effects of a microgravity environment on humans, which is necessary to take into consideration when planning future human exploration missions to Mars.

In addition to the heart cells experiment, scientists from the University of Minnesota are currently flying an experiment on the space station that will study the long-term effects of bone loss in a microgravity environment. Osteoporosis is a degenerative disease that often affects those who are bedridden for long periods of time. New drug therapies and physical therapy techniques are currently under

Figure 4. Astronaut Terri Virts working exchanging a media exchange during a stem cell experiment
Image credit: NASA

investigation and bone stem cell research could provide insight into how cellular changes occur under reduced gravitational loading.

Twenty-four single-well plates have been flown frozen to ISS until the crew thaws them and performs a media exchange. All of the cell plates will then be placed in an on-orbit incubator that will precisely control temperature and carbon dioxide levels for the mission duration. Ultimately, scientists will be looking for changes in cell differentiation and gene expression.

Previous experiments aboard ISS have included studies of bacterial resistance to antibiotics, drug metabolism pathways and even the ability for spiders and other living organisms to adapt to the microgravity environment. All of these experiments provide powerful insight into problems that are experienced on Earth as well as those that will inevitably occur during human missions beyond low-earth orbit.

Space Mishaps

The technical design and implementation of the ISS has not been without issues and challenges. Any great engineering feat has unknown unknowns, and the space station is no exception. The following three examples highlight a few technical mishaps that have occurred in the space station's lifetime and highlight the extensive capability of operators and engineers on the ground to drive towards the solution.

Solar Array Deployment Tear

As mentioned in the Subsystems section, the solar arrays are a crucial aspect of the space station's power source. They take sunlight from millions of kilometres away and convert it into up to 120 kilowatts of electrical energy to be used on-board (Wright, 2013d). The solar arrays were launched in a folded configuration due to volume constraints for the Space Shuttle then unfolded (or deployed) on orbit after being integrated to the station. During the deployment of one of the solar arrays the material substrate holding the photovoltaic cells tore (Mosher, 2007). The engineers and operators at Johnson Space Center planned an emergency EVA in less than two days (relative to the normal several months' timeframe) to plan a solution to the issue. It was then up to astronaut Scott Parazynski to snip off the frayed wires that had ripped the solar array and install five straps to reinforce the damaged area (Stein, 2007). The EVA lasted over seven hours and saved the delay of installing the next module to grow the station, the European laboratory.

Ammonia Leak During EVA

Ammonia is the extremely toxic fluid used in the ISS coolant loops to transport heat from internal components on-board to the space station's radiators that, in turn, radiate that heat to deep space (National Aeronautics and Space Administration, 2010). Ammonia, although toxic, is selected as the fluid of choice due to its high capacity for storing heat and low freezing point. In a 2001 spacewalk, astronaut Capt. Robert Curbeam Jr. was conducting an EVA to rewire the electrical system of the space station to draw power from a new set of solar arrays when Curbeam and astronaut Thomas Jones saw a "silvery, fluttering stream of snowflakes jetting out" from the space station (Schwartz, 2006). A defective valve released five percent of the ammonia coolant supply to space (Georgakas, 2012). This escaping ammonia froze onto Curbeam's spacesuit as he closed the leak using a device built into the valve in the case it began to leak. It was then up to the team of operators in mission control to decide how to get both astronauts back inside without also bringing in the toxic ammonia that now coating one of the spacesuits. Through quick thinking and planning, the mission control team was able to give Curbeam direction to position himself while spacewalking in sunlight during the space station's orbit, causing the ammonia to be in direct contact with sunlight, warm up, and evaporate off of his space suit (Georgakas, 2012). Due to the extensive training of the astronauts and the quick thinking of the support team, a potential life-threatening toxic leak contaminating the crew habitat was averted.

Space Suit Water Leak

Space suits have been critical in the development of the ISS. The astronauts that travelled to the station to help assembly its pieces frequently donned space suits and performed Extra-vehicular Activities (EVAs) to build and maintain the modules, components, and consumables only accessible externally. On July 16, 2013 an incident occurred 44 minutes into an EVA where a crew member reported water inside his helmet and both he and the ground team supporting the crew member were unable to determine the source of the fluid leakage (*International Space Station (ISS) EVA Suit Water Intrusion High Visibility Close Call*, 2013). The EVA continued, but the water continued leaking into the astronaut's helmet and eventually began to pool around his face. This event caused the immediate end of the EVA and the crew member was brought back inside the station's airlock and his suit was removed in an expedited fashion. Water leaking onto the face or neck of someone on Earth is not usually perceived as a life-threatening scenario, but due to the microgravity environment and closed surroundings of the spacesuit, the water pooled around the eyes, nose, and mouth of the crew member. Water tension plays an higher relative role of water movement in microgravity because gravity cannot counteract the tension has on a surface, meaning that water "sticks" to surfaces far more in space (Motil & Singh, 2004). Due to the physics of water in microgravity, the water pooling around the EVA crew members face lead to partial blockage of the breathing pathway and a potential life-threatening scenario. In this instance, the protocols of the operations team were able to safely route the astronaut back into the space station and to safety, but this situation represents how something simple on Earth can be much more technically complex when encountered in space.

SOLUTIONS AND RECOMMENDATIONS

Throughout its lifetime the ISS has highlighted the technical challenges that may be overcome through international coordination. The technical achievement of transporting 420,000 kilograms of mass into Earth's orbit, assembling components built from different parts of the globe in space, and creating an artificial atmosphere for humans to live and work safely is a paramount achievement of humankind. As denoted in the Space Mishaps section, there are challenges yet to face. When moving forward it is important to continue the international collaboration and focus on solutions that will benefit the goal of traveling farther out into the solar system, to the Moon and Mars.

FUTURE RESEARCH DIRECTIONS

The ISS is the first long duration human space habitat built in history and will continue to be a guiding light for persistent human presence in space. The ISS provides the equipment and facilities to support advanced research for both Earth-based and space-based scientific questions, in addition to serving as a developmental platform for future technologies.

As the utilization of the ISS increases for scientific research, experiments will continue to answer some of science's largest questions. Examples of ongoing and future experiments that are or will fly on the ISS include: musculoskeletal research performed on rodents, protein crystal growth, DNA sequencing, and human psychological ramifications during long-duration spaceflight. Technological demonstrations will include small satellite deployments, additive manufacturing in microgravity, and advanced life support system (International Space Exploration Coordination Group (ISECG), 2018).

The ISS has taught humankind many things about spaceflight in both the technical system development as well as the technical architectural approach for space station design. Many commercial companies are using the lessons learned by the ISS and its members to develop future private space stations. With international focus shifting to the Moon and Mars, the design, assembly, and operation of a large Low Earth Orbit (LEO) space station is not in the plans (International Space Exploration Coordination Group (ISECG), 2018). Programs like commercial cargo and commercial crew have paved a route for a new technical architecture approach to replenish supplies and send crew to the ISS. This approach is being applied the design of new space stations in the implementation of companies like Bigelow Aerospace and Axiom Space.

The technical lessons learned from the ISS will be applied to the design and manufacturing of the Deep Space Gateway, the next planned international space station which will orbit the Moon (International Space Exploration Coordination Group (ISECG), 2018). This space station will be significantly smaller that the ISS and will differ in a major technical paradigm. The Deep Space Gateway, or DSG, will be "crew tended" as opposed to "crew inhabited" like the ISS. The technical design of the systems will be required to have much larger mean time between failure (MTBF) rates and reduced maintenance needs because there will not always be a human there to troubleshoot and correct issues. An additional major technical difference is the distance DSG will be in space relative to the Earth. Whereas the ISS orbits Earth in Low Earth Orbit (LEO) approximately 400 km in altitude, the DSG will orbit in Lunar vicinity approximately 400,000 km from Earth.

CONCLUSION

After more than 15 years of continuous habitation, the International Space Station is now the first stepping stone in the world's exploration of the final frontier. New astronauts and experiments continue to arrive at the orbiting research facility. Since the beginning of the assembly, systems have continuously been upgraded to improve efficiency and performance, and new technologies have continuously been tested. For example, Bigelow Aerospace is testing their new kind of pressurized space habitat denoted in Table 1, one that is inflatable to decrease launch volume. Furthermore, new exercise techniques and eating regimens have decreased the rate at which astronauts lose bone and muscle tissue while in microgravity. And lastly, long-standing scientific questions, like the nature of dark matter, are starting to be better investigated and understood through experiments like the Alpha Magnetic Spectrometer (AMS-02). These examples are just a few of many that show the evolution of the space station's capabilities and future trends of what we can expect to see in future space exploration missions.

Given the continued use of the ISS throughout more than two decades, the end of lifetime of the ISS is coming closer. At one point the ISS will have to be terminated, just like any other satellite or space station. What the exact date of termination will be is yet unclear. The original date was set to be 2016, but all State Parties have agreed to an extension first through 2020 and, with a December 2016 affirmative decision by the ESA Council of Ministers, now through 2024. There is a possibility of a third "life extension" until 2028 (Malik, 2014; Nimon, 2014).

Although plans for decommissioning are not complete, the end of ISS's mission will be a spectacular show because its modules will re-enter the Earth's atmosphere and burn up upon re-entry. In the future, international partners are targeting missions to the Moon and, eventually, to Mars (International Space Exploration Coordination Group (ISECG), 2018). The lessons that have been and will be learned on the ISS will likely prove to be crucial in the coming decades and humankind's exploration of space. The systems that will be needed to venture farther into space include human health and behavioural research, high-bandwidth communication systems, bio-regenerative life support systems, advanced in-space manufacturing techniques, on-orbit servicing, and in-situ resource utilization technologies. These advancements will lead to the ultimate goal of planetary settlements all having used the ISS as a proving ground (*NASA's Journey to Mars: Pioneering Next Steps in Space Exploration*, 2015).

The International Space Station is not only one of the largest technical endeavours created by mankind in outer space. It is also an outstanding legal achievement. Since this very complex project consists of an international partnership between 15 Member States and their five space agencies, reciprocal cooperation had to be fostered and many topics had to be legally regulated. The following chapter will give an overview of how rights and obligations between the parties were codified, to ensure the safe conduct and use of mankind's first permanent home in space.

REFERENCES

Center for the Advancement of Science in Space. (2015). *History and Timeline of the ISS*. Retrieved January 12, 2015, from http://www.iss-casis.org/About/ISSTimeline.aspx

Dinerman, T. (2008). *Business on the ISS beyond space tourism*. The Space Review.

Eckart, P. (1996). *Spaceflight Support and Biospherics*. Microcosm Press, Kluwer Academic Publishers Incorporated.

European Space Agency. (2010). *Environment Control and Life Support System*. ECLSS.

European Space Agency. (2013). *Cupola*. Retrieved January 12, 2015, from http://www.esa.int/Our_Activities/Human_Spaceflight/International_Space_Station/Cupola

Garcia, M. (2016). *About the Space Station: Facts and Figures*. Retrieved January 1, 2016, from https://www.nasa.gov/mission_pages/station/main/onthestation/facts_and_figures.html

Garcia, M. (2018). *Visiting Vehicle Launches, Arrivals and Departures*. Retrieved from https://www.nasa.gov/feature/visiting-vehicle-launches-arrivals-and-departures

Georgakas, S. (2012). *Contamination Risk During EVA*. *Space Safety Magazine*.

Harrison, A. A. (2001). *Spacefaring The Human Dimension*. Berkeley, CA: University of California Press.

Intergovernmental Agreement (IGA) on Space Station Cooperation (1998).

International Space Exploration Coordination Group (ISECG). (2018). *The Global Exploration Roadmap*. Author.

International Space Station. (2008). *Environmental Control and Life Support System*. National Aeronautic and Space Administration.

International Space Station (ISS) EVA Suit Water Intrusion High Visibility Close Call. (2013).

Jacobson, A., & Hume, A. (2016). *NASA Communications Network to Double Space Station Data Rates*. Retrieved from https://www.nasa.gov/feature/goddard/2016/nasa-communications-network-to-double-space-station-data-rates

Kauderer, A. (2013). *Canadarm2 and the Mobile Servicing System*. Retrieved January 12, 2015, from http://www.nasa.gov/mission_pages/station/structure/elements/mss.html

Lafrance, A. (2015). The Internet in Space? Slow as Dial-Up. *Atlantic (Boston, Mass.)*.

Larson, W. J., & Pranke, L. K. (2000). *Human Spaceflight: Mission Analysis and Design* (W. J. Larson & L. K. Pranke, Eds.). McGraw-Hill Higher Education.

Malik, T. (2014). *International Space Station gets Life Extension through 2024*. Space.com.

May, S. (2015). *Working in Space*. Retrieved January 12, 2015, from https://www.nasa.gov/audience/foreducators/stem-on-station/ditl_working

Mosher, D. (2007). *Space Station Surgery: Astronauts Mend Torn Solar Wing*. Space.com.

Moskowitz, C. (2009). *Space Station Population Hits Record High.* Retrieved January 12, 2015, from http://www.space.com/7003-space-station-population-hits-record-high.html

Motil, B., & Singh, B. (2004). NASA's microgravity fluid physics strategic research roadmap. *42nd AIAA Aerospace Sciences Meeting and Exhibit*, 123.

NASA's Journey to Mars: Pioneering Next Steps in Space Exploration. (2015).

National Aeronautics and Space Administration. (2010). *Reference Guide to the International Space Station.* Washington, DC: NASA.

National Aeronautics and Space Administration. (2013). *Audit Report of the Office of Inspector General, July 2013, NASA's efforts to maximise research on the international space station.* NASA.

National Aeronautics and Space Administration. (2015). *ISS Reference Guide.* Author.

National Aeronautics and Space Administration. (2016). *Year Long Mission.* Retrieved June 11, 2016, from https://www.nasa.gov/1ym

National Aeronautics and Space Administration. (2018). *International Space Station Utilization Statistics.* Retrieved November 11, 2018, from https://www.nasa.gov/sites/default/files/atoms/files/iss_utilization_statistics_brochure_expeditions_0-52_final.pdf

New Member States. (n.d.).

Nimon, J. (2014). *Space Station 2024 Extension Expands Economic and Research Horizons.* Retrieved from https://www.nasa.gov/mission_pages/station/research/news/2024extension/

Rainey, K. (2016). *Space Station Research Experiments.* Retrieved January 1, 2016, from http://www.nasa.gov/mission_pages/station/research/experiments_category/index

Schwartz, J. (2006). A Toxic Leak Haunts the Shuttle Crew. *The New York Times.*

Stein, R. (2007). Space Station's Damaged Panel Is Fixed. *Washington Post.*

Tkatchova, S. (2011). *Space-Based Technologies and Commercialized Development: Economic Implications and Benefits.* Academic Press.

Walter, E. (2012). The privatisation and commercialisation of Outer Space. In C. Brünner & A. Soucek (Eds.), *Outer Space in Society, Politics and Law* (p. 500). Springer Verlag.

Wertz, J. R., & Larson, W. J. (Eds.). (1999). Space Mission Analysis and Design (3rd ed.). Microcosm Press, Kluwer Academic Publishers Incorporated.

Wright, J. (2013a). *Cupola Observational Module.* Retrieved January 12, 2015, from https://www.nasa.gov/mission_pages/station/structure/elements/cupola.html#.Vu7oYeIrKCq

Wright, J. (2013b). *Kibo Laboratory.* Retrieved January 12, 2015, from https://www.nasa.gov/mission_pages/station/structure/elements/jem.html#.Vu7ocOIrKCq

Wright, J. (2013c). *Quest Airlock.* Retrieved January 12, 2015, from https://www.nasa.gov/mission_pages/station/structure/elements/quest.html#.Vu7oaeIrKCp

Wright, J. (2013d). *Solar Arrays*. Retrieved January 12, 2015, from https://www.nasa.gov/mission_pages/station/structure/elements/solar_arrays.html#.Vu7omOIrKCp

ADDITIONAL READING

Canadian Space Agency. (n.d.). Canadian Space Agency (CSA). Retrieved from http://www.asc-csa.gc.ca/eng/iss/

Dick, S. J., & Launius, R. D. (2007). *Societal Impact of Spaceflight*. NASA History Office.

European Space Agency. (n.d.). European Space Agency (ESA). Retrieved from http://www.esa.int/esaHS/iss.html

Japan Aerospace Exploration Agency. (n.d.). Japan Aerospace Exploration Agency (JAXA). Retrieved from http://iss.jaxa.jp/en/

National Aeronautics and Space Administration. (2015). *ISS Reference Guide*.

National Aeronautics and Space Administration. (n.d.-a). International Space Station. Retrieved from www.nasa.gov/station%0D

National Aeronautics and Space Administration. (n.d.-b). Station Science. Retrieved from www.nasa.gov/iss-science

Russian Federal Space Agency. (n.d.). *Russian Federal Space Agency (Roscosmos)*.

KEY TERMS AND DEFINITIONS

Center for the Advancement of Science in Space (CASIS): Manager of the International Space Station (ISS) U.S. National Laboratory with the goal to ensure the ISS is fully utilized to benefit life on Earth.

Environmental Control and Life Support System: The collection of components designed and built to measure and maintain the necessary habitat to support life in human spaceflight.

Human Spaceflight: Space travel with crew or passengers aboard the spacecraft.

International Space Station: A habitable artificial satellite built by member nations that conducts testing for future space technologies and microgravity research for Earth-based science.

Journey to Mars: Plan that outlines goals in the bipartisan NASA Authorization Act of 2010 and in the U.S. National Space Policy, also issued in 2010 to develop the capabilities needed to send humans to Mars in the 2030s.

Life Science: A branch of science that focuses on the study of organisms. Research in this area occurs in space to determine the response of organisms in the microgravity environment.

Microgravity: The condition in which people or objects appear to be weightless.

Space Mishaps: Accidents or incidents that have occurred during human spaceflight.

Chapter 7
The International Space Station: Legal Aspects

Sarah Katharina Germann
Austrian Federal Ministry for Europe, Integration and Foreign Affairs, Austria

ABSTRACT

The International Space Station is certainly one of the most astounding achievements of humankind in space. Especially from a legal point of view, the creation of the Intergovernmental Agreement (IGA) and its sub-instruments specifically for the ISS was a major success: the IGA was the first grand-scale multi-national legally binding space-related treaty drafted, ratified, and implemented by the major space faring nations since the drafting of the five UN Space Treaties roughly 20 years before. And still today, the legal framework of the ISS is a stand-alone legal system which can serve as model for other missions, as it refines and develops in an innovative way the rules laid out in the five UN Space Treaties and at the same time manages to coordinate and organize management, utilization, and financing between all the partners. This chapter intends to shed light on the complex legal system governing the ISS and to point out the novelties of the IGA-Structure in comparison with the conventional body of international space law.

INTRODUCTION

All views, thoughts, and opinions expressed belong solely to the author, and do not necessarily reflect the official policy or position of the Austrian Foreign Ministry.

All activities in outer space are governed by a legal regime. This applies to the placement of satellites, launching of space vehicles, conducting manned missions as well as building and operating space stations. The International Space Station (ISS), like all other space activities, is governed by public international law and international space law. However, there is an additional set of rules applicable to the ISS, which was created solely for this unique project and milestone in international cooperation in space.

The legal core piece, which was specifically drafted for the ISS mission is the Intergovernmental Agreement (IGA), a treaty under international law.1 The ISS started out as a NASA-project at the end of the 1970s, to which the US under President Reagan invited interested nations to join, in order to set up a permanently manned Space Station (Farand, 1998). Accordingly, an agreement was then negoti-

DOI: 10.4018/978-1-5225-7256-5.ch007

ated between the governments of the United States, Canada, Japan, and eleven European countries and signed in January 1988 (1988 IGA-Agreement). As a result of the political and structural collapse of the Soviet Union and the end of the Cold War, Russia was formally invited to join the partnership in 1993. The agreement was thus renegotiated into the so-called '1998 version' (1998 IGA-Agreement, hereinafter: IGA) which entered into force in 2001 (Von der Dunk, 2009, 149-150). The parties to the IGA are Canada, the "European Partner" (Belgium, Denmark, France, Germany, Italy, the Netherlands, Norway, Spain, Sweden, Switzerland, the United Kingdom), Japan, Russia and the United States. This version of the agreement is still in force today. Even though the original partners started negotiating the legal structure already in 1988, the technical in-space construction of the ISS did not start until 1998.

BACKGROUND

As already mentioned, for a complex mission such as the ISS, there are multiple sets of legal and regulatory instruments applicable. Primarily, as the ISS is operated in outer space, an area beyond national jurisdiction where multiple sovereign nations interact with each other, public international law is applicable. This includes the more specific framework of international space law, relevant to activities in outer space. The IGA refers to these instruments *expressis verbis* (Article 5, 1998 IGA-Agreement). On a second layer, the IGA sets up a three-tier framework in itself for the ISS: The IGA features a number of substantial rules and was specifically tailored to cover the basic legal structure of the ISS. Additionally, it refers to two levels of subsidiary instruments, namely the memoranda of understanding (MOUs) and the implementing arrangements (IAs) (Article 4/2, 1998 IGA-Agreement). The IGA stipulates that these instruments shall provide more details where the IGA provides for the basics, but more precise rules on the implementation are necessary (Article 4/2, 1998 IGA-Agreement) (for details on MOUs and IAs see chapter below). On a third layer, finally, there is the respective national law of the partners to the mission, which can play an important role on their respective modules and in other regards.

Legal Framework Applicable to the ISS

Public International Law and International Space Law

Article 1 of the IGA relegates to the whole corpus of public international law (Article 1, 1998 IGA-Agreement). This means that all sources and norms of international law (see Art. 38 of the ICJ-Statute) are to be applied to the ISS and its missions. These include international treaties (e.g. the UN Charter), customary international law (e.g. the prohibition of force) and general principles of international law (such as the principle *pacta sunt servanda,* which means that legally binding agreements must be performed). Following the rule of *lex specialis derogat legi generali* (meaning that a more specific rule precedes the more general rule), only in cases where more special and specific norms are available, such as the IGA, these rules precede general public international law in regard to the ISS.

International space law, as a subcategory of international law provides for the legal principles concerning space activities and is laid out in the UN Space Treaties[2]. Those principles include, *inter alia,* the principle of free use and exploration of outer space (Article I, Outer Space Treaty 1967, hereinafter OST), the principle of peaceful use of outer space (Preamble and Article IV, OST) and the principle of due regard to corresponding interests of other states in outer space (Article IX OST). The UN Space

Treaties do not only set the basic framework but also the legal boundaries for space related agreements between states, such as the IGA.

At this point it has to be noted that public international law, of which the outer space treaties are a part, would be binding to the parties and apply to the ISS in any case, even without specific references in Article 1 of the IGA. This is, as all parties to the IGA are also parties to the UN-Charter, hence the ICJ-Statute and most other major treaties of public international law, as well as the main UN Space Treaties, and are subject to other sources of international law, such as customary international law and the general principles of international law. Still, those references are often used and are meant to show recognition of the hierarchy of sources in all the applicable legal instruments. In this tradition, also the OST stipulates that all activities in the exploration and use of outer space are to be carried out in accordance with international Law, including the UN-Charter (Article III, OST).

The question might arise, why the need for a special set of rules for the ISS, if general rules apply which regulate all space activities. The answer is simply that not all aspects which needed to be regulated for the specific nature and functioning of the ISS-project were sufficiently regulated in the necessary detail by the then existing framework. In regard to the main UN Space Treaties it has to be mentioned that their rules are rather broad and general. Furthermore, they have been negotiated and drafted between the 1960s and 80s, have not been amended since and until today no major multilateral space treaty followed. As space flight and technological possibilities have evolved tremendously since then, there are certain aspects in new space endeavours that may not be addressed sufficiently by the UN Space Treaties as they were not envisioned by the time of their drafting. This creates the situation that specific space activities are not regulated sufficiently, or that it is unclear how existing international space law applies.

With the IGA, the drafters had the possibility to enhance certain existing aspects and to clarify legal uncertainties regarding the operation of the ISS. However, as state parties to the UN Space Treaties, the drafters at the same time had to respect the already existing principles. As a result, only such areas were regulated, which were not fully covered by the existing law and where existing principles could be refined (Sharpe & Tronchetti, 2015, 625). As a result, some unique solutions have been found for legal issues ranging from liability to criminal jurisdiction, which are all in accordance with existing space law and public international law.

The IGA

As stated before, the IGA establishes basic goals, rights, and obligations for the partners. It is a treaty under international law such as the UN Space Treaties. However, it is binding for a smaller number of states in comparison to the main UN Space Treaties (only to the Partners to the ISS as the IGAs state parties) and is applicable only to the ISS and its missions. As mentioned before, the IGA presents clarifications to existing legal uncertainties and controversies and creates additional solutions for legal problems that had not existed prior to the ISS. The IGA includes procedural norms (e.g. amendments, withdrawal), technical norms (e.g. utilization, evolution, operation), but also typical provisions dealing with issues like liability, criminal jurisdiction and intellectual property. It furthermore builds the legal foundation for the MOUs and the IAs and therefore acts as a base (often also regarded as "over-arching umbrella") for the multi-layered structure.

MOUs

At the second level there are four MOUs established between the partners. The MOUs were signed in 1998 and are in effect today. All the MOUs were drafted, negotiated and signed by NASA on the one hand and another partner space agency on the other (see e.g. MOU NASA/JAXA, 1998), to cover cooperation regarding a specific field relevant to the ISS mission. This means the MOUs are not multilateral (i.e. signed by more than two parties, e.g. all of the partners like the IGA) but bilateral. NASA being a constant party to the MOUs is determined by the IGA, which allocates the overall program management to NASA (Article 7/2, IGA 1998). The MOUs are all constructed the same way, which means they feature virtually the same content in the same articles. Hence, they all look and read the same, with the exception of the Preambles and certain articles, which are tailored to the respective parties of the MOU at hand (e.g. which space station elements are provided or the respective responsibilities).

The MOUs describe the roles and responsibilities of the partners in an even more detailed way than the IGA and they govern the management bodies necessary to regulate the design, development and now operation and utilisation of the station (ESA, International Space Station Legal Framework). The content of the articles is more technical, compared to the IGA. For example, apart from the necessary procedural norms and projected goals and milestones, the space station itself is described in detail, the space station elements and modules are listed, and it is stipulated which partner had or has to provide which elements. Furthermore, access, use and utilization rights are determined and a long list of specific responsibilities (e.g. for costs, management and safety) is established in the MOUs.

IAs

At the third level, the IAs are small-scale instruments to implement the MOUs by distributing concrete guidelines and tasks among the national agencies whenever the need arises. (ESA, International Space Station Legal Framework). Just to name an example, one of these numerous IAs is the so-called Balance Agreement between NASA and Roscosmos from 1996 (Balance Agreement, 1996). It mainly concerns the mutual contributions between the two space agencies for the ISS and enumerates all hardware and software items that have (or had) to be contributed by one of the two partners. Additionally, it features a range from more general rules (e.g. setting English as the operating language or assuring that each agency provides sufficient food, supplies and personal items for its astronauts) to highly technical norms such as detailed instructions for workflow during initial assembly.

Barter Agreements

Often the legal structure of the ISS is referred to as a "three-tier structure". The "three-tiers" refer to the IGA, the MOUs and the IAs, which are the core instruments. However, in this context, also the so-called barter agreements have to be acknowledged as part of the legal system governing the ISS. They are mentioned in Article 9/2 of the IGA, which gives the partners the right to barter or sell any portion of their respective allocations. The terms and conditions of any barter or sale are negotiated by the individual partners to the transaction.

A barter is defined as a formal exchange of goods or services between the agencies "without a corresponding financial transaction, i.e. without an exchange of monetary funds." (ESA, International Space Station, Barter Agreements). This way, one partner receives goods or services, which he would otherwise have had to provide for himself, from another partner, who produces them regularly, without paying money in return. A typical barter would be one party delivering goods (hardware), while the other party is providing services or access opportunities to its ISS facilities, as it happened for example concerning the launch of the Columbus Orbital Facility, an ESA module (ESA, International Space Station, Barter Agreements). The barter was concluded between NASA and ESA in 1997 (as amended in 2000 and 2001) and secured that NASA would launch the Columbus module and its initial payload on its Shuttle quasi for free, but in return for hardware, such as, for example, Modules and Freezer equipment, as well as software and support for software development (Veldhuyzen, & Grifoni, 1999). A major incentive for such agreements is the reduction of financial risks (ESA, International Space Station, Barter Agreements). These barters are used to balance possible adjustments of obligations and contributions stipulated in the legal framework if they have cost implications, so that no additional money needs to be paid and no amending or renegotiating of legal instruments needs to be undertaken (Balance Agreement, 1996).

Other Agreements

In Article 11/2 the IGA mentions verbatim a Code of Conduct, which was developed by the Multilateral Crew Operations Panel in 2000 and represents a set of non-binding guidelines for crew members (Farand, 2005) (see below). Even with IGA being silent on further instruments, there is room for additional agreements and arrangements between the partners of the ISS. Examples of such agreements are the "MOU concluded in 1990 (and amended in 1997) between NASA and the Italian Space Agency (ASI) for the development by ASI of the Mini Pressurised Logistics Module (MPLM)" as well as an "arrangement signed in 1996 between ESA and the RSA for the delivery by ESA of a European Robotic Arm (ERA) to be used on the Russian segment of the Space Station" (Farand, 1998).

National Law in Outer Space

In an environment like outer space, where big projects can be conglomerates of partnerships and co-operations of various nations, the question of who has control over space objects in a legal sense arises. This specific form of control, called jurisdiction, has to be differentiated from ownership in the sense of civil law. Jurisdiction means the power to legally regulate, ergo to create laws (Shaw, M. N, 1997, 469). According to Article 5 of the IGA (which refers to the more general norm in Article VIII of the OST) jurisdiction on the ISS is retained by the partner who registers the respective space object with the UN registry (Article 5, 1998 IGA-Agreement). Consequently, the partner who registered has jurisdiction. Hence, this partner's national law applies to the said space object. However, the national law only applies *to* and *on-board* of the space object, which is surrounded by deep space, where only international law can apply. Theoretically speaking, in this special case, jurisdiction is "extended from the state's territory to the module of the station" (Soucek, 2012, 391). This concept can also be found in maritime law, where (speaking simplified) national law applies to and on the vessel itself, but regarding the surrounding high seas only international law applies.

Article 5 IGA requires each partner to register the space objects it provides and redirects to Article II of the Registration Convention of 1976. In practice this means that a NASA Module will be registered by the US and hence US national law applies. Likewise, for a module registered by Japan, Japanese national law is applicable. For Europe, the situation is more difficult, as 11 countries are involved within the ESA mission that is part of the ISS. ESA registers space objects on behalf of the European partners

(Balsano & Wheeler, 2006, 67). Each of the States can then claim sovereignty and jurisdiction (Walter, 2012, 551; Balsano & Wheeler, 2006, 67). As to how conflicting claims over jurisdiction and control would be settled, the framework offers no solution (Cogliati-Bantz, 2015, 256-257). From a strictly legal perspective, a multi-national space agency like ESA does not have the competency to create "national" law, i.e. carry out jurisdiction. In practice however, ESA is mandated to do exactly this, more precisely to act on behalf of the European state who claimed jurisdiction.

Specific Regulations Within the ISS Framework

Ownership and Utilization Arts. 6 and 9 IGA

In a legal sense, ownership means the strongest civil right to exclusive possession, the fullest right of dominion over property (Mattei, 2000, 77ff). Ownership therefore has to be differentiated from concepts like jurisdiction and control (Article 5/2, IGA), which were explained earlier.

According to Article 6/1 IGA, each partner owns what he respectively provided for the station (1998 IGA-Agreement). Even though the owner would have the most exclusive right to utilize his own module solely, by signing the IGA the partners have decided upon certain fixed shares for collective utilization, for the sake of the project (Article 9/1, IGA). This is because only some partners provide User Elements, and others provide resources to operate the Station. The specific allocations are determined in the MOUs and Implementing Agreements. (Article 9/1, IGA).

As an example, Article 8/3/b of all the four MOUs (see e.g. MOU NASA/ESA, 1998) lists:

- NASA has 97.7% utilization rights on its own modules and accommodation sites for external payloads, 46.7% on ESA and JAXA's modules and user accommodations.
- ROSCOSMOS has 100% utilization rights on its own modules and accommodation sites.
- ESA and JAXA have each 51% utilization rights on their own modules.
- CSA has 2.3% utilization rights of the user accommodations provided by NASA, ESA and JAXA.

Concerning crew time and resources that remain for utilization, 76.6% are allocated to NASA; 12.8% to JAXA; 8.3% to ESA; and 2.3% to CSA (Articles 8/3/b and 8/3/c MOU NASA/ESA, 1998).

Regarding all the capabilities of the Station that remain beyond these allocations, the partners can arrange specific agreements and use them with respect to these arrangements (Article 4/2, MOU NASA/ESA,1998).

Organizational Structure: Art. 7 IGA

Article 7/2 IGA holds that NASA is mainly responsible for the overall programme management and coordination of the Space Station, as mentioned above.

The partners to the IGA coordinate their activities and are each represented in the Multilateral Coordination Board (MCB), which is the highest coordination body established by the legal framework (Balogh, 2012, 213). The MOUs merely state that MCB comprises representatives of all partners to the ISS (Article 8.1.b MOU, 1998). In a US/Russian Interim Agreement it was specified that that the MCB

Figure 1. Utilization diagrams for crew time and resources
Source: Author's Elaboration

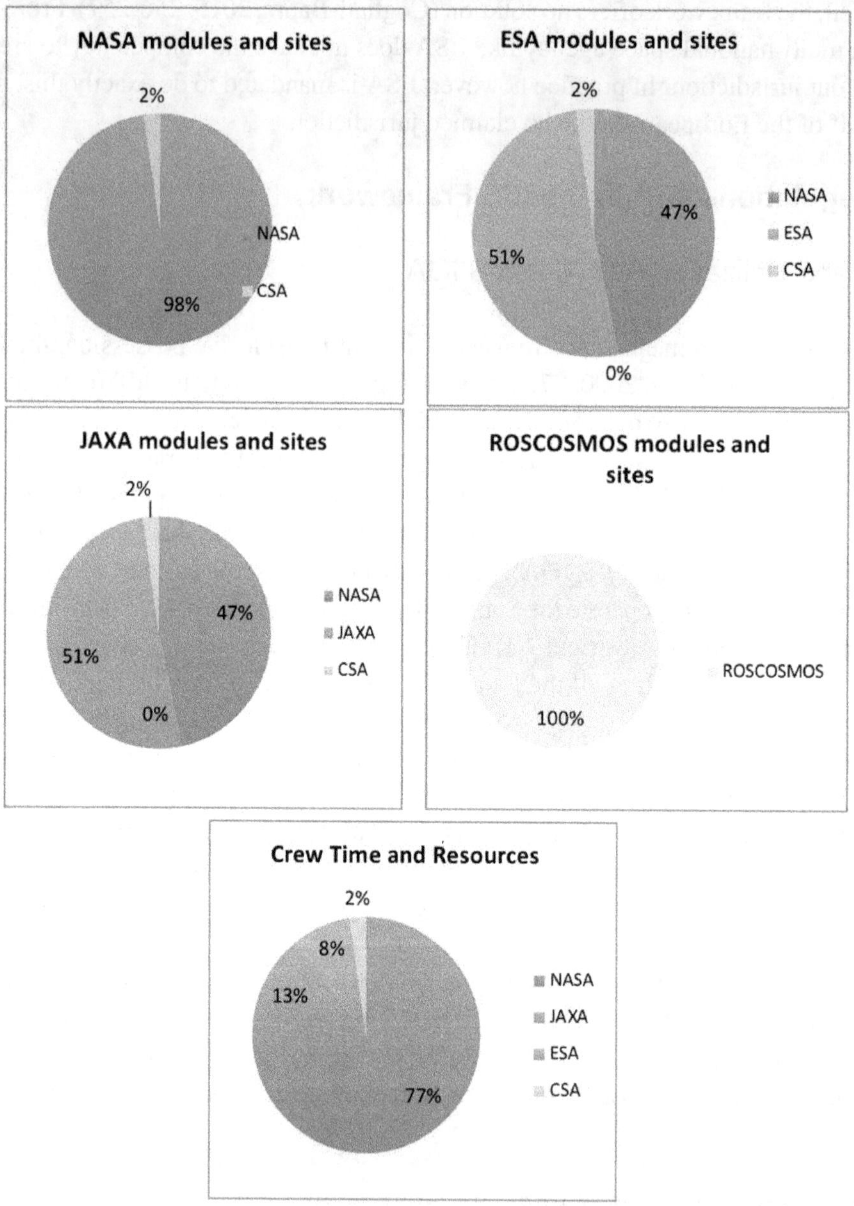

shall be comprised of the NASA Space Station Program Director, the Roscosmos Deputy General Director, the STA Director-General of the Research and Development Bureau, the ESA Columbus Programme Department Head and the CSA Vice President for Human Space Flight (Article 3, Interim Agreement, 1994). Decisions are reached by consensus, which means that a decision is adopted, as soon as no one from the negotiating parties opposes the decision reached. Following the rule that NASA leads the overall management, the NASA Space Station Director chairs the MCB. If a decision cannot be reached by consensus, the Chairman will decide.

Liability: Arts. 16-17 IGA

The UN Space treaties regulate liability in outer space. Article VII OST stipulates the basic rule that each state party to the OST is liable for damages its space objects inflict to another state party. This principle is refined in the Liability Convention of 1972 (hereinafter LC), which sets up an absolute liability regime for damages occurring on Earth and a fault-based liability regime regarding damages occurring in outer space (Article II-III, LC). The OST and the LC do apply to the ISS just as much as the other Space Treaties do, and expressively through Article 17 IGA. The IGA, however, goes further and beyond what the OST and the LC provides for.

Many terms within the LC are not legally defined in an exhaustive way, and are thereby open for interpretation: so, for example "space object", "property" or "damage". The LC defines "damage" to be compensable if it is either physical damage to natural persons or damage to property of states, natural persons or international organizations (Article I lit. d, Liability Convention, 1972). This is considered an exhaustive enumeration (Gorove, 1978). However, the whole terminology is very vague, and it is not clear, e.g. which kind of damaged property is meant to be included. Typically, the term "property" is interpreted in this context to mean only physical property (Carpanelli & Cohen, 2014). It has since been a hot topic between space lawyers if immaterial, economic or consequential damage (summarized as "indirect damages") is or should be included (Von der Dunk, 2015, 84-86; Lee, 2012 149; Carpanelli & Cohen, 2014). Indirect damages are all damages that do not occur directly to a physical possession but affect either one's psyche, bank account or other property only subsequently connected to the original damage (Christol, 1982, 95-97). For example, loss of profit due to a damaged space object which is used commercially would not constitute a material damage but a consequential damage.

The IGA introduces a clarification for the just mentioned problem: Article 16/2 lit c features a broader definition of "damage" than the one in the LC and does include indirect damages verbatim[3] (Article 16/2/c, 1998 IGA-Agreement).

The LC does not feature specific norms regarding alleviations of the conventional liability regime damages regarding cooperating partner states or related entities in a specific undertaking. In a project like the ISS, which is highly dependent on the friendly cooperation of the partners, it would be an unfavourable outcome, if the partners were constantly liable to each other in case of damages. This is especially true in an environment like outer space where damages can occur very easily.

Hence, Article 16/3 of the IGA presents a system of cross-waivers. This means, in general, that the partners will not claim compensation for damage resulting from certain operations between themselves or between one of the parties and a "related entity" (e.g. subcontractors) or the employees of the related entities[4]. In return, also these related entities, have to waive their possible claims (Article 16/2/f, 1998 IGA-Agreement). This facilitates cooperation between the partners.

Intellectual Property: Art. 21 IGA

Ownership cannot only be obtained over physical property, e.g. a space object itself, but also over intangible property like thoughts, ideas and inventions, or for example the industrial design of said space object. These properties are also called Intellectual Property (IP) and are defined as including "*the rights relating to literary, artistic and scientific works; (…) inventions in all fields of human endeavour; scientific discoveries; industrial designs; trademarks, service marks, and commercial names and designations*" (Article 2, WIPO Convention, 1967, as amended). It is important that IP is protected as an incentive to research and an attraction to investors. This is equally true for matters regarding the ISS, as many experiments regarding human physiology, medicine, biology, science and technology are carried out on board of the ISS and the results need to be legally protected (ESA, Patents and Space-Related Inventions). Otherwise, the experiments would not be financially viable for the conductor and investments from the private sector would cease (Maskus, 2000, Preface ix).

The question here in connection to the ISS is: who can claim ownership rights over IP (IPRs) created on board of the ISS and where can they be registered, following which law? The question arises because patent law is generally only applicable on the territory of a state (WIPO, 2004, Paras 1 and 40), whereas outer space does not underlie any national sovereignty and cannot be appropriated by any means (Article II, Outer Space Treaty, 1967). Again, similar to what was explained before - regarding the concept of jurisdiction and control - the IGA stipulates that territory is extended from earth to the respective module on the ISS (Article 21/2, 1998 IGA-Agreement). This means that standard national IP protection applies. Hence, there is not one homogenous, specific ISS intellectual property regime applicable, but different national IPR regimes co-exist on the ISS as a "*Quasi-Territorial Approach*" (Von der Dunk, 2009, 155). Practically this means that not the person who created the IP will necessarily be able to register it following his national law. The patent law of the Partner State, who registered the module on which the invention took place, should be applicable.

However, from a practical view, there is legal uncertainty if, without explicit statutory provision in their national laws, national patent law can be applied at all in outer space and will be exercised by the courts in case of an infringement (WIPO, 2004, Para 44). The US took the simple path to solve this problem. Primarily, NASA holds the opinion that experiments that have already been planned on the ground and are then performed on the ISS still follow the initial pre-ISS inventorship determination (ISS US National Laboratory, 2010). This means they are dealt with as if the invention took place on earth. For cases where an experiment is planned and takes place solely on board of the ISS, the US made its Patent Act applicable to space objects and inventions under its jurisdiction and control through an explicit statutory provision (US Patent Law, 2007; Balsano & Wheeler, 2006, 66). For the situation concerning IPRs in the US in particular, it has to be noted, that on the ground of US legislation, NASA, as a Federal Agency is required to retain certain IP Rights (even when experiments are conducted by universities or the industry) and additionally, if inventions are created using NASA resources and funds, NASA is authorized to "take-title" (which in effect means claim ownership) (NASA, 2013; US Code on National and Commercial Space Programs, 2010). Hence, US private actors and universities see it as a risky venture to conduct experiments on the ISS (NASA, 2013).

The European partners, on the other hand, face a different problem, as there are 11 States under one space agency. The IGA states that regarding an invention on board of an ESA module, any European Partner States may claim IP rights (Article 21/2, 1998 IGA-Agreement; Balsano & Wheeler, 2006, 69).[5] However, only Germany (ESA, Patents and Space-Related Inventions) and Italy modified their national patent laws to be applied to space inventions, as the US did (Balsano & Wheeler, 2006, 67). This means they added an explicit statutory provision to their national patent codes to include space objects and inventions to be protected as any other inventions. For the other countries there is still legal uncertainty if, without explicit statutory provision in their national laws making them applicable to space objects or activity, national patent law can be at all applied in outer space and will be exercised by the courts in case of an infringement (WIPO, 2004, Para 44). But there is still another problem: Potentially up to 11 European national patent laws could be applicable to inventions being made in ESA modules. Similar to what has been stated for jurisdiction and control, ESA registers modules on behalf of the European

States, but one of the States has to finally register the patent. A EU harmonization of intellectual property law would circumvent problems arising from this multitude of applicable laws. However, EU Patent law is one of the least harmonized areas in Community Law (Dimopolous, 2015, 59).

Another question regards consequences in case of an infringement of intellectual property. This is the case, for example, if someone uses or sells intellectual property, which he does not own. According to Article 16/3/d/4 IGA, the cross-waivers for liability do not apply to intellectual property claims, therefore it is possible that partners will claim redress for IP infringements between each other. Similar to what is explained above, the respective national laws will be applicable for the claims. As for European partner states, in case of infringement, the owner of the intellectual property can choose one of the potentially 11 applicable patent laws of the European partner states in order to get protection (Balsano & Wheeler, 2006, 68). Seeking redress in more than one European partner country however is not possible due to Article 21/4 IGA. (Balsano & Wheeler, 2006, 68). Moreover, all European Partner States have to recognize licences and patents that are enforceable at least in one European Partner State, which includes the recognition of reimbursement in case of infringement (Article 21/5, 1998 IGA-Agreement).

Code of Conduct for ISS Crews: Art 11 IGA

The ISS is now permanently inhabited and serves as a precursor to a "first city in space" (Marshall, 1998). This "city", like every other coexistence of human beings needs to be (or traditionally is) governed by a set of rules. The activities of human beings on the ISS are regulated partly through national law, directly through the IGA or the MOUs as well as the Code of Conduct.

National Law does not automatically apply in outer space. However, through the concept of jurisdiction and control, which was mentioned before, the partners can apply their national law to their nationals on board the Space Station (Art 5/2 1998 IGA-Agreement). This means, for example, that Russian nationals on the ISS are still bound to the usual set of rules they would be bound to on Russian territory.

But the daily life of crew members on the ISS is further regulated. On the one hand some provisions of the IGA and MOUs are directly relevant to astronaut activities (Farand, 2005). Additionally, Article 11/2 IGA introduces a code of conduct for the space station crew, which was developed by the Multilateral Crew Operations Panel in 2000 (Farand, 2005). From a legal point of view, this code of conduct is originally not a binding treaty, but a mere recommendation or so-called guidelines, which the crew members should abide by. However, the partner states have a legal obligation through the IGA to ensure that the crew members observe the code. Therefore, the code of conduct itself does not bind crew members, but the partners made crew members sign the code either through the terms and conditions of employment or invited them to sign it individually (Farand, 2005). In effect, the rules of the code of conduct are binding to crew members, nevertheless.

Topics covered by the code of conduct are: establishment of a chain of command, disciplinary regulations, security guidelines and safety procedures (Article 11.6 MOU, 1998; Section I/A, COCC, 2000). The code shall provide for a harmonious atmosphere, exclude preferential treatment, private financial gain, and disclosure of protected data (Section II, COCC, 2000; Farand, 2005).

Violations of the code of conduct are addressed by a specific ISS disciplinary policy, which was developed by the Multilateral Crew Operations Panel and the Multilateral Coordination Board (Section IV, COCC, 2000). Besides the legal consequences provided for in said ISS disciplinary policy, the partners apply relevant laws and procedures pursuant to their own personnel policy, which leads to a two-step approach and possibly also to a differentiated treatment, depending on which space agency the astronauts belong to (Farand, 2005).

For activities not on board of the ISS but e.g. pre-flight, in flight and post-flight, separate sets of rules are applicable: These are the ISS flight Rules and ISS program Rules, as well as operational directives and management policies (Section II/B COCC, 2000).

Criminal Jurisdiction: Art. 22 IGA

The Space Treaties have no explicit regulations concerning criminal jurisdiction. When the UN treaties were drafted, no one realistically considered that human existence in outer space could one day be so casual and permanent that crimes might be committed. The OST merely states in Article VIII that a state has jurisdiction and control over the personnel of a space object it registered. This means, that in principle the national (criminal) laws of the state apply to the personnel on board of the spacecraft or modules they registered – noting, that said personnel doesn't necessarily need to consist of citizens of the state of registry. In the case of the ISS this would probably lead to chaos and/or legal uncertainty, as five (and more) different national laws would apply, differing from module to module, depending on who registered it.

The IGA therefore provides specific provisions for the case of criminal misconduct in orbit (Article 22, 1998 IGA-Agreement). It changes (and simplifies) the rule of Article VIII OST slightly and states that partner states have primary criminal jurisdiction over their respective nationals. However, in certain cases if one of the partners to the ISS is the victim of said misconduct (the "affected state"), it can, as a secondary means, exercise criminal jurisdiction itself, as an addition to the previously mentioned rule, to safeguard that the case is dealt with effectively. In case that the "affected state" suffers damages to proprietary flight elements or to its nationals through the action of a national of another partner state ("perpetrator's State"), it can exercise criminal jurisdiction if two requirements are met: i) consultations have been held and ii) the "perpetrator's State" either agrees to it or fails to submit the case to its competent authorities for prosecution. So far, Article 22 IGA has never been invoked.

It is still an unsolved problem, who would serve as the arbiter in the event that two partner countries were entitled to exercise criminal jurisdiction, and both chose to do so (Ratner, 1999).

There are other means for exercising criminal jurisdiction, namely prosecution under the Uniform Code of Military Justice (only for members of the Military) and under the IGA as a legal basis for extradition if an extradition treaty does not exist (Frost, 2014).

SOLUTIONS AND RECOMMENDATIONS

As can be seen from the aforementioned, the IGA provides innovative and ingenious solutions to legal uncertainties to the existing legal framework, tailored to the ISS mission. The implementation of the legal framework has shown its success, especially in adapting to the different situations and in responding to the specific needs (UNCOPUOS, 2013). "The IGA and the four MOUs have been flexible enough to provide an adequate legal framework for the functioning of the partnership, without the need for amendment, while the Implementing Arrangements, and other arrangements of programme instruments have been developed as needed (UNCOPUOS, 2013)." However, there are still certain aspects that have been demonstrated in this chapter, that are in need of further conclusive clarification. A relatively new topic that might will at one point will possibly give rise to new treaties and regulations are increasing commercial use of space and specifically space tourism. For example, according to news outlets, NASA is currently exploring "policy or legal or programmatic changes" in order to broaden possibilities for "pure for-profit activities" on board of the ISS, such as advertising, marketing and "fly trinkets" (Wall, 2018).

All these issues should be closely reviewed and taken into account as "lessons learnt" for any future successor mission and their respective framework.

Additionally, in light of technological advancement and a new era of private actors in space, and in absence of additional international binding rules, there will be a rising need for multi-national binding agreements such as the IGA, regulating specific projects and missions like the ISS.

FUTURE RESEARCH DIRECTIONS

Every human-made object in outer space has an expiry date. The ISS was originally built to last around fifteen years, but it's life span has since been extended since, currently until 2024 (Bolden, 2014), according to media reports with an outlook to 2028 (Washington Post, 2016), if funding can be kept up until then. With the inevitable expiration of the ISS an outstanding legal project will reach its end. The specific legal framework (the IGA, MOUs and IAs) will also have fulfilled its purpose and therefore be terminated, hence lose its binding force or will need to be adjusted to meet future needs. However, as just mentioned, the IGA structure can act as a model framework for future space stations. Because even though the ISS might be terminated, with the current projects in planning, it can already be foreseen that the ISS was not the last international space station in history. One of the latest plans for a new space station was announced by a US private venture called Axiom Space LLC: to launch the first module of the private commercial ISS successor in 2020 (David, 2017). Axiom Space has NASA-approved plans to attach their first module to the ISS to work alongside it during the rest of the ISS's lifespan before being detached in order to build up and assemble the new Axiom Space Station (David, 2017). Another private venture, Las Vegas-based Bigelow Aerospace, has already launched BEAM, the Bigelow Expandable Activity Module, which has been docked to the ISS since 2016 and is planned to be further developed in order to one day replace the ISS (Wall, 2016; Davis, 2018). The space activities envisioned for the next decades will undoubtedly represent even bigger leaps for all of humankind and should be conducted within a detailed and comprehensive legal framework.

CONCLUSION

The ISS was a major step forward in many ways. As for the legal community, the IGA and its sub-instruments in the 90s were the first grand-scale multi-national legally binding treaties drafted and actually implemented by the major space faring nations in the past 20 years. Additionally, the project finally sent a positive signal towards major international cooperation in space, but it also united superpowers in space during the cold war, a time when cooperation between them on earth was very rare. More recently, in the light of the crisis in Ukraine, political tensions between the US and Russia spilled over to space matters and the ISS. According to various media reports, Russia threatened in 2014 that it would deny the use of the ISS[6] to the US after 2020 as a response to US sanctions (Bora, 2014). However, by the end of the year 2014, it was announced that irrespective of the political tensions, the cooperation in space would continue until the planned end of mission (Zak, 2015). Therefore, the ISS can and should be seen as an international peace project.

In conclusion it can be said that the ISS is certainly the most astounding achievement of humankind in space – so far.

REFERENCES

Agreement among the Government of Canada, Governments of Member States of the European Space Agency, the Government of Japan, the Government of the Russian Federation, and the Government of the United States of America concerning Cooperation on the Civil International Space Station, Washington, 29 January 1998, entered into force 27 March 2001 (1998 IGA-Agreement)

Agreement among the Government of the United States of America, Governments of Member States of the European Space Agency, the Government of Japan, and the Government of Canada on Cooperation in the Detailed Design, Development, Operation, and Utilization of the Permanently Manned Civil Space Station, Washington, done 29 September 1988, entered into force between the Government of Japan and the Government of the United States 30 January 1992, (1988 IGA-Agreement)

Agreement Governing the Activities of States on the Moon and Other Celestial Bodies, entered into force 11 July 1984, 1363 UNTS 21, (Moon Agreement, 1984)

Agreement on the Rescue of Astronauts, the Return of Astronauts, and the Return of Objects Launched into Outer Space, entered into force on 3 December 1968, 672 UNTS 119, (Rescue and Return Agreement, 1968)

Balogh, W. (2012). Institutional Aspects. In *Outer Space in Society, Politics and Law*. Wien: Springer Verlag.

Balsano, A. M., & Wheeler, J. (2006). The IGA and ESA, Protecting Intellectual Property Rights in the Context of ISS Activities. In *The International Space Station: Commercial Utilisation form a European Legal Perspective*. Leiden: Martinus Nijhoff Publishers. doi:10.1163/ej.9789004152564.i-267.41

Bolden, C. (2014). Obama Administration Extends International Space Station Until at Least 2024. *NASA Blogs*. Available at https://blogs.nasa.gov/bolden/2014/01/08/obama-administration-extends-international-space-station-until-at-least-2024/

Bora, K. (2014). Russia Threatens to Ban US from using International Space Station, Bars Rocket Engines for military Use over Ukraine Crisis. *International Business Times*. Available at http://www.ibtimes.com/russia-threatens-ban-us-using-international-space-station-bars-rocket-engines-military-use-1584065

Carpanelli, E., & Cohen, B. (2014). Interpreting "Damage Caused by Space Objects" under the 1972 Liability Convention. In *Proceedings of the International Institute of Space Law 2013*. The Hague: Eleven International Publishing.

Charter of the United Nations, entered into force 24 October 1945, 1 UNTS XVI (UN-Charter)

Christol, C. (1982). *The Modern International Law of Outer Space*. New York: Pergamon Press.

Cogliati-Bantz, V. (2015). *Means of Transportation and Registration of Nationality: Transportation registered by International Organizations*. Abingdon, UK: Routledge.

Convention Establishing the World Intellectual Property Organization. done at Stockholm on 14 July 1967, as amended September 1979, 828 UNTS 3, (WIPO Convention)

Convention on International Liability for Damage Caused by Space Objects, entered into force 1 September 1972, 961 UNTS 187, (Liability Convention, 1972)

Convention on Registration of Objects Launched into Outer Space, entered into force 15 September 1976, 1023 UNTS 15, (Registration Convention, 1976)

David, L. (2017). *Private Space Station Coming Soon?* Available at https://www.scientificamerican.com/article/private-space-station-coming-soon-company-aiming-for-2020-launch/

Davis, J. (2018). Goodbye, ISS. Hello, private space stations? *The Planetary Society*. Available at http://www.planetary.org/blogs/jason-davis/2018/20180220-bye-iss-hello-private-stations.html

Dimopolous, A. (2015). An Institutional Perspective II: The Role of the CJEU in the Unitary (EU) Patent System. In *The Unitary EU Patent System*. London: Hart/Bloomsbury.

ESA. (n.d.a). *International Space Station Legal Framework*. Available at http://www.esa.int/Our_Activities/Human_Spaceflight/International_Space_Station/International_Space_Station_legal_framework

ESA. (n.d.b). *International Space Station, Barter Agreements*. Available at http://m.esa.int/Our_Activities/Human_Spaceflight/International_Space_Station/ESA_s_International_Space_Station_barter_agreements

ESA. (n.d.c). *Patents and Space-Related Inventions*. Available at http://m.esa.int/About_Us/Law_at_ESA/Intellectual_Property_Rights/Patents_and_space-related_inventions

Farand, A. (1998). The Space Station Cooperation Framework. *ESA Bulletin*, 94.

Farand, A. (2005). *Astronauts' behaviour onboard the International Space Station: regulatory framework*. available at http://portal.unesco.org/shs/en/files/8480/11224719601Farand.pdf/Farand.pdf

Frost, R. (2014). *Here's What Happens If You Commit A Crime in Outer Space*. Available at http://www.businessinsider.com/heres-what-happens-if-you-commit-a-crime-in-outer-space-2014-1?IR=T

Interim Agreement Between the National Aeronautics and Space Administration of the United States of America and Russian Space Agency for the Conduct of Activities Leading to Russian Partnership in Detailed Design. (1994). *Development, Operation and Utilization on the Permanently Manned Civil Space Station, done at Washington, 23 June, 1994*. Interim Agreement.

ISS US National Laboratory. (2010). *Reference Model for the International Space Station*. Available at https://www.nasa.gov/pdf/487816main_10_09_22%20ISS%20National%20Lab%20ProOrbis%20Reference%20Model.pdf

Lee, R. (2012). *Law and Regulation of Commercial Mining of Minerals in Outer Space*. Dordrecht: Springer. doi:10.1007/978-94-007-2039-8

Marshall, T. (1998). 15 Sign Pact for Space Station. *Washington Times.*

Maskus, K. E. (2000). *Intellectual Property Rights in the Global Economy*. Washington, DC: Institute for International Economics.

Mattei, U. (2000). *Basic Principles of Property Law: A Comparative Legal and Economic Introduction. Legal Studies No 93*. Westport, CT: Greenwood Publishing Group.

Memorandum of Understanding between the National Aeronautics and Space Administration of the United States of America and the Canadian Space Agency concerning Cooperation on the Civil International Space Station, done at Washington D.C., Jan 29 1998. (n.d.). Available at http://web.archive.org/web/20091029094938/http://www.nasa.gov/mission_pages/station/structure/elements/nasa_csa.html

Memorandum of Understanding between the National Aeronautics and Space Administration of the United States of America and the Government of Japan concerning Cooperation on the Civil International Space Station, done at Washington, Feb 24 1998. (n.d.). Available at http://web.archive.org/web/20091029094939/http://www.nasa.gov/mission_pages/station/structure/elements/nasa_japan.html

NASA Advisory Council Recommendation, Research Subcommittee of the Human Exploration and Operations Committee (HEOC-01). (n.d.). Available at https://www.nasa.gov/sites/default/files/files/NASA_Responses_Package_1.pdf

Protocol including terms, conditions and assumptions, summary balance of contributions and obligations to international space station (ISS) and resulting rights of NASA and RSA to ISS utilization accommodations and resources, and flight opportunities, done at Moscow, Jun 11 1996. (n.d.). Available at https://archive.org/stream/nsia-25923192035014/25923192035014_djvu.txt

Ratner, S., (1999). Establishing the Extraterrestrial: Criminal Jurisdiction and the International Space Station. *Boston College International and Comperative Law Review, 22*(2).

Sharpe, C., & Tronchetti, F. (2015). Legal aspects of public manned spaceflight and space station operations. In *Handbook of Space Law*. Cheltenham, UK: Edward Elgar Publishing Limited. doi:10.4337/9781781000366.00020

Shaw, M. N. (1997). *International Law* (7th ed.). Cambridge University Press.

Soucek, A. (2012). *International Law. In Outer Space in Society, Politics and Law*. Wien: Springer Verlag.

Treaty on Principles Governing the Activities of States in the Exploration and Use of Outer Space, including the Moon and Other Celestial Bodies, entered into force 10 October 1967, 610 UNTS 205, (Outer Space Treaty, 1967)

UNOOSA. (2013). *The Legal Framework of the International Space Station*. Retrieved from http://www.unoosa.org/pdf/pres/lsc2013/tech-05E.pdf

U.S. Patent Law, 35 U.S.C. §§ 1 et seq., Consolidated Patent Laws (2007), (US Patent Law, Inventions in Outer Space) National and Commercial Space Programs, 51 U.S.C. Pub. L. 111–314, § 3, 124 Stat. 3328. (2010). Available at https://www.gpo.gov/fdsys/pkg/USCODE-2010-title51/pdf/USCODE-2010-title51.pdf

Veldhuyzen, R., & Grifoni, E. (1999). *No Exchange of Funds – The ESA Barter Agreements for the International Space Station*. Available at http://www.esa.int/esapub/bulletin/bullet99/veld99.pdf

Von der Dunk, F. (2009). *Space Law in the Age of the International Space Station, University of Nebraska – Lincoln: Paper 6*. Space and Telecommunications Law Program Faculty Publications.

Von der Dunk, F. (2015). International Space Law. In *Handbook of Space Law*. Cheltenham, UK: Edward Elgar Publishing. doi:10.4337/9781781000366.00011

Wall, M. (2016). *Private Space Stations May Take Flight in 2020*. Available at http://www.space.com/34377-private-space-stations-may-take-flight-in-2020.html

Wall, M. (2018). *NASA Looking at Ways to Commercialize International Space Station*. Available at https://www.space.com/39880-nasa-commercialize-international-space-station.html

Walter, E. (2012). The privatisation and commercialisation of Outer Space. In *Outer Space in Society, Politics and Law*. Wien: Springer Verlag.

Washington Post Live. (2016). NASA Administrator: We used to buy the rocket. Now we buy the service. *The Washington Post*. Available at https://www.washingtonpost.com/blogs/post-live/wp/2016/04/07/meet-nasa-administrator-charles-bolden/?utm_term=.c5f4b719f9c4

WIPO. (2004). *Intellectual Property and Space Activities*. Paper prepared by the International Bureau of WIPO. Available at http://www.wipo.int/export/sites/www/patent-law/en/developments/pdf/ip_space.pdf

Zak, A. (2015). A Rare Look at the Russian Side of the Space Station. *Air & Space Magazine Smithsonian*. Available at http://www.airspacemag.com/space/rare-look-russian-side-space-station-180956244/#rPCY1SgwIRFktBbq.99

ADDITIONAL READING

Agreement among the Government of Canada, Governments of Member States of the European Space Agency, the Government of Japan, the Government of the Russian Federation, and the Government of the United States of America concerning Cooperation on the Civil International Space Station, Washington, 29 January 1998, entered into force 27 March 2001 (1998 IGA-Agreement)

Balsano, A. M., & Wheeler, J. (2006). The IGA and ESA, Protecting Intellectual Property Rights in the Context of ISS Activities. In *Von der Dunk & F., Brus, M., The International Space Station: Commercial Utilisation form a European Legal Perspective*. Leiden, Boston: Martinus Nijhoff Publishers. doi:10.1163/ej.9789004152564.i-267.41

ESA. International Space Station Legal Framework, available at http://www.esa.int/Our_Activities/Human_Spaceflight/International_Space_Station/International_Space_Station_legal_framework

Farand, A. (1998). The Space Station Cooperation Framework. *ESA Bulletin*, 94.

KEY TERMS AND DEFINITIONS

IA: Implementing Arrangements, in the context of the ISS are subordinated instruments connected to the MOUs that implement what is provided for in the MOUs.

IGA: The Intergovernmental Agreement is an International Treaty between Canada, Governments of Member States of the European Space Agency, Japan, Russia and the US establishing specific rules applicable to the ISS. It is the core piece of the legal framework of the ISS and serves as an over-arching umbrella. It defines the rights and obligations of the partners.

Intellectual Property: Is ownership over intangible property like thoughts, ideas and inventions, or for example the industrial design of a space object.

International Space Law: Is the body of rules that governs activities in Outer Space and includes inter alia the five UN Space Treaties.

ISS: The International Space Station is a manned mission in outer space which is permanently inhabited by a crew who conducts scientific research and maintenance of the mission.

Jurisdiction: Means the power to legally regulate, ergo to create laws.

Liability: Refers to the state of having legal—often financial—obligations towards someone as a result of a specific action, in the context of space law due to damages occurring. In Space Law the terms liability and responsibility need to be strictly differentiated.

MOU: A Memorandum of Understanding is a—traditionally non-binding—agreement between partners. In respect to the ISS, four MOUs have been established, each between NASA and one of the other partners to regulate in more detail what the IGA provides for. However, these MOUs have been drafted to entail legal obligations.

Public International Law: Is the combined set of rules that govern the relations between States and International Organizations among themselves and with individuals. Sources of International Law are defined in Article 38 of the ICJ Statute and includes treaties, customary law, general principles but also judicial decisions and teachings of the most highly qualified publicists.

ENDNOTES

1 At this point it has to be noted that some scholars (see Loosch, R., (1990), The International Space Station – the Legal Framework, in Les Stations Spatiales Habitées - Aspects Juridiques. Manned Space Stations - Legal Issues. Proceedings of the Colloquium held 7-8 Nobember, 1989 in Paris, 56) view the IGA to be more politically binding than a legally binding treaty in regard to the US, as the US handled the IGA as an "executive agreement" which does not require ratification by the US Senate; see further Farand, 1998.

2 The Outer Space Treaty of 1967, the Rescue Agreement of 1968, the Liability Convention of 1972, the Registration Convention of 1976 and the Moon Agreement of 1984 (the last is not mentioned in the Preamble of the IGA).

3 "The term 'damage' means: (1) bodily injury to, or other impairment of health of, or death of, any person; (2) damage to, loss of, or loss of use of any property; (3) loss of revenue or profits; or (4) other direct, indirect or consequential damage."

4 However, there are exceptions, *inter alia,* regarding claims for damage caused by wilful misconduct or claims by a natural person for bodily injury, impairment of health or death. (Article 16/3/d IGA).

5 "(…) except that for ESA-registered elements any European Partner State may deem the activity to have occurred within its territory";

6 The Russian module can exist independently from the US module, but not *vice versa*, and until today, the only way to reach the ISS is via the Russian Soyuz Spacecraft.

Chapter 8
Mega-Constellations:
Technical Aspects

Arthur K. Lacombe
ISAE-SUPAERO, France

ABSTRACT

This chapter focuses on satellite mega-constellations, as mega-constellations raise several technical issues from telecommunication to launching these huge sets of satellites. Through an overview of basic signal science (periodic waves, types of modulation) and a discussion about telecommunication systems, actual 4G and future 5G networks, the chapter aims at giving without prior scientific background, sufficient material to better understand mega-constellation from its basic functioning to sophisticated satellites systems on orbit providing wide communication services all around the Earth. Major topics are evoked about such constellations be it on launching, space debris, challenges, and technologies used from a technical point of view.

INTRODUCTION

As of 2017, 54% of the world population has been connected to the Internet (Internet World Stats, 2017). Since its beginning almost 40 years ago, the Internet has grown, and more people have been linked together. However, connecting the whole planet - by building ground infrastructures and shelving millions of kilometres of cables - would be too costly and take over decades to reach isolated areas. Yet, a simple way of accomplishing it exists; new-born private companies entering into the space industry have serious projects that are disrupting settled telecommunication satellite operators. These new companies are thriving on a new concept called mega-constellation. Mega-constellations are composed of hundreds or thousands of low-cost satellites orbiting all around the Earth and covering the entire planet, providing access to the Internet even in the most desert regions. The telecommunication sector has always been the leader in the privatization of space use and continues to do it. Mega-constellations raise numerous technical challenges, especially for signal/frequency utilization or on constellation architecture (number of satellites, launches, orbits…) that would be discussed in this chapter. Because mega-constellation fascinates people, an overview of basic science and engineering about signal, telecommunication and

DOI: 10.4018/978-1-5225-7256-5.ch008

communication satellite is necessary to comprehend the whole set of problems that are raised by large constellation.

BACKGROUND

Telecommunication satellites are mainly placed in geostationary orbit at 36.000 km from the Earth. However, they are large and heavy (about 2 to 6 tons), thus the launch is extremely costly (up to $62 million for a Falcon 9 launch). Mega-constellation concepts in low Earth orbit (between 300 to 1400 km mostly) using lighter satellites at lower cost, are on the verge of becoming reality. A serious debate of such constellation's usefulness is taking place for filling the digital gap to provide Internet access for isolated areas. Mega-constellations, composed of hundreds and thousands of orbit satellites around the Earth, are under harsh debate in most space debris mitigation discussions due to the increasing risk of collision that they imply as several space agencies have warned. Companies has been claiming that they will take a proper care in deorbiting systems to mitigate the risks. Also, at the edge of the emerging 5G network, a battle for frequencies is on-going. Satellite operators and ground telecommunication operators are applying for the same frequencies. This chapter will give an overview of the governing physical principle behind telecommunication: signals; followed by their use in satellites. Once satellite functioning is developed, background on yet existing constellations will introduce the trending topic of mega-constellation and its raising concerns.

THEORY OF TELECOMMUNICATION

Telecommunication comes from '*tele*' meaning 'far' and '*communicare*' meaning 'sharing', it brings the idea of an exchange of information between remote/distant protagonists. According to the article 1.3 of the International Telecommunication Union (ITU), telecommunication is defined by any transmission, emission or reception of signs, signals, writings, images and sounds or intelligence of any nature by wire, radio, optical or other electromagnetic systems (ITU Radio Regulation, 2016).

The section addresses different notions dealing with actual issues such as frequency bands, the sharing of the radio frequency spectrum, telecommunication networks and the 4G/5G transition.

Signal, the Origin of Telecommunication

The very basis of communication is the signal, which is the way of transmitting and receiving information. Telecommunication uses specific signals to send data called periodic signals, allowing us to build recognized symbols that can be interpreted by decoders into readable messages. Those signals have been described for many years by waves and mathematical functions.

What Is a Signal?

A signal is a mean to convey and to remotely transmit information (Priemer, R., 1997). Several kinds of signals exist and are used in our daily lives such as visual signs, light, electrical signals or sounds. Typically, people and animals are using different kind of signals for orienting, moving or sensing. For instance: bats emit ultrasounds signals to detect obstacles and preys in the dark (Knörnschild, M., Jung, M., Nagy, M., Metz, M., Kalko, E., 2008); traffic lights signals make people understand when they need to stop on the road; and, even while surfing at home on the Internet with your smartphone, the latter communicates with operator's equipment with Wi-Fi signal.

This chapter focuses on specific signals that are used by systems creating artificial waves, unlike natural waves that are created by natural events in the Universe, without any electromagnetic man-made devices. Basically, a wave is a disturbance travelling through a medium. Earthquakes, tsunamis, light and sound are all examples of natural waves we can encounter. The medium then can be the soil for earthquake, water for tsunamis or even vacuum for sunlight.

Thus, telecommunication science aims at developing systems and methods to convey information efficiently with artificial waves. For this purpose, it uses signal processing, which is a means to operate on a signal in order to analyse, synthesize and modify the information it is holding. Such signal processing is possible through the use of periodic signals.

The following section will explain the basic techniques, vocabulary issues that arise today when working on communication, satellite or the Internet.

What Is a Periodic Signal?

A signal is said to be periodic when it repeats in a regular time period. Such signals are represented mathematically using periodic functions. The first function to study is the sine function which is the most basic periodic function used in signal processing.

Simple Representation of a Periodic Wave

When it comes to basic periodic signal, the first word attached to this, is sinusoid. The sinusoid or sine wave is simple wave which fluctuates with the time. A sine wave is represented on Figure 1. It can be observed also while playing with a rope by giving impulses from one side. Thus, the sine wave will propagate along the rope to end at the other side.

On Figure 1, the sinusoid consists in a wave varying from a range value U to –U (its opposite) alternatively. The U can be for instance the maximal height of the wave propagating along the rope.

Three main characteristics are basically identified: the amplitude, the phase (or the shift) and the frequency (or the period).

- Amplitude U

The amplitude measures the magnitude of the change of the signal values over one period of time. Most of the time, especially in electronics, the unity used to describe the amplitude is the Volt noted V.

- Phase φ

The phase of the signal corresponds to the shift of the sine function, considering that if the phase is zero the sine function begins its cycle at value 0 then increasing to reach the value U. The unity of the phase is the radian or the degree (°). Some examples of different phases and the shifts associated are represented in figure 2.

Figure 1. Description of a sine wave and its parameters

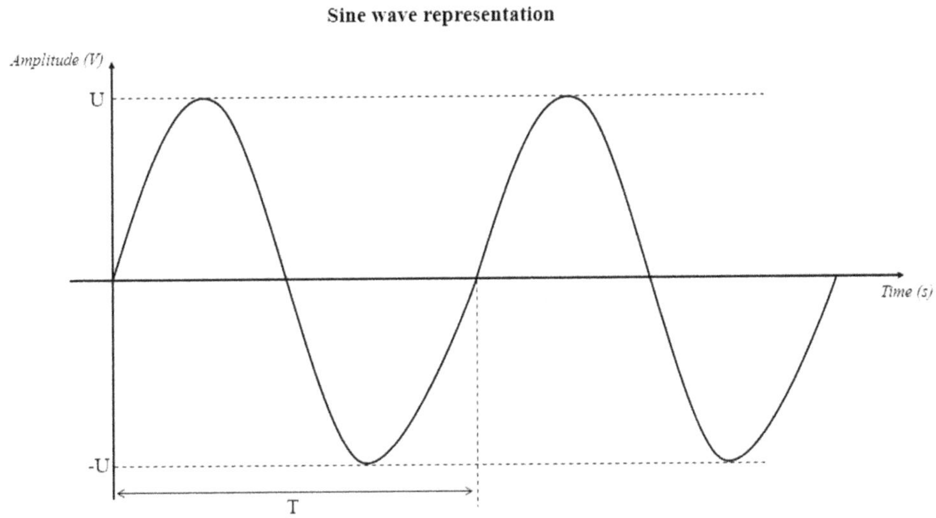

Figure 2. Sine waves captured with different phases

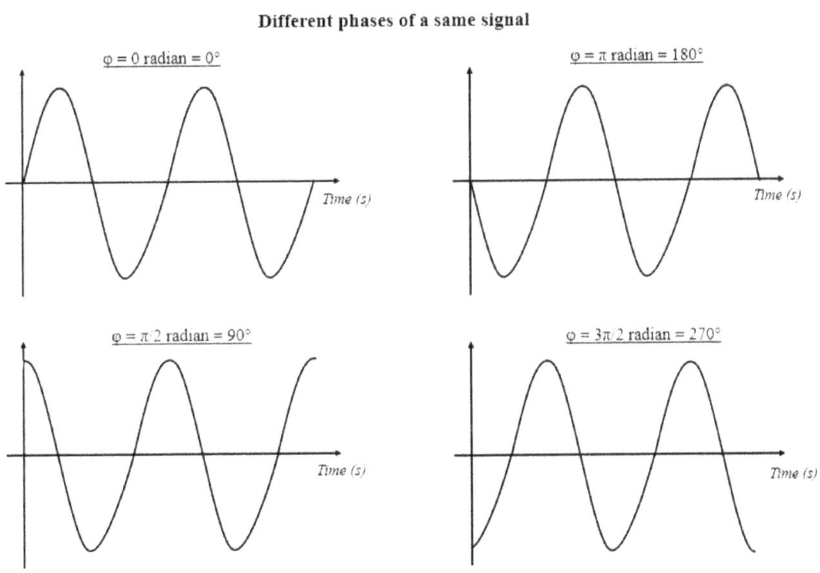

- Frequency f / Period T

The frequency is the number of cycles that the signal repeats each second. If the cycle of the signal lasts a period of time T then the frequency is: f = 1/T. The unity is the Hertz noted Hz. Frequency can go from 0 Hz (no signal) to infinity (or at least to the maximum actual measurable with instruments). If somebody gives an impulse every second to the rope, the corresponding frequency is 1 Hz. If somebody gives 1,000,000 impulses every second to the rope, the frequency is 1,000,000 Hz noted 1 MHz.

Example: If we have a signal with a period of 2 seconds then:

- The period T is 2 seconds meaning that 1 cycle lasts 2 seconds.
- The frequency is f = 1/2 = 0.5 Hz meaning 0.5 cycle per second (equivalent to 1 cycle in 2 seconds).

In literature, special symbology allows scientists to reduce the number used. For 1,000 Hz, it will be written 1 kHz (read kilohertz). For 1,000,000 Hz, it will be written 1 MHz (read megahertz). For 1,000,000,000 Hz, it will be written 1 GHz (read gigahertz). Those correspond to the most used symbols in telecommunication and mega-constellation.

Non-Sinusoidal Periodic Waveforms

In electronics, sine wave has a key role. Electronic components and devices, when in contact with a power source, will emit electrical waves such as a sine wave. Other interesting waves are observed and used in electronics, such as square waves or triangle waves that are represented on figure 3.

Carrier Wave and Modulation/Demodulation

In telecommunication, the wave conveying the information is called carrier wave. It consists in a waveform modulated meaning that it has been modified to contain the message. This modification is called modulation. Basically, modulation modifies the signal characteristics (amplitude, phase, frequency) transmitting specific patterns such as binary digits that are massively used for electronic system communication. It consists in communicating using only 2 values identified as 0 and 1. Modulation will create binary words composed of numerous digits such as 10010. Several modulations are used be it on

Figure 3. Sine, triangle and square waveforms

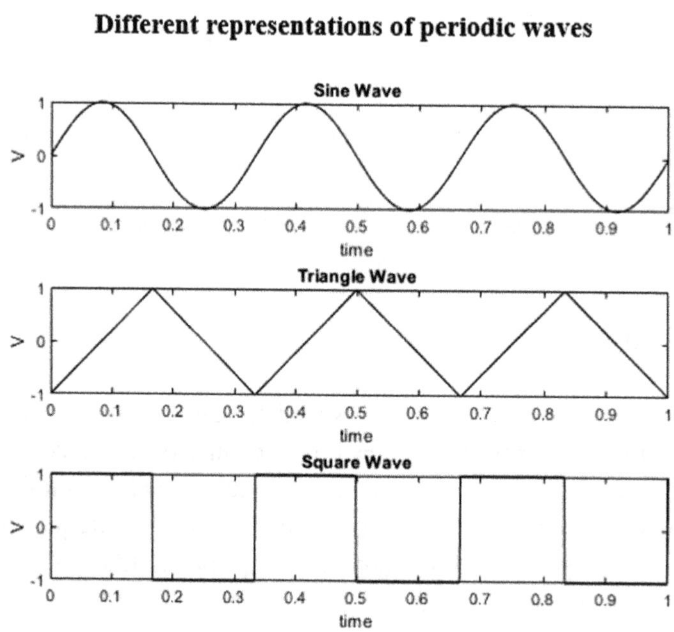

amplitude (AM), on frequency (FM or Frequency Shift-Keying) or on phase depending on technologies and applications (Vlajic, N., 2010). Different types of modulation are used to encode the binary word: 10010 on figure 4. Once such signal is received it needs to be decoded from the carrier wave. This operation is called demodulation.

Telecommunication satellite are complexed spacecraft using sophisticated modulation/demodulation techniques, but they are based on the same principal (AM, FM, PM). Satellites can emit several

Figure 4. Frequency Shift-Keying example

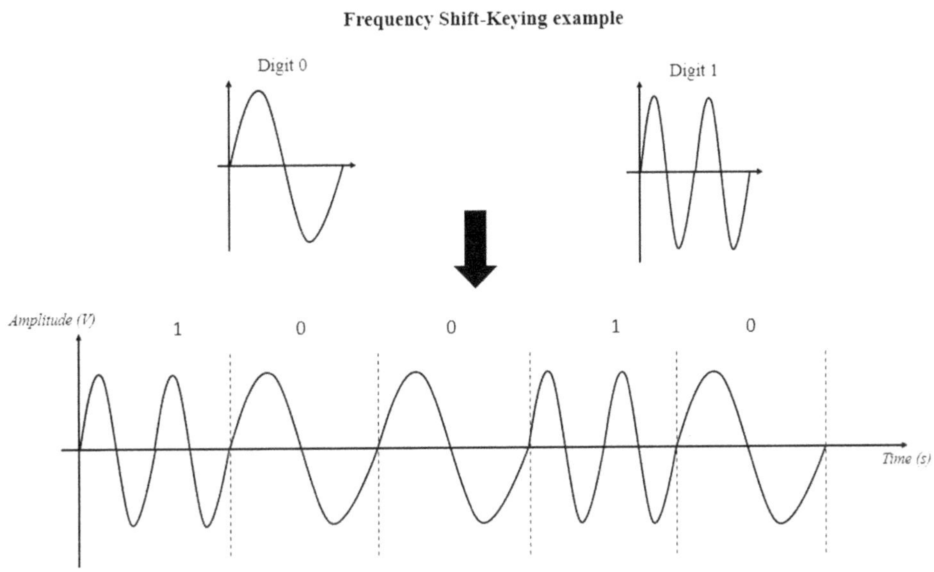

Figure 5. Amplitude modulation example

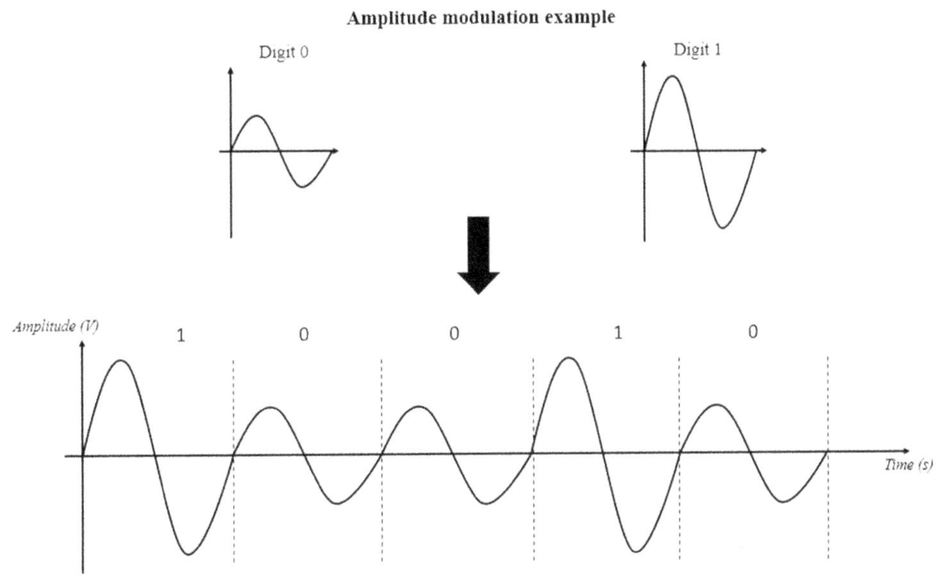

Figure 6. Phase Shift-Keying example

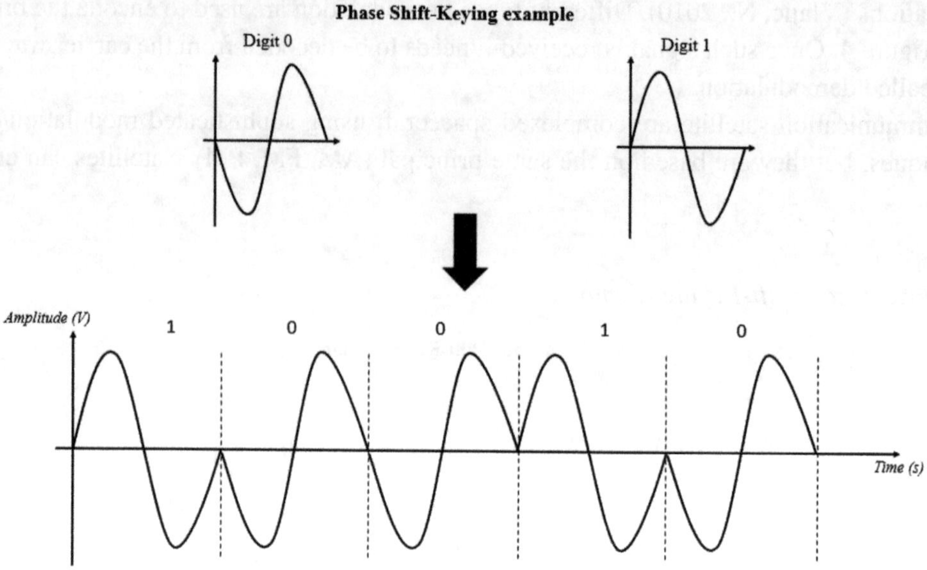

signals at the same time using different frequencies or use a common signal to emit different messages at different times.

Frequency Bands

Electromagnetic Spectrum

Discovered by Maxwell (1831-1879), a famous British mathematician and physicist, electromagnetic waves are created when electric and magnetic fields fluctuate together (Purcell, E.& Morin, D., 2013). They are propagative waves meaning that they propagate within a medium. Electromagnetic waves have particular properties due to their nature and they find application in various fields such as telecommunication and medicine. Electromagnetic waves can propagate whether in the vacuum of the outer space or in the matter except metals. This propagation property makes electromagnetic waves appealing for human applications and explains their extensive and wide range of utilization over the previous decades to the present day. In our daily lives, these waves are everywhere. Our environment is immersed within electromagnetic waves (light, Wi-Fi, radio…).

Since the 1890s, many scientists have worked on developing theories on electromagnetic waves. It has led to the development of systems using such waves for communication, medical or even cooking purposes. The most prominent scientists being Maxwell, Fourier, Bell and many others.

Here is a non-exhaustive list of electromagnetic waves (Maxwell, J. C., 1865):

- X rays,
- UV,
- Light,
- Microwaves,

- Radio waves.

Those waves differ in their frequencies and their properties. Some of them are useful for medical application like X-rays for checking the state of your bones whereas other can travel through high distance to allow communication like radio waves.

The wavelength λ (in meters) is related to the frequency f by the following relation: $\lambda = c/f$ where c is the speed of electromagnetic waves, mostly called speed of light (because light is an electromagnetic wave and this speed has been determined studying light). C is a constant value of approximately 300,000 km per second.

Thus, the longer the wave, the lower the frequency.

The whole electromagnetic spectrum on figure 7, displays the different kind of waves and their origins/applications.

Radio waves are often called long waves due to their lower frequencies compared to other electromagnetic spectrum waves. Frequencies are from 0 Hz to 100 GHz.

Gamma rays have the highest frequencies found in the electromagnetic spectrum with frequency above 10^{20} Hz. Gamma rays are produced within the universe during powerful events like solar flares or collapse of stars (NASA, 2016). On Earth, they are produced by nuclear reactions.

The lowest frequencies are produced by the functioning of some systems like magnets or electricity transportation in high voltage line (around 50 Hz).

Most of the radio frequencies are used for telecommunications. Waves are produced by emitting tower's antennas, phones or satellites.

Constantly, the Earth is being hit by light, UV and X-rays coming from our star, the Sun.

Radio Spectrum

For telecommunication application, the use of radio waves is more convenient, as less power is needed to be emitted. Today, most human activities requiring connectivity, mainly for transmitting data, employ systems emitting a specific allocated frequency of the radio spectrum. It means that a license, specifying

Figure 7. Representation of the electromagnetic spectrum and zoom on radio spectrum applications

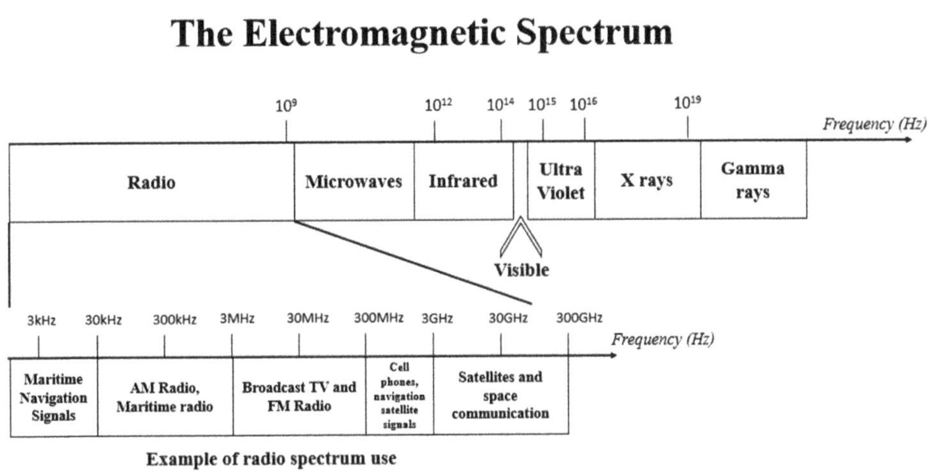

the frequency the system is allowed to use, is required. This will be discussed in detail in the legal part of this chapter. To prevent conflicts and to manage this rare resource (frequencies are not unlimited and the same frequency cannot be allocated on the same region for two different applications due to interference), the ITU governs over the spectrum to coordinate international policy and insure the access to the resource (ITU Radio Regulations, 2016). These rules aim also at preventing from interferences that could be critical for some applications (rescue for instance). That is why a licensing regime has been implemented.

On figure 7, the radio spectrum is represented by the different applications and their frequency range. Maritime and aeronautical sectors are deeply linked with radio emissions for navigation. Rescue and emergency services are also dependent of such emission for their efficiency. Numerous actors are encountered in the radio spectrum as scientists for measurements and wild life tracking, transportation companies (airlines, maritime companies), fishermen, armies for surveillance and defense systems, radio show, TV broadcasters, satellites companies. Therefore, even if ITU rules do exist to avoid interference with other signals, interference may happen and cause your application to stop running well.

But What Is Interference?

An interference is an alteration of a signal by another signal having a closed frequency range. The altered signal is difficult to decode, then it becomes unusable. There is no special means to prevent it except avoiding using the same frequency than another operator in the same area.

Equipment

How to Create Artificial Waves?

Lot of electronic systems emit waves such as phones, satellites or radio stations. They all create waves with a common functioning principle. In fact, each system emitting waves has a frequency source, an electronic oscillator generating a radio signal.

A modulator device follows the oscillator, to adapt the signal with the information to transmit by modulating it, meaning that it will apply modulation on the signal as explained on figure 8.

After being modulated, the signal is too weak and is not able to reach remote areas, therefore it passes through an amplifier that will empower the signal without changing the frequency.

Finally, the antenna radiates the signal (Stoehr, M., 2012). Depending on the technology used, the antenna can be directive, meaning that it points to specific zones where a receiver is predicted to be, or non-directive meaning that it broadcasts the signal in a large region or all over the antenna.

The receiving pattern is composed as well of an antenna, an amplifier and a demodulator. A huge amount of energy is lost in the environment while travelling through the vacuum or in the atmosphere, thus the amplifier is needed before demodulating the signal.

Some noises altered the signal at the emission and the reception that is why systems have implemented filters to clean the signal and limit the effect of the noise in the interpretation of the signal. The overall systems both emitting and receiving are represented in a block diagram on figure 8. Figure 8 displays the evolution of the signal, its way and main systems operating on it.

Figure 8. Simplified block diagram of emitters and receivers

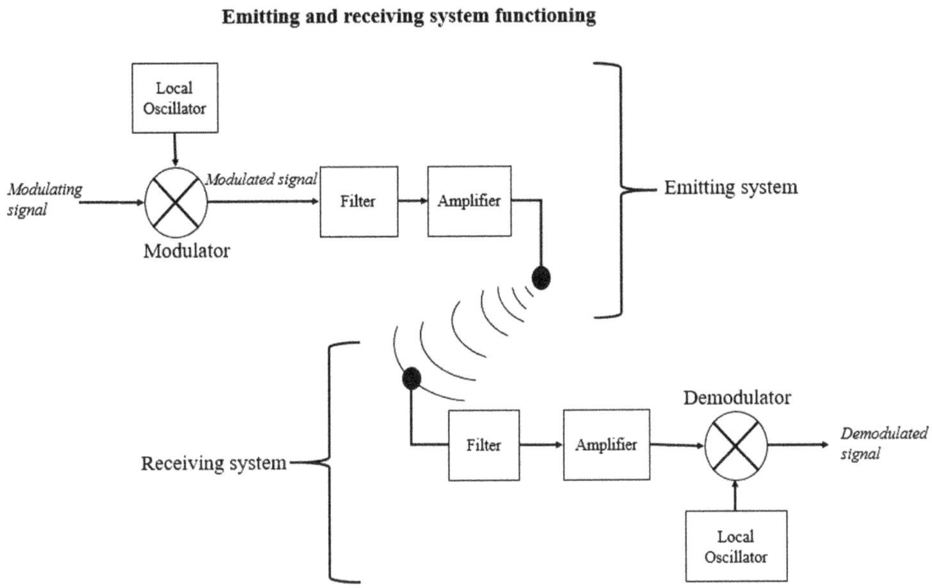

Bandwidth

A frequency band is an interval in the frequency domain delimited by a lower and an upper frequency. For instance, the electromagnetic spectrum is divided into bands (radio band, UV band, Infrared band, X-ray band), as highlighted by figure 7, based on physical features and behaviours of the waves. The radio band is also divided in frequency bands by IEEE (Institute of Electrical and Electronics Engineers) technical norms (IEEE, 2002) but the ITU established another division of the radio spectrum, represented on figure 7 (below the whole electromagnetic spectrum), in order to manage more efficiently the spectrum's use. However, even if the ITU divided the radio spectrum into wide band, a band can be divided into several other bands and so on. Most of the time, a narrow band is allocated to an operator and not only a unique frequency. For instance, in the 3-30 GHz band, an operator can obtain a license for exploiting the 24.5-25.5 GHz band. Instead of using a unique carrier frequency, systems can use different frequencies working at the same time and preventing from causing interference between each other. Thus, allowing a band to provide the system a margin of error upon the frequency used. It is *de facto* a technical method to prevent also interference. The bandwidth is the difference between an upper and a lower frequency used in the band. Several criteria exist to determine the corresponding frequencies as explained further.

The bandwidth used by device is deeply linked with the data rate (amount of data transmitting per second, in bits per second) that it can offer. The Shannon's formula linked both data rate and bandwidth:

$$C = Q \log_2 \left(1 + \frac{S}{N}\right)$$

C is the binary rate in bits per second, Q is the bandwidth (in Hz), the ratio S/N (Signal/Noise) determines the quality of the signal (Shannon, C. E., 1949). The average power of the signal S and of the noise N are measured in Watts.

The wider the bandwidth Q, the higher the data rate C. An important pattern to have in mind is no matter the frequency used, you may have a high frequency but still have a low data rate if the bandwidth is too narrow. For instance, 2 systems function with 2 different sets of frequencies and bandwidth: one having a central frequency of 20 GHz with a bandwidth going from 18 GHz to 22 GHz (Q = 22-18 = 4 GHz); and the second having a central frequency of 30 GHz with a bandwidth going from 29 GHz to 31 GHz (Q = 31-29 = 2 GHz). Then if the system works with the same signal quality (S/N), the rest of the equation (part on the right of Q) is considered equal for both, and the only difference remaining is the bandwidth (Q). The first system uses a lower central frequency than the second one, but its bandwidth is wider (4 GHz) than the second one (2 GHz), meaning the data rate is higher with the first system.

The bandwidth is determined by a criterion called the 3 dB (decibel) bandwidth which corresponds to a mathematical criterion, that evaluates when the distant frequencies are too weak to contain information. In fact, in most systems, a central frequency is used, and upper and lower frequencies are emitted with decreasing energy compared to the central one. When the energy is too low for those frequencies, it can't be distinguished from the noise. The 3 dB criterion or other x dB are chosen to calculate the bandwidth available and take into account the decreasing energy of frequencies as evoked.

Figure 9 is an illustration of bandwidth determination on a frequency spectrum using the 3 dB criterion. Frequency within the 3 dB band drawn on the figure 9 is comprised on the bandwidth. The criterion is the same with another amount of dB. The dB band is determined from the top of the frequency curve.

Figure 9. Bandwidth determination using frequency spectrum and 3dB criteria

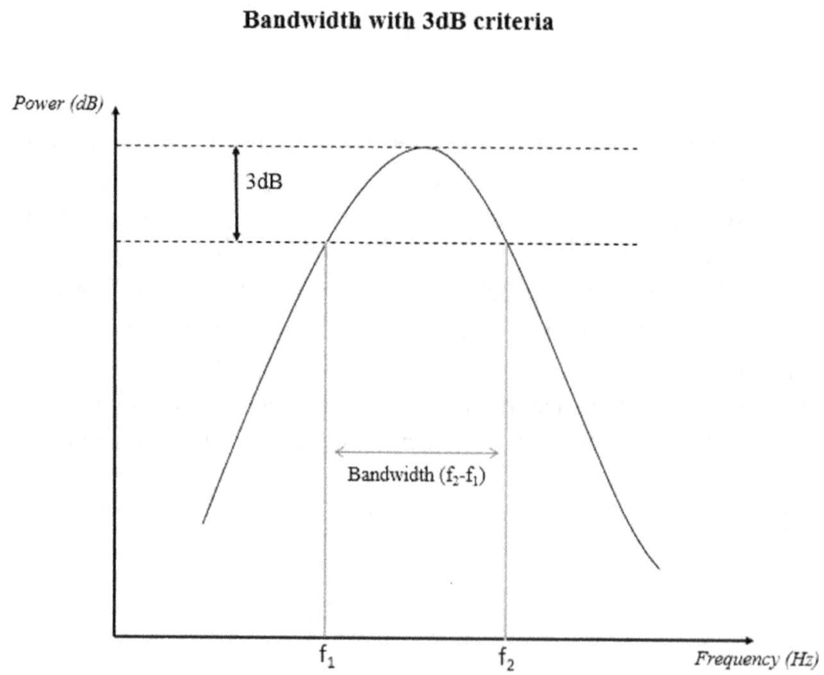

4G / 5G

4G, 5G are words that overflow everywhere because companies are undertaking to launch the first 5G within a few months. But how many can relate what is it about? In fact, G stands for generation. These generations correspond to technical international standards for electronic devices allowing them to use the network. Standards are developed by international standardization groups (3GPP, ITU, IEEE, etc.). Devices (cell phones, laptops, connected watches) shall respect these technical norms to connect with the corresponding generation network and access the service (a cell phone compatible only for 3G network cannot use the 4G network). Generally, a new network generation implies a higher data rate, a better quality of services and new features (elaborated in the following section). Actually, the 4G is widespread and developed all over the world, yet engineers are working on the new generation. Succeeding in managing different types of devices connected at the same time requires rigorous technical rules and standardization. When talking about a phone compatible with 4G for instance, it means that it does respect technical norms established by a standardization agency and is then able to connect the network using a 4G terminal. Let us have a more detailed look at these standards.

4G

Actually, technical norms for the 4G or Long Term Evolution Advanced (LTE-Advanced) was set by 3GPP (3rd Generation Partnership Project) and the ITU. LTE norms corresponds to a transition between 3G and 4G.

In theory, a device can reach a data speed rate of 300 Mb/s and can even reach 1 Gb/s for certain category under the 4G in downlink. The speeds are different in downlink and uplink. In uplink the data rate is lower than in downlink. Most of the time, 10 times lower.

Depending on the region, frequencies allocated for the 4G devices differ. In Asia and Europe, the 800 MHz and the 2600 MHz frequency bands were allocated whereas the 700 MHz and the 1700-2100 MHz are used in the Americas (ARCEP, 2018). Have you ever experienced this issue while looking for 4G abroad? That might be a valuable reason why. Some online websites give an insight of the possible compatibility of your personal device according to national norms.

As the data rate depends on the bandwidth, it appears clearly here that the higher the number of users, the lower the data rate. The main source of decrease is the share of the bandwidth or slots between the different users.

5G

5G is a trending word, heard more and more nowadays as the future of 4G. Nevertheless, the standards and norms are not fixed yet even if the 3GPP and ITU are working on it, to release them by 2020 (Bertenyi, B., n. d.). It does not prevent states from anticipating its arrival and prospecting for distributing frequency bands. Actually, in France, the first test will be held within the 3.5 GHz and the 26 GHz bands.

In theory, the 5G should bring data rate to unequalled value by now. The goal is to ensure a data rate of 10 Gb/s uplink and 20 Gb/s downlink for the very first tests. The 5G will be able to target 50 Gb/s in the future.

New generation often rimes with new technologies. In fact, the 5G will need new antennas and new algorithms to undertake the challenge of supporting thousands of connected devices on the same access point that can be dealt by the actual 4G.

With the 5G, the connected world will know dramatic changes. It will allow the free development of smart grids that are smart networks specialized in a specific field (energy, health, and banking). For instance, an energy smart grid would be an electrical grid that is able to measure energy consumption and production in the whole network to adapt the production of the different energy production units (solar arrays, thermal reactor, dams, etc.). These networks will be able to adapt the electric production with people's consumption for instance; or connect hospitals with their patients. Cloud computing (access to shared resources from different devices be it cell phones, computers, tablets, etc.) and the Internet of Everything (multiplication of connected objects: computers, phones, refrigerators, watches, printers, etc.) will come to a new golden age with the new capacities of 5G networks.

5G networks would be adapted for specific applications depending on the architecture of the network. For instance, some networks will be structurally designed to sustain mass connections whereas others will be designed to maintain a constant data rate and high-quality connection (precisely for health application and vital organisms).

Ensuring mass connectivity is a challenge that needs new algorithms and technologies to be developed.

TELECOMMUNICATION SATELLITES

Nowadays, it is relatively easy to communicate with someone wherever he or she is located in the world. This is possible thanks to networks that have been built over decades to connect each point of the planet, allowing the transfer of information. Several kinds of networks have been developed, ground networks using cables connecting point to point were historically first. However, during the 90s, with improvements in telecommunication and satellites, satellite-based network appeared. Those networks are used by TV broadcasters all over the world and recently companies announced the arrival of global satellite-based internet networks. Such networks are based on communication satellites that this section will explore.

Table 1. Overview of generations from 1G to future 5G

Generation	1G	2G	3G	4G (LTE-Advanced)	5G
Period	1980-1990	1990-2000	2000-2010	2010 - 2020	2020-2030
Data rate	2 kb/s	64 kb/s	144 kb/s – 2 Mb/s	100 Mb/s – 1 Gb/s	1 Gb/s – 10 Gb/s
Technology	Analog cellular	Digital cellular (GSM)	EDGE	LTE, WIFI, LTE-Advanced	
Application	Voice call	Text message, image and MMS (Multi Messaging Service)	Web based applications	Web surfing and streaming	Mass connectivity, Internet of Things/Everything

Architecture of Communication Satellites

Sophisticated systems are required to perform TV broadcast or data transmission via satellites. As it is impossible for now to perform in-orbit repairing, satellites need to be robust and have redundant systems.

A telecommunication satellite is composed of 2 main parts: the bus spacecraft and the payload (Bousquet, M., 2018).

- *The bus* spacecraft is more or less a standardized architecture that will fulfil the positioning and the life support mission. To succeed in the mission, the architecture consists of various subsystems. First, the hardware architecture in which every subsystem is fixed. Secondly, to position the satellite, thrust engines are used allowing movement and keeping the antennas pointing to the right area. Such engines can use either solid propellant or liquid propellant, and currently electric thrusters are developed. These engines receive orders from the control and command subsystem that consists of a certain number of actuators and sensors to orientate the thrusters and an on-board computer to determine the right move with the right information from sensors. To supply each component, solar arrays transform solar' radiations into electricity that is stored in batteries. The batteries are discharging when no solar rays reach the satellite, and are charging when enough radiations hit the solar arrays. It can be also called a solar generator. Solar arrays and batteries are necessary to permit the satellite functioning even if it has been damaged by a space debris for instance.
- *The payload* is a load carried by a spacecraft that is necessary to the purpose of the mission. In the case of telecommunication, payload consists of emitting and receiving systems similar to those previously described.

A satellite alone cannot work out its mission. In order to carry on correctly, ground infrastructures are required. On the ground, a control station receives and communicates with the satellite adapting its coordinates, controlling its flight parameters and reprogramming the satellite if needed. Also, some satellite-based network may be connected to ground networks using what are called gateways. Gateways are antennas that receive signals from satellites and send them through the classic on-ground network.

Satellite-Based Communication Networks

Numerous companies are operating satellite-based network for telecommunication purposes such as SES (Luxembourg), Intelsat (US), Eutelsat (France), Telesat (Canada) or SKY Perfect/JSAT (Japan). These companies mostly possess satellites in Geostationary Earth Orbit (GEO). The GEO is the most used orbit for telecommunication because it presents the huge advantage of keeping the satellite above a fixed point on Earth. The demand for having access to this orbit is high and regulated by ITU (positions and frequency slots). The GEO is about 36 000 km from the Earth's surface; and the satellite placed within GEO has the same rotating period around the Earth than the Earth around itself. That is why the satellite is constantly above a fixed point. It makes telecommunication secure for always having a satellite to send and transmit data. However, due to the large distance from the Earth, there is a delay in transmitting information, about 0.25 seconds, which is limiting some applications in terms of speed such as internet (Bousquet, M., 2018). Such satellite is used to carry out communication missions for TV broadcast or data transmission.

LEO and MEO (Medium Earth Orbit) are also used for communication networks. Iridium and Globalstar provide a communication service from constellation in LEO whereas SES operates satellites from MEO with its O3B constellation. Their functioning is further elaborated in the mega-constellation section, in the basic communication constellation subsection.

Internet by Satellite Networks

Some operators provide internet via satellite-based networks such as Eutelsat and SES. A subscriber to an internet satellite service uses a specific device composed of an antenna and a decoder to receive and emit signals to the satellite. The satellite is connected to a ground station providing internet service. The subscriber can then be linked with this network thanks to the device which emits the signal to the satellite and the satellite that repeats the signal to the ground station. The satellite is a "bent pipe" analogue to a signal repeater. Such satellite network can offer a data rate from 2 Megabits per second (upstream) to 20 Megabits per second (downstream) which is close to the ADSL data rate (called high speed) but quite far from the newest technologies using fibre optic allowing data rate from 30 Mb/s to 1 Gigabits per second (called very high speed).1 Gigabits per second represent a speed that is one thousand times faster than 1 Megabits per second. Indeed, internet satellites are actually placed in GEO orbit and are limited due to transmitting delays about 0.2 seconds. However, the great advantage of using such a network is that people that are not connected to ground networks in isolated areas can access to the Internet thanks to high speed data connection. In France, the main operator Orange has concluded an agreement with Eutelsat to provide internet via satellite to areas non-connected via ground telecommunication networks. It aims at reducing connectivity inequalities between regions, bridging the so-called digital divide. As a satellite in GEO is fixed upon a certain point on Earth, it will maintain a permanent connection to the Internet.

GEO orbit is limiting the speed of the internet that can feed such a network. However, a project emerged a few years ago, using mega-constellation in LEO to provide high quality internet with less delay. The last part will focus on mega-constellations.

MEGA-CONSTELLATIONS

Miniaturization of technologies, new electrical engines, launching costs dropping, all these events put together has led to the concept of mega constellation. Or, more precisely, has pushed the space industry to lower even further their costs. The specificity of the space industry was low number of products leaving the manufacturer keeping it in a cutting-edge artisanal industry. Nonetheless, mega constellation will significantly transform the space industry by undertaking massive production of satellites. They will allow new services but they will also raise a number of challenges for launching, operating and servicing that are discussed in this section. Mega-constellations are not a contemporary revolution since ideas has emerged since the 1990s, especially with the Teledesic project. It consisted in 288 LEO satellites for providing telecommunication services. Pharaonic cost and technical issues had stopped the project. Nowadays, many companies plan to launch mega-constellation, however only OneWeb and SpaceX have secured sufficient amount of money to launch their project with a high level of certainty. This section is often referring to OneWeb and SpaceX in discussing technical points because their projects are sufficiently advanced and well documented. Nonetheless, they should be considered as a broad example of common issues related to mega constellation.

Basic Communication Constellation

Before the arrival of mega-constellation doing telecommunication, companies such as Iridium and Globalstar launched their own constellation with fewer satellites, between 40 and 70, to provide connection to isolated areas such as the sea or desert (making it useful for explorers, scientists, or isolated population). Today, they still function, and they work in two different ways, further explained in this subsection.

Iridium

Iridium is composed of 66 active satellites orbiting at 780 km from Earth (Maine, K & Devieux, C & Swan, Peter., 1995). They are distributed on 6 different orbital planes (11 satellites per orbital plane) on *quasi* polar orbit. An orbital plane is a geometric plane that includes the circular or elliptic orbit of the satellite. Each satellite passes by the North and the South poles at each rotation. The constellation covers the entire world and is extremely useful for exploration near the poles, where geostationary satellite signals can hardly cover. Figure 10 sketches the orbital plans of an Iridium type polar constellation.

Iridium uses a point to point system passing through only Iridium satellites, both points need to be equipped with an Iridium compatible phone. The scheme of the call is the following:

1. A scientist, located in Antarctica, wants to call his laboratory in the Japan. As no "local" operator provides a service, he uses his Iridium-compatible phone. The laboratory is also equipped for receiving the call thanks to the phone also compatible with Iridium.
2. The scientist makes the call. The signal of the phone is sent to an Iridium satellite orbiting above Antarctica.
3. The satellite is looking for the receiver but does not find it. Then, it sends the signal to another satellite. This is called an inter-satellite link.
4. The new satellite looks for the receiver but does not find it. Then, it sends the signal to another satellite.

Figure 10. Polar orbital plan scheme for Iridium constellation

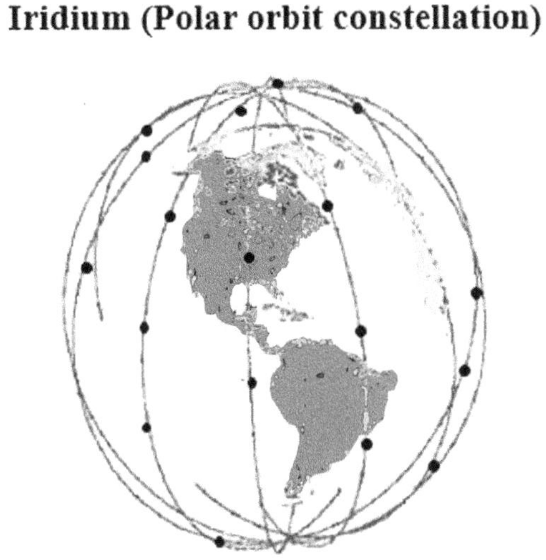

5. The following satellite finds the receiver. It sends the signal to the Iridium-compatible phone in Japan. A link is then established. The communication can then start.

The signal then goes from the emitter to the satellite and passes through other satellites of the constellation if the receiver is not under coverage of the same satellite. Then the signal is sent back to Earth for the receiver.

Most boats and polar scientists are equipped with an Iridium-compatible phone because the Iridium constellation provides a service worldwide even above the ocean unlike Globalstar.

Globalstar

Globalstar is composed of 48 active satellites orbiting at 1,400 km from Earth. They are distributed in 8 different orbital planes (6 satellites per orbital plane) each having a 52° inclination with the equator. The constellation covers the zone between the two polar circles very well, but the poles are barely covered. Figure 11 represents the orbital planes of a Globalstar type inclined constellation.

The Globalstar system works differently from the Iridium one. Globalstar satellite also uses ground antennas to connect to the normal network. The functioning is as follows (Globalstar, 2000):

- Scientists, in the Utah desert (USA) wants to call his laboratory in Argentina. As there is no available network where they are, they pass the call with his cell phone thanks to a Globalstar device (looking like a Wi-Fi modem) connected to the Globalstar network.
- The signal is sent from the cell phone to the modem and from the modem to a Globalstar satellite above the Utah desert.
- Instead of sending the signal to another constellation satellite, the signal is emitted back to the ground to a ground station called gateway. A gateway is an antenna sizing more than 10 meters' diameter (up to 13.5 meters' diameter for a ViaSat Earth station).
- Once the signal has reached the gateway, it is sent to the ground network until the receiver is found.
- The signal is thus received by the laboratory.

Figure 11. Inclined orbital plan scheme for Globalstar constellation

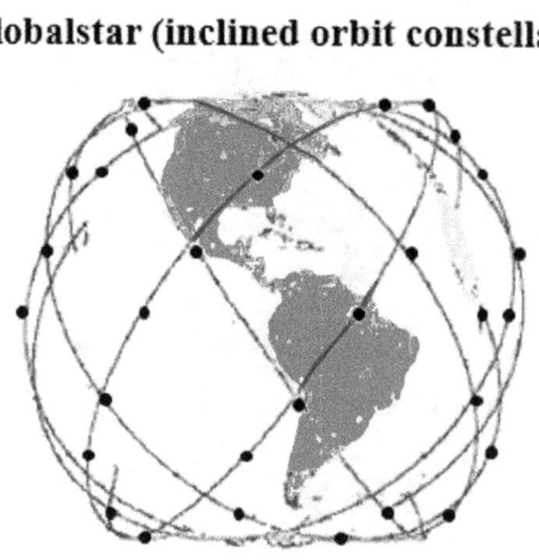

Satellites in these schemes are just "bent pipe" which means they just repeat the signal they received to a near ground station unlike Iridium satellites that look and maintain communication by themselves. 40 ground stations are necessary all around the world. There are only a few Globalstar services above the ocean due to the lack of ground stations. Globalstar is useful when going to or living in isolated areas such as jungles, deserts and steppes.

O3b Constellation

O3b ("Other 3 billion") is an SES telecommunication constellation in MEO at an altitude of about 8,000 km. It consists in 16 satellites (to be extended to 20 in 2019) orbiting in an equatorial plane. O3b constellation aimed to deliver internet connection to more than 40% of the population (the "Other 3 billion"), especially to non-connected areas. As MEO satellites are closer to Earth than in GEO, the signal delay is reduced from 250 ms (in GEO) to 50 ms. SES planned to build seven O3b mPower satellites, interoperable with the rest of the constellation to provide high speed internet.

Characteristics of Mega-Constellation

Definition

As defined previously, a constellation is a group of satellites working together to fulfil a mission. The term mega-constellation appeared recently to name projects foreseeing hundreds / thousands of satellites for covering the whole planet (Henry, C., 2017). Examples of mega-constellation projects among many others are OneWeb and SpaceX. These two main companies have announced such projects with respectively 900 and 4,400 satellites planned to be launched into LEO. They aim at connecting the planet to the Internet via a satellite-based network. Those mega-constellations architecture make a big leap compared to previous known constellations above mentioned. Indeed, the philosophy of conceiving satellites has completely changed. For instance, a loss of an instrument/system was quite critical for the mission of the GEO satellites, therefore engineers had to put twice the same instrument/system on-board whereas mega-constellation satellites can have just one to lower the cost and reduce the weight of the satellite.

Advantages

One major characteristic of these mega-constellation projects is that a loss of several satellites has a limited impact on the constellation itself and then the service keeps working. For that reason, companies designed "low cost" satellites that can be made quickly. The industry needs to sharply increase their manufacturing facilities to produce satellites at an unprecedented rate so far. This is causing incredible changes within the space industry. The cost of components is dramatically reducing due to higher production. The shape of the space industry is now evolving thanks to the actors of the so-called New Space (private companies that make business within the space industry).

Sending satellites to GEO is really costly and not adapted for covering the whole planet, especially the polar zones of both hemispheres. The idea is then having a large coverage by sending myriads of small satellites (smallsats), with an average mass each of 150 kg for OneWeb and 380 kg for SpaceX (Henry, C., 2017), compared to the classic telecommunication satellite around 2 to 6 tons in GEO orbit.

Thanks to the limited impact of the loss of several satellites, the satellite network is robust. This fail-safe conception ensures the good functioning of the service without being jeopardized by satellite failures. Even more, launching another satellite should not take 2 years (a classic contract duration from building to launching a telecommunication satellite) but less (a few months to a year) and will have a more reasonable cost. In that way, we may envision mega-constellation as a resilient network able to work and to be repaired faster than ever.

Orbital Mechanics

Covering the whole planet is not as easy as it seems: several parameters need to be taken into account for positioning satellites like the coverage of antennas beam (the zone covered by one satellite), the altitude, the number of satellites and the desired performance for data transmission, among others. Launchers can inject in orbit several satellites per launch since LEO and MEO satellites are usually smaller than GEO satellites.

Launch

OneWeb satellites are small and light. Micro launcher projects are thriving everywhere however just a few are yet in operation to launch one smallsat at a time (Electron rocket only by now). Launching of satellite clusters (several satellites mounted on a common disposal system) has become a common thing for providing launches with Soyuz, Proton or Ariane V.

How Is It Possible?

Such disposal is possible thanks to the last stage of the rocket. This last stage has the crucial role to manoeuvre with the cluster of satellites and inject them at the right point in orbit. The easiest way is to place all the satellites of the cluster within the same orbital plane (same inclination). In fact, changing an orbit inclination is a really costly manoeuvre, fuel-efficiently speaking. Only Russian Soyuz vehicle launched 73 CubeSats in 3 different orbits, more for technological demonstration than real market expectation. The mission architecture for satellites to place in the same orbital plane is the following:

- The upper stage will bring the whole clusters of satellites to the designated circular orbit. Then it will release the satellites that need to be placed (one or more) keeping the others attached to the upper stage.
- The upper stage will brake to reduce its speed changing its trajectory and going to an elliptical orbit that will create a shift with the first satellites released (the upper stage will take less time to make a complete rotation).
- When the upper stage reaches again the apogee, it will ignite the motor again to circularize the orbit and make it coincident with the orbit of the previous satellites. It will then release other satellites.
- The upper stage will reiterate the operation until all satellites are placed.

How Many Launchers Are Necessary?

Depending on the launcher contracted to send satellites in orbit, the number differs. For instance, Soyuz that is considered as a medium-class launcher will place in orbit between 32-36 satellites per launch (Clark, S., 2015). For now, OneWeb has contracted with Arianespace for 21 launches with Soyuz carrying the near 700 satellites constellation into space. It also has a contract for 39 missions with Blue Origin (Henry, C., 2017) and 11 launches with Proton (Zolik, 2018), a heavy launcher.

Obviously to put the Starlink constellation of SpaceX in LEO (about 4,400 satellites), it will require 220 Falcon 9 launches (20 satellites per launch vehicle) to complete the constellation. The company has to launch 2,200 satellites by 2024 for keeping the license delivered by the FCC. In practice it means that, SpaceX needs to make 18 Falcon 9 flight per year. SpaceX would have the choice also using Falcon 9 heavy to put a higher number of satellites on orbit (64 per launch) and thus reducing the number of launches required.

Why Not Using Smaller Rockets?

First of all, just a few micro launchers are available or on the road to be operative on the market (see the Electron rocket of Rocket Lab) and their capacities are very limited. Therefore, it would be a one satellite per launch for most micro launchers. Building a constellation within a short frame time may not be possible by now using micro launchers due to the very limited offer.

Secondly, the cost per kilogram is much cheaper with a bigger launcher. The cost of flying with an Electron launch vehicle is about $40,000 per kilogram, nearly twice as much as the Soyuz one.

Nevertheless, micro launchers may be essential in the future for managing the constellation and responding in a shorter amount of time to replace a failed satellite or to launch an active satellite removal to deorbit debris for instance.

Positioning

Positioning satellites is key to covering the whole planet. By taking into account the technical characteristics of the satellite antennas (power, pointing accuracy and angle of the antenna) and the altitude of the satellite orbit, the number of orbital planes and satellites per plane can be determined.

Most of mega-constellation satellites are to be positioned into a specific LEO orbit called polar orbit (or quasi-polar orbit). This orbit passes through both North and South poles (or at least close to it for quasi-polar orbit). As mentioned above, OneWeb has divided its constellation into 18 polar orbital planes, each of them populated with 40 satellites. Each orbit is separated by 10° from the other and each satellite of each orbit would be distant by 9° from another satellite.

SpaceX constellation, Starlink, has been designed with 83 different orbital planes. However, the first 1,600 satellites are planned to be placed upon 32 orbital planes inclined by 53° with respect to the equator. Fifty satellites would orbit each plane. The very first salvo will be able to provide internet services by 2019.

A mega-constellation bears back-up satellites also which are satellites that can be used for replacing a deteriorated one. OneWeb planned to build more than 250 spares satellites, representing more than 10 spare satellite per plane. Dealing with this special feature and the high number of satellites per orbit, the constellation ensures a minimum service and redundancy of critical systems.

Space Debris

Mega-constellations are raising concerns about space debris especially in LEO. Making numerous constellations significantly increases the risk of collisions. Satellites, when hit by debris at relative speed near 10 km/s, may explode into multiple pieces. Thus, the loss of control of one of the satellites would be threatening for the rest of the constellation. With the growing demand for launching in LEO, the orbit is likely to be congested. This demand is linked to the advantages of the position for Earth observations, CubeSats, space stations and now communication mega-constellations. Such situations have pushed OneWeb to call for deorbiting systems with a high level of reliability. Each satellite will be equipped with ion electric propulsion enabling deorbiting at any time even in case of malfunctioning satellites. It will help to avoid congesting the orbital plane by failed/old satellites.

Operational active debris removal (ADR) is not foreseeable in the near future. Only a few concepts do exist by now but companies may invest in the future to prevent incidents and to possess a new fleet management tool.

Space Traffic Management

As of 2017, about 1,100 operating satellites (Union of Concerned Scientists, 2017) in LEO were listed which represents an increase of 40% since 2016. Indeed, the proliferation of CubeSats (very light satellites that can be built at a very low cost to realize multiple missions) has participated significantly to such an increase. Nevertheless, with the total number of satellites that SpaceX and OneWeb plan to put in LEO, at least 7,000 satellites will be in LEO by 2025. It raises challenging issues dealing with space traffic management. Some people have been arguing that space needs a traffic management like aviation has its own. This idea made its way in the International Civil Aviation Organization's debate. Nonetheless, moving in space is quite different than flying into the air and is very fuel-consuming. Space Agencies and private companies are working on new concepts to offer to the space world tools for managing space traffic, while international organizations are debating on legal regimes and needs for space traffic management.

Nowadays, private companies such as MAXAR TECHNOLOGIES are moving to build on-orbit servicing spacecraft to provide support to GEO satellite such as refuelling or movement assistance, whereas space agencies such as the European Space Agency (ESA), backed by industrial company's expertise, are developing the Space Rider. This vehicle is a small version of the Intermediate eXperimental Vehicle (IXV) aiming at providing satellite support in LEO such as launch, deorbit, or safe re-entry. A safe re-entry will allow a company to take a satellite in orbit back to Earth, repair it and send it to space once again. Nevertheless, a cheaper way may be to send a new satellite in the case of mega-constellation due to the operation cost. Robotic technologies are evolving quickly, and first prototype of on-orbit repair services should appear before the end of the next decade.

Launching and Deorbiting

Mega-constellation management will evolve thanks to the arrival of a new class of launchers called micro-launchers. Due to their mass capacity they are only able to send between 50 kg to 300 kg in LEO. Companies have planned to launch several times per month and even several times per week. This launch rate revolution should reduce the delay from contract to launch date at about a couple of months in the following years.

Most mega-constellation satellites are planned to have a small electric engine, thus allowing avoidance manoeuvre and the end-of-life re-entry mandatory for LEO within 25 years. Each satellite should then be able to deorbit by itself in case of damages or failures impeding operational utilization.

Functioning

How Does Mega-Constellation Aim to Provide the Internet?

As explained in the telecommunication satellite section, several space segment architectures are possible (satellites send data to a gateway or to a terminal). The first generation of OneWeb satellite is based upon the Globalstar architecture scheme, meaning that user needs to have a OneWeb device (actually, it looks like a small shoebox) that will directly communicate with above satellites. The user can connect its phone to the device. The satellite will receive the signal and send it back to a specific OneWeb Gateway connected to the overall Internet Network providing then access to the Internet. Nonetheless inter-satellite links are not planned for the first constellation generation; thus the satellite will need to find a gateway under its path to be able to send the signal. Yet, OneWeb is working on the second generation with optical laser communication between OneWeb satellites that will limit the necessity of gateway coverage.

How to Cover the Entire Earth?

A massive production of OneWeb receivers has begun under a contract with Hughes Network Systems to build gateways. The gateway will be having several antennas features to communicate with the LEO constellation. Multiple antennas are needed due to the satellite speed. The satellite rotates around the Earth in about 1 to 2 hours depending on its altitude. Having several antennas allows the gateway to track following satellites while the first one is disappearing from the focus of the gateway. Such system ensures the continuity of the service. The gateway will be able to handle over 10,000 hand-offs/connections with satellites per second, which is a real breakthrough for now.

OneWeb has decided to, either provide direct access meaning that you can have one OneWeb terminal devices installed in your home and benefit from it thanks to 3G/4G network access and Wi-Fi mode, or to install terminals which can be accessed by operators to extend their coverage.

As of October 2018, SpaceX has not revealed its constellation architecture.

What Is the Coverage Offered?

Each point of the world will be overhung by more than 20 satellites for the Starlink constellation. Thus, even if certain satellites failed, the number of satellites in visibility of ground stations is high enough to ensure services.

However, some criticism arose for the coverage from the OneWeb first generation mega-constellation. For instance, since there is no satellite gateway on the Saint-Helena Island, the access to OneWeb services will not be possible because the Island is too isolated and too far away from a continental ground network. Only inter-satellite links can achieve providing a regular access to the Internet to the Saint-Helena Island, which is also true for other archipelagos or isolated areas. Another option would be to build a ground station on the Island but the cost of such installation can be too excessive compared to the revenue that can generate the low demand.

Frequency Management

Performances

OneWeb has already planned to offer the first peak speeds of 500 Mb/s everywhere and the second generation would be able to provide speeds beyond 2,5 Gb/s by 2021.

The first generation would be able to handle about 7 Terabits/s (one million times more than 7 Megabits/s) of capacity while the second one will reach 120 terabits/s followed by the target of 1,000 Tb/s of capacity by 2025 with the third generation (Henry, C., Oct., 2017).

SpaceX thinks it can get a latency of 25 milliseconds by 2019. This represents a big improvement compared to the 600 millimetres latency of current communication satellites. Compared with ground systems and DSL connections especially, this latency is superior. It still represents an opportunity for unconnected areas. SpaceX targets speeds of 1Gb/s by 2019 (Brodkin, J., 2017). The higher the number of satellites, the higher the capacity.

By now, ground technologies depending on whether copper cables are used with DSL technology, or optical fibre, can provide speeds of 500 Mb/s to some few Gb/s. However, since ARPU (Average Revenue Per User) is too low, major telecommunication companies have neglected covering the whole population in isolated areas.

The 3GPP explained that satellite-based network is the key to cover the whole Earth with 5G and is working closely with satellite operators to adapt technical recommendations. For instance, the satellite network will enable airlines to broadcast 5G to their on-board customer during the whole flight thanks to the permanent presence of satellites above planes (Reals, K., 2018).

Use of Frequencies

Albeit OneWeb satellites are conceived to transmit and to receive signals from the Ku band (12-18 GHz frequency range) of the radio frequency spectrum which is often use for satellite communications and broadcasting television, OneWeb has been asking an authorization to the Federal Communication Commission (FCC, the United-States telecommunication regulator agency) for using the V band (40-75 GHz frequency range) of the radio frequency spectrum for its future generation. This band has not been intensively operated by now. There has been only a few uses for specific telecommunication satellite mainly.

Starlink satellites are conceived to transmit and to receive signals from Ku band, Ka band (26,5-40 GHz frequency range) and like OneWeb, SpaceX has asked for using the V-band also.

In order to prevent interferences between satellites of the different constellations, satellites need to work with sufficiently distant frequencies to limit signals alteration, therefore it is mandatory to receive an authorization from ITU and the National Telecommunication Regulation Agency. Granting the right to access frequencies is currently still under debate for constellations. Before such projects exist, launching a unique satellite using a particular frequency was sufficient to keep the license, whereas the FCC agreed to give SpaceX the operating license if it could launch its full constellation (1,600 satellites) within 9 years starting in 2018.

A main issue with LEO constellations is that they may interfere with signals emitted by GEO satellites. Indeed, they are placed between GEO satellites and the Earth and they operate in the same frequency band (remember that interference happens when frequencies are close). To prevent such events, OneWeb implemented a specific procedure consisting in progressively turning off signals when passing over the equator. The interference may occur when the LEO satellite is near the Equatorial plane (GEO satellites are *de facto* in this plane). By switching off the LEO satellites when crossing an area close to the equatorial plane, interferences are thus avoided in principle. It remains to be proved that such a procedure is indeed effective in space.

Other Type of Mega-Constellations

Many other mega-constellation projects exist around the world. Boeing, Telesat, Samsung are the other main companies aiming at developing such a capability. Limited information is available about these contemplated constellations. Even the government of China has also planned to launch a 300-Hongyan-satellite communication constellation in LEO to provide internet access to its people and beyond its borders to the rest of the world. Fleet Space Technologies, an Australia new space comer, planned to launch more than 100 smallsats for Internet of Things (IoT) purposes such as connecting and tracking connected objects for resource management in agriculture or mining industry. Telecommunication is not the only benefit that mega-constellation of satellites can provide. Earth observation is a major activity in LEO also. Companies like EarthNow, a start-up that recently raised funds from the biggest players (Airbus, OneWeb, Bill Gates and Softbank) (Henry, C. (2018), plans to build an impressive Earth monitoring system composed of 500 satellites orbiting in LEO. This mega-constellation will provide a-real-time video of any point on the Earth within 20 milliseconds and with 20 frames per second. Indeed, such a mega-constellation is raising interesting challenges. Due to the huge amount of data that need to be collected and analysed, it would be really hard to send the raw data to the ground and process it. It would require numerous heavy super-calculator and databases. The idea is therefore to directly process images captured by satellite on-board sensors and only send down to Earth the relevant processed data. In doing so, the need for database storage is drastically reduced and the method is time saving. Applications resulting from the Earth Observation mega-constellation are various like: change-based detection (the system will observe different areas and identify changes such as a road being built), on real-time tracking to study traffic jams or maritime channels, or intelligence support. Planet Labs, a California based company, has yet launched near 300 small satellites (only 150 still actives) for its Flocks constellations with the goal of providing a complete image of Earth per day.

Table 2. Overview main advantages/limitations of mega-constellation and GEO satellite network

	Mega-constellation	GEO satellite network
Advantages	- Low signal delay (high transmission rate) - Global coverage including polar regions - Fail safe mode (resilient to satellite failure) - Lower cost per satellite	- Wide coverage with 2-3 satellites - Constantly over the same areas - A few ground stations needed - No inter-satellite link
Limitations	- Dramatically increasing number of objects in LEO - Long deployment phase (numerous launches needed) - New ground stations or inter-satellite links needed - Possible interference with GEO satellite	- Polar region not deserved - Satellite failure can severely affect the service - Signal delay (250 ms) - Cost per satellite - In-orbit positioning requires more fuel

CONCLUSION

Mega-constellations are a turning point for telecommunication networks. Bridging the digital gap will be possible in a short time with mega-constellation. Humanity can benefit very much from such projects, while people need to be conscious that mega-constellation projects are also a threat to safe, secure and sustainable space activities in LEO. Indeed, the risk of collision is increasing due to the thousands of new satellites in LEO bringing a serious concern to space agencies about the possibly to getting closer to the Kessler syndrome that would cause the proliferation of space debris generated by debris-debris collisions. The existing processes and frameworks need to be revised as they were drafted and issued without consideration of mega-constellation activities. Such initiative is crucial to not threaten space access and its use in the future. Reconsiderations of international regimes related to LEO and other orbits will be a keystone to implement effective policies, regulations and frameworks. Coordination and cooperation of all space actors (space fairing nations, developing countries, space agencies, manufacturers, satellite operators, …) should be deepened to reinforce coherent and fair space traffic management. The shift from state activities to private entrepreneurs in space is accelerating through the burgeoning mega-constellation projects around the world. Further articulation between state agencies and private actors will be beneficial to the overall space economy and stability. ITU, UNOSA, COPUOS are all long lasting international agencies that shaped the outer space treaty activities. With the new challenges they are facing, profound changes and reforms of recommendations are necessary to enhance Humanity benefits from space and it is a unique opportunity to build a new joint vision of the future of space.

REFERENCES

Autorité de Régulation des communications électroniques et des Postes (ARCEP). (2018). *4G*. Retrieved from https://www.arcep.fr/index.php?id=9971

Bat Conservation Trust. (2018). *Echolocation.* Retrieved form http://www.bats.org.uk/pages/echolocation.html

Bertenyi, B. (n.d.). *3GPP system standards heading into the 5G era.* Retrieved from http://www.3gpp.org/news-events/3gpp-news/1614-sa_5g

Bousquet, M. (2018). ISAE-SUPAERO lecture: Introduction to satellite communication systems. Toulouse, France: ISAE-SUPAERO.

Brodkin, J. (2017, March). With latency as low as 25ms, SpaceX to launch broadband satellites in 2019. *Ars Technica.* Retrieved from https://arstechnica.com/information-technology/2017/05/spacexs-falcon-9-rocket-will-launch-thousands-of-broadband-satellites/

Clark, S. (2015, July). OneWeb launch deal called largest commercial rocket buy in history. *Spaceflight Now.* Retrieved from https://spaceflightnow.com/2015/07/01/oneweb-launch-deal-called-largest-commercial-rocket-buy-in-history/

Cobley, P. (n.d.). *Communication.* Retrieved from https://semioticon.com/seo/C/communication.html

Dirk Krebs, G. (2017). *Syncom 1, 2, 3.* Retrieved from http://space.skyrocket.de/doc_sdat/syncom-1.htm

European Space Agency. (2017, January). *Space Debris by Numbers.* Retrieved from https://www.esa.int/Our_Activities/Operations/Space_Debris/Space_debris_by_the_numbers

Globalstar. (2000). *Description of Globalstar System.* Retrieved from https://gsproductsupport.files.wordpress.com/2009/04/description-of-the-globalstar-system-gs-tr-94-0001-rev-e-2000-12-07.pdf

Henry, C. (2017, March). Blue Origin signs OneWeb as second customer for New Glenn reusable rocket. *Space News.* Retrieved from http://spacenews.com/blue-origin-gets-oneweb-as-second-new-glenn-customer/

Henry, C. (2017, October). SpaceX, OneWeb detail constellation plans to Congress. *Space News.* Retrieved from http://spacenews.com/spacex-oneweb-detail-constellation-plans-to-congress/

Henry, C. (2018, April). Startup with Softbank, Airbus investment planning video constellation with several hundred satellites. *Space News.* Retrieved from http://spacenews.com/startup-with-softbank-airbus-investment-planning-video-constellation-with-several-hundred-satellites/

IEEE. (2002). *IEEE Std 521-2002 (Revision of IEEE Std 521-1984).* IEEE Standard Letter Designations for Radar-Frequency Bands.

International Telecommunication Union. (2016). *ITU Radio Regulations.* Retrieved from https://www.itu.int/pub/R-REG-RR/en

Internet World Stats. (2017, December). *Internet Users in the World by Regions.* Retrieved from https://www.internetworldstats.com/stats.htm

Knörnschild, M., Jung, M., Nagy, M., Metz, M., & Kalko, E. (2008). Bat echolocation calls facilitate social communication. *Proceedings of the Royal Society, Biological sciences.* Retrieved from http://rspb.royalsocietypublishing.org/content/279/1748/4827

Maine, K., Devieux, C., & Swan, P. (1995). *Overview of IRIDIUM satellite network.* doi:10.1109/WESCON.1995.485428

Maxwell, J. C. (1865, January). A Dynamical Theory of the Electromagnetic Field. *Philosophical Transactions of the Royal Society of London, 155*, 459–512. doi:10.1098/rstl.1865.0008

National Aeronautics and Space Administration. Science Mission Directorate. (2010). *Gamma Rays.* Retrieved from https://science.nasa.gov/ems/12_gammarays

Pancaldi, F., Vitetta, G., Kalbasi, R., Al-Dahir, N., Uysal, M., & Mheidat, H. (2008). Single-carrier frequency domain equalization. Signal Processing Magazine, 37–56.

Priemer, R. (1991). *Introductory Signal Processing.* Hackensack, NJ: World Scientific.

Purcell, E., & Morin, D. (2013). *Electricity and Magnetism* (3rd ed.). New York: Academic Press.

Reals, K. (2018, March). OneWeb Looks To Disrupt Inflight Connectivity Market With LEO Network. *Aviation Week.* Retrieved from http://aviationweek.com/aircraft-interiors/oneweb-looks-disrupt-inflight-connectivity-market-leo-network

Shannon, C. E. (1949, January). Communication in the Presence of Noise. *Proceedings of the Institute of Radio Engineers*, 447-457.

Stoehr, M. (2012, December). *Getting Started With A Short Range Radio Design.* Retrieved from http://www.electronicdesign.com/communications/getting-started-short-range-radio-design

Union of Concerned Scientists. (2017, November). *USC Satellite Database*. Retrieved from https://www.ucsusa.org/nuclear-weapons/space-weapons/satellite-database#.WL2lyPJCjNE

Vlajic, N. (2010). *Analog Transmission of Digital Data course*. Stanford University. Retrieved from https://web.stanford.edu/class/ee102b/contents/DigitalModulation.pdf

Zolik. (2018, April). *Europe buys from Russia 11 rocket "proton Average"*. Retrieved from http://handofmoscow.com/2018/04/12/europe-buys-from-russia-11-rockets-proton-average/

KEY TERMS AND DEFINITIONS

Frequency: Number cycle repetitions that occur within one second.

Internet: Interconnection network born in the 60s that is connecting people through electronic devices using internet protocol.

Mega-Constellation: A constellation that is composed of several hundreds and thousands of satellites orbiting the Earth.

Modulation: A means to encode or decode a signal in the purpose to convey information, most often binary words.

Outer Space: The area above the Von Karman line (approximately between 80 to 100 km).

Satellites: A system placed in orbit around a celestial body, mainly Earth, carrying instrumentation in the purpose of fulfilling a determined mission such as telecommunication, Earth observation, or science.

Signal: A signal is a means to convey and to remotely transmit information.

Telecommunication: The means to convey information using emitting and receiving features be it using writings, signs, waves, or other transmission way.

Chapter 9
Mega-Constellations: Legal Aspects

Alberto Rueda Carazo
Université Paris-Sud XI, France

ABSTRACT

As a reflection of the fast-paced world we currently live in, a new era in satellite technology begins to emerge. This sector, of vital importance for our day-to-day lives, is in a period of significant change, characterized by the use of large satellite constellations. These mega-constellations aim to provide, among other things, worldwide internet coverage. Since 2015, satellite internet providers such as Space X and OneWeb have been working on the idea of mega-constellations and intend to launch circa 4.000 and 882 satellites respectively, thus forcing satellite manufacturers to lower the costs drastically. However, many legal issues may arise since the vast amounts of satellites will significantly increase the risk of collision with other satellites or spacecraft.

INTRODUCTION

According to the United Nations Office for Outer Space Affairs (UNOOSA), in 2018 there are 2075 satellites orbiting around the Earth. (Online Index, 2018). Soon, this figure will be greatly increased due to the incorporation of large satellite constellations formed by hundreds or even thousands of satellites. It is obvious that the increase in the number of satellites will be accompanied by an increase in risk of collision and a number of legal issues that have not been addressed so far. The current tendencies will change the outer space environment and its traditional principles in a radical way. (Vila, 2016).

In 2016, Elon Musk asked for prior permission to the Federal Communications Commission (FCC) to launch a constellation of 4,425 satellites during the next ten years with the purpose of offering a global satellite Internet service. (De Selding, 2016). Needless to say that the Space Surveillance and Tracking segment of the Space Situational Awareness (SSA) Programme will gain more importance due to the thousands of new satellites launched into outer space. On the bright side, companies like OneWeb aim to use these macro-constellations to provide communication service to anywhere on Earth, hence facilitating the access to communications of developing countries.

DOI: 10.4018/978-1-5225-7256-5.ch009

The number of launches every year is expected to be constant even if the complete constellation is deployed in its entirety, since these types of satellites will have to be constantly replaced due to its relatively short life span. (De Selding, 2016). Logically, the number of collisions will significantly increase because of the high concentration of functional and non-functional satellites.

As a response to the imminent risk posed by macro-constellations, the next question must be discussed: How do space actors, public or private, decide which orbit or which frequency they will use in order to not interfere with the activities of other actors? This chapter tries to answer this question by explaining what is the role played by the International Telecommunication Union (ITU or the Union). Interestingly, the ITU, which is the oldest international organisation, is the one in charge of managing and coordinating this state-of-the-art technology. The Union is a specialised agency of the United Nations and is in charge of promoting and making accessible communications systems to both developed and developing countries. Therefore, it is from paramount importance to understand the structure, the current legal framework of the ITU, and how the process of orbit/spectrum allocation works. This chapter intends to give the reader a broad insight of the legal framework of the ITU in order to better understand the legal complexities lying behind macro-constellations.

Finally, the chapter will analyse and discuss the main problem arising from macro-constellations in light of the current international legal framework. The growing number of space debris in such important resources as the geostationary orbit or the low-Earth orbit presents one of the major drawbacks and threats posed by large satellite constellations. (Leinberg, 1989). As a result of the broadly drafted *Corpus Iuris Spatialis*, and due to the legal uncertainty it creates, states and international organisations have adopted a number of legal guidelines to promote space debris mitigation. (De Man, 2013). These mitigation guidelines are essential, albeit not sufficient. Their non-legally binding character makes it difficult to implement them thus requiring a compelling series of international legislation that can address this important problem. The most important recommendations will be discussed.

BACKGROUND

In order to provide a complete background to the reader, this part will focus on three main aspects that will help to understand the relevance of the ITU in light of the advent of macro-constellations. These three sections are the history, the structure and the legal framework of the Union.

Firstly, from a historical perspective it is worth noting that, since its conception, telecommunications have morphed from a military application into an essential global network able to connect people around the world. Due to its nature at the moment of its inception, there was a lack of international cooperation and a general sentiment of reluctance towards connecting the networks, since it could be seen as detrimental to the sovereignty of the States. There were numerous national telegraph networks that were not interconnected amongst each other and, therefore, it required a vast amount of separate agreements between countries to regulate the different aspects of telegraphy, an essential cornerstone for telecommunications. (Coddding & Rutkowski, 1982) (Bernhardt, 1984). It is important to remark the first two main organisations that laid the foundations for the future ITU. In 1850, Prussia, Austria, Bavaria and Saxony established the Austro-German Telegraph Union. It was soon followed by the creation of the West European Telegraph Union on 29 December 1855 by Belgium, France, Sardinia, Spain and Switzerland. Despite the great uniformity across the continent, there were still two separate actors regulating the same activity. As a solution for the absolute dissonance and for the sake of establishing harmonised international standards that would facilitate interconnection, European States[1] founded the International Telegraph Union on 17 May 1865. (Standage, 1998). This Union was intended to strengthen cooperation between the States, being the first international organisation with universal vocation. From a treaty perspective, the idea was to have a Convention, which laid down the main principles, and technical annexes (regulations), which would be able to be modified periodically. Both the Convention and the technical annexes had treaty value.

The beginning of the XX century was marked by the arrival of wireless telegraphy. The 1906 Berlin Conference instituted the International Radiotelegraph Union following the same structure as the International Telegraph Union.

The ITU as we know it nowadays, was the result of a merger set up at the 1932 Madrid Conference. The objective of such conference was to integrate the conventions from the International Telegraph Union (1865) and the International Radiotelegraph Union (1906) thus creating a united international organisation. (Sands, Klein & Bowelt, 2009). On September 16, 1946, and under the request of the United States of America, the ITU invited its members to the Atlantic City Conference. As an outcome of this conference, the International Frequency Registration Board (IFRB) was created with the purpose of regulating the use of frequencies. Finally, in 1947 the ITU became the first specialised agency of the United Nations (UN). (White, White & White, 1988) (Jakhu, 1983).

Secondly, a look in the legal instruments of the Union is crucial in order to comprehend its structure. As enshrined in article 1 of the ITU Constitution, its purposes encompass a wide range of activities. From maintaining and promoting international cooperation among its member states to rendering assistance to developing countries in the field of telecommunications. In contemplation of these general purposes, article 1 further explains what are the actions to be taken in order to achieve such goals. The ITU will facilitate the standardisation of telecommunications and coordinate the allocation and allotment of frequencies and orbital positions in order to avoid the harmful interference between countries. (ITU Constitution, 1992).

By virtue of a substantial process of modernisation implemented during the Additional Plenipotentiary Conference held in Geneva in 1992, the ITU was provided with a new structure that would allow it to face the challenges posed at the time. (Tchikaya, 1998). This long-awaited reform gave the ITU the capacity to adapt itself to the changes that the telecommunication sector was undergoing. (Savage, 2016). The result of this restructuring was the formation of three Sectors in charge of managing the different activities.

- **The Radiocommunication Sector**'s (ITU-R) main objective is to ensure the rational, equitable, efficient and economical use of the frequency spectrum by all radiocommunication services. In doing so, the ITU-R is in charge of allocating frequency bands, the allotment of frequencies, the registration of frequency assignments, and any associated orbital positions in the geostationary-satellite orbit and the coordination of efforts to avoid harmful interferences. The world radiocommunication conferences (WRC) are held every three to four years and aim to revise Radio Regulations, the international treaty governing the use of the radio-frequency spectrum and the geostationary-satellite and non-geostationary-satellite orbits.
- **The Telecommunication Standardisation Sector** (ITU-T) focuses on the adoption of technical standards at an international level in order to ensure the consistency among equipment and services. It is the oldest activity of the ITU. There are 13 study groups, which develop technical recommendations; around 200 recommendations per year approved by the World Telecommunication Standardisation Assembly, which meets every 4 years. It is worth noting that despite the legally

non-binding nature of these recommendations, as they are *soft law*, States comply with them to avoid a possible technical isolation.

- **The Telecommunication Development Sector** (ITU-D) promotes international cooperation by giving technical assistance and by developing and expanding telecommunication networks and services with a particular focus in developing countries. This new system allowed the ITU to rapidly aboard any problematic case arising from new technological developments, addressed in the conferences organised periodically.

The main structure of the ITU, among which are these three sectors, is envisioned in article 7 of its Constitution providing that the Union shall comprise: the Plenipotentiary Conference, which is the supreme organ of the Union; the Council, which acts on behalf of the Plenipotentiary Conference; world conferences on international telecommunications; the Radiocommunication Sector, including world and regional radiocommunication conferences, radiocommunication assemblies and the Radio Regulations Board; the Telecommunication Standardization Sector, including world telecommunication standardization assemblies; the Telecommunication Development Sector, including world and regional telecommunication development conferences; and the General Secretariat. (ITU Constitution, 1992)

Lastly, a general knowledge of the legal framework of the Union will indicate the reader what are the sources of law of the Union and what are its main characteristics. The Secretary-General, who acts as a legal representative of the Union, organises together with the Coordination Committee, the activities of the Union.

The legal framework of the Union encompasses both its Constitution (CS) and the Convention (CV). Both texts have treaty status and therefore are binding upon member states. Both, the Constitution and the Convention are complemented by the Administrative Regulations– Technical rules regularly modified within the framework of world conferences. Radio Regulations and International Telecommunication Regulations, also binding on Member States, are an essential part of such Administrative Regulations.

The Union fully recognises "the sovereign right of each State to regulate its telecommunications". This principle provides that states do not transfer their sovereignty to the ITU for the regulation of telecommunications, considering that the Union is an organisation for international cooperation and not an organisation in charge of managing or regulating telecommunications. The ITU therefore acts only in the interest of sovereign states, and in no case it is meant to substitute them.

In light of the principle of equality among States, Members States shall participate in the different conferences on a "one State, one vote" basis. The quorum required to amend the Constitution by the Plenipotentiary Conference is a two-thirds majority while amendments to the Convention only require a majority of the delegations that are authorised to vote.

Generally speaking, States have the right to make reservation to treaties in order to "exclude or modify the legal effects of certain provisions of the treaty in their application to that State".(VCLT, 1969) However, in accordance with article 19 of the Vienna Convention of the Law of the Treaties (VCLT), amendments to the Constitution or the Convention shall enter into force as a single amending instrument and, as enshrined in the ITU Constitution, in article 55(6) and 42(6) respectively, "ratification, acceptance or approval of, or accession to, only a part of such an amending instrument shall be excluded". On the other hand, reservations to the Administrative Regulations are permitted. Nonetheless, it is presumed that States will not exercise this right due to the tremendous setbacks and technical issues that may originate from this option. Due to the strategic nature of telecommunications and its vital importance for the international community, even if the resolutions or recommendations are not legally binding, compliance with them is a condition *sine que non* for States to avoid complete chaos and ensure the efficient functioning of international telecommunications. (Frowein, 2004) (Detter, 1996)

Regulatory Aspects of the ITU

The analysis of the regulation set by the ITU is from paramount importance when speaking of macro-constellations. The main characteristic of these types of constellations is the large amount of satellites using the same radiofrequency and orbits. When operating large satellite constellations, it is important to beware that both radio frequencies and orbits are a finite resource and therefore must be used adequately and in accordance with the provisions of the Radio Regulations (RR). (ITU Constitution, 1992) The Radio Regulations provide an exhaustive procedure with the objective of protecting the different operators from interferences which "endangers the functioning of a radio navigation service or of other safety services or seriously degrades, obstructs, or repeatedly interrupts a radio communication service operating in accordance with" such regulations. (ITU Radio Regulations, 2015)

Article II of the Outer Space Treaty encapsulates the principle of non-appropriation stating that "outer space, including the moon and other celestial bodies, is not subject to national appropriation by claim of sovereignty, by means of use or occupation, or by any other means." (OST, 1967) This means that neither States nor private companies have the capacity to own radiofrequencies or orbital slots. Instead, the ITU provides an exclusive use for a particular operator with the aim of ensuring the correct functioning of global communications as stated in its Constitution. (ITU Constitution, 1992)

There are two processes to access orbital and associated spectrum resources. These processes can be found at the Radio Regulations of the ITU and are referred to as "first come, first served" principle and the *a priori* approach. The "first come, first served" finds its roots in the cooperation and coordination among states in an attempt of international recognition and protection from harmful interferences. As a solution for a possible monopolisation of the orbit/spectrum by industrialised, spacefaring nations, the *a priori* plan intends to make an equitable distribution of the different frequencies and orbital slots among the members of the Union. (Jakhu, 1983) These plans assure future access to developing countries by reserving orbital resources and bandwidth.

For this purpose, the allocation of frequencies, the ITU divides the world into three different regions as it can be seen in Figure 1. These regions are structured in the following manner: region 1, which comprises Europe, Middle East and Asia; region 2, comprising the American continent; and region 3, including Asia and Australasia.

The first general plan for Broadcasting Satellite Services (BSS) adopted in 1977 by the World Administrative Radio Conference (WARC) for Broadcasting Satellite Services set up a Plan for regions 1 (11,7- 12,5 GHz) and 3 (11,7- 12,2 GHz). Region 2 (12,2- 12,7 GHz) adopted its plan a few years later at the 1983 Regional Administrative Conference (RARC) for the Planning of the BSS. (Allison, 2014) Although the WARC ORB-85 and 88[2] was to analyse the access to the geostationary orbit it ended up setting up a partial plan for fixed-satellite services (FSS). The different representatives managed to find a middle ground between the interests of the developing countries, in guaranteeing space access, and the interests of the developed countries, in retaining the current demand-based process.

The basic common phases that need to be fulfilled in order to obtain access are described in the Radio Regulations. Before initiating the procedure, it is important to remember that the International Telecommunication Union is an international organisation and thus, individual operators cannot interact directly with it. Alternatively, the governmental department designated by the states will request radiofrequen-

Figure 1. ITU Regions
Source: ITU

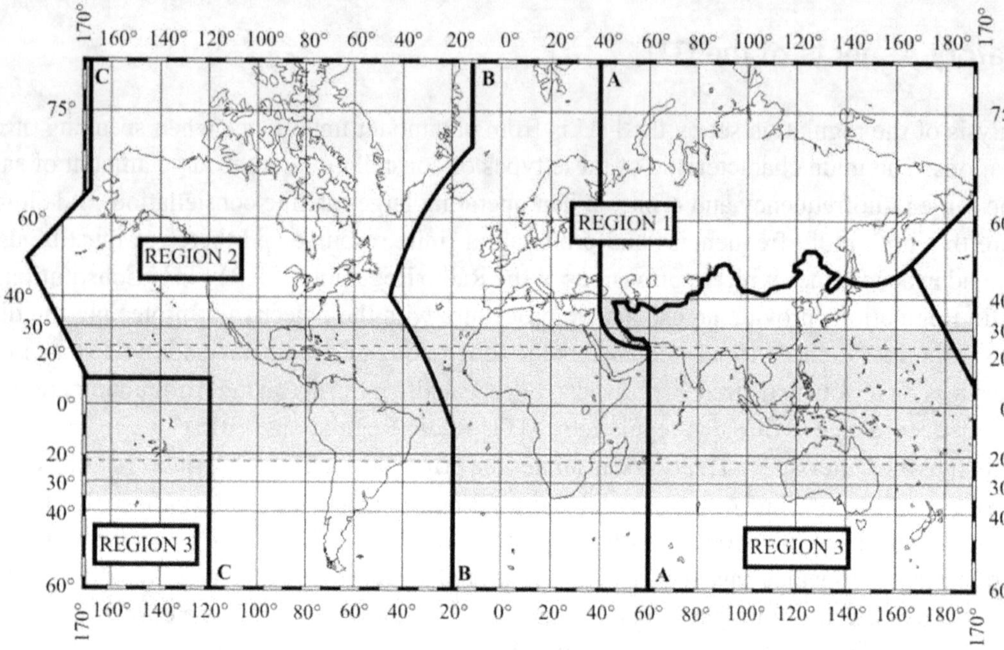

cies and orbital slots on behalf of them. These governmental departments or services "responsible for discharging the obligations undertaken in the Constitution of the ITU" will be referred as "*administration*". (ITU Radio Regulations, 2015)

The procedure starts with a report containing a general description of the network or system that will be sent to the ITU Radiocommunication Bureau (Bureau) by one administration. This report is the so-called Advance Publication Information (API) and is meant to be filed no earlier than 7 years and no later than 2 years before the date of bringing into use the satellite network. The International Frequency Information Circular (BR IFIC) will publish this information before the planned date of bringing into use said network or system. This information includes, *inter alia*, the identity of the identity of the satellite network, the geographical area in which the station is located, the date of bringing the frequency assignment into use and general information about the orbit where the satellite will be operating. (ITU Radio Regulation, Appendix 4, 2015). If the country representing the operator fails to complete the project in a period of 7 years since the bureau's receipt of the API, the project will be cancelled and will not receive protection from harmful interferences.

A coordination procedure must be carried out, in particular, with a view to avoiding harmful interferences with existing networks or systems. If there are indications suggesting that the proposed system may interfere with an existing one, consultations with the registering administration may be undertaken. (ITU Constitution, 1992). Subsequently, the registering state or administration will notify the ITU Radiocommunication Bureau so as to safeguard its assignment recording it in the Master International Frequency Register. (ITU Radio Regulation, 2015). The national administration should submit the coordination request between two and six years before the assignment is brought into use.

After the Bureau has examined whether the notice is in conformity with the Table of Frequency Allocation, the pertinent provisions of the RR and provided that such examination leads to a favourable finding, the Bureau will record the requested assignment in the Master International Frequency Register thus granting it with international protection. The satellite operator has now the duty to bring into use the frequency assignment within the required period or the Bureau will be forced to cancel the entry after duly informing the administration three months in advance before the end of the prescribed period. (ITU Radio Regulation, 2015)

As previously mentioned, private operators do not have the authority to directly initiate the procedure before the ITU. Accordingly, they must request the national authority[3] to ask as an intermediary in order to allocate frequencies and orbital slots. Taking as an example the last case, the FCC would be the selected agency in charge of determining whether or not the operator is worthy of a license to provide commercial satellite services within the United States of America. (United States Communication Act, 1934)

Macro-constellations may be facing difficulties in the future since, as already discussed, getting cleared frequencies and orbital slots requires a lot of bureaucracy not only at ITU level but also at a national level. In addition to this long administrative process it must not be forgotten that due to the unparalleled amount of satellites forming these types of constellations, the procedures may prolong themselves in time. As it happened during the late 1990's, the demand of spectrum/orbit will dramatically increase, hampering the coordination of satellite networks.

To this can be added the increasing number of applications for satellite frequencies and orbital slots aiming to reserve a frequency or an orbital slot without actually using it. This phenomenon based on the "first come, first served" principle, challenges the procedures of the ITU impairing its ability to rightfully manage space telecommunications resources. This practice of "overfilling" triggered by the competition between private companies in order to secure their access to these space resources is also known as "paper satellites". The changes that the satellite industry experimented during that time are comparable to those of today. The reduction of costs to access space, the introduction of a significantly higher number of private actors and the great revenues behind these new technological trends, will make impossible to meet the demands of all of the actors. (Allison, 2014) This could generate a new wave of massive amounts of paperwork sent to the Union that will never be used.

The landmark case of the remote Pacific island state of Tonga who filed to the Union sixteen prime orbital locations over the Pacific, Asia, and America. A former official of Intelsat who established a joint venture with the Kingdom of Tonga, called Tongasat, leaded this initiative. Matt Nilson, the former official, was familiar with the ITU procedures and initiated the filings with the aim of selling the acquired positions for large quantities of money. (Aoki, 2013). As could be expected, Intelsat was opposed to this initiative and considered it an abuse of the procedure. Despite the fact that the conduct could have meant a breach of Tonga's obligations under article 44 of the ITU Constitution, since it completely disregarded the fact that radio frequencies and orbits are a limited natural resource to be used in a rational, efficient and economical way, the International Frequency registration Board (IFBR) had no legal underpinnings to not register the satellite network. (Thompson, 1996) (Delzeit & Beal, 1996). Needless to say, this case served as an example to other countries that also wanted to secure orbital positions. These countries realised the rising value of this finite resource and so they filed a series of applications to later sell these rights to interested countries.

At the moment, it was evident that there was an imperative need to take appropriate measures in order to mitigate this practice. Satellite operators did not have the certainty that their networks were going to receive protection from harmful interferences. The 1994 Plenipotentiary Conference held in Kyoto adopted resolution 18 to try to come up with a solution for the paper satellites. It was up to the national administrations to implement a process of due diligence in which they would make sure that the notified systems would eventually be put into use. (Lyall, 1996). The ITU adopted various methods to try to seek a solution and, finally, succeeded in solving the processing backlog without causing any harm to the administrations. However, efficient use of orbit/spectrum resources still remains a major challenge nowadays since it is crucial to continue the work of the recording frequency assignments and making sure that regardless the issues macro-constellations may bring up, the rational, equitable, efficient and economic use of frequencies and orbital positions is ensured.

Mega-Constellations and Space Debris

Having now a broad insight in the procedures and different steps that need to be taken in order to put into orbit a large satellite constellation or mega-constellation, this section aims to give the reader an understanding of the possible drawbacks that deploying such amount of satellites could have. Among the most notorious issues there is one that is particularly important to address, the generation of space debris. It is going to be challenging to increase the current number of space objects orbiting around the Earth bearing in mind the level of space debris. Environmental issues are by far the most relevant and serious issue when speaking about macro-constellation.

As discussed in previous chapters of this book, the United Nations General Assembly has adopted, since 1967, five treaties dealing with space activities. Concerning space debris issue there are only two of them that are applicable, the Outer Space Treaty (OST) and the Liability Convention. The definition of the notion "space debris" given by the Inter-Agency Space Debris Coordination Committee (IADC) does not clarify whether it can be considered a "space object" since it broadly refers to it as "all man-made objects including fragments and elements thereof, in Earth orbit or re-entering the atmosphere that is non-functional." The great majority of scholars do believe that even if a satellite has not broken up, but it is a non-functioning space object falls indeed within the category of space debris. Inactive or non-functioning satellites are space objects and therefore their fragments in case they break up will also be considered space objects. In this case, the most comprehensive definition has been given by UN COPUOS in the realm of its Scientific and Technical Sub-committee. It defines it as "human-generated, non-functional objects in Earth orbit or re-entering the atmosphere." The report continues explaining that approximately, 60% of the catalogued space debris objects are fragments of spacecraft and rocket bodies. The report gives a list of the main objects that contribute to the increase of space debris such as "satellites that have reached their end-of-life and are left in orbit; Upper states of launchers which had been used to place satellites in orbit; operational debris, which are objects intentionally released during the mission; the result of fragmentation either by a collision between two objects in orbit or from a space object accidentally or intentionally exploding". (UN COPUOS, 2011).

Companies such as Space X and OneWeb and Boeing, are leading the way to flooding Earth's orbits with satellites. It is estimated that in the next years the number of satellites operating around the Earth will be tripled. (Weeden, 2016) As we have previously explained, radio frequencies and orbital slots are finite resources that are progressively becoming scarce. The congestion propitiated by these constellations will significantly increase the risk of collision, which would have atrocious consequences. One collision can generate up to 5,000 pieces of debris that would stay in orbit for around 100 years. (Davenport, 2016).

In 1978, Donald Kessler seemed to have foreshadowed what may be a reality in a few years. A cascading process in which fragments will collide with other objects generating in turn more fragments that will collide and generate even more space debris. (Kessler & Cour-Palais, 1978). Private companies need to be aware of this fact since, most likely, they will have to indemnify the Sate that is supervising their activities. Article VI of the Outer Space Treaty establishes that "States Parties to the Treaty shall bear international responsibility for national activities in outer space" even if such activities are carried out by non-governmental entities. These activities would require the authorisation of and constant supervision of the correspondent state. Because a state can be held liable for damages even if it did not partake in the activity that caused the damage, it will include an indemnification provision in its contract with the satellite operator or private entity wishing to carry out activities in outer space. (Kerrest & Smith, 2009). This mechanism will ensure that non-governmental entities comply with the provisions set out in the Outer Space Treaty and in general public international law.

Although generally speaking United Nations General Assembly Resolutions are non-legally binding, some of them have such an important content that national legislations adopt their core principles into their laws in order to be able to implement them. This is the case of the UN COPUOS space debris mitigation guidelines. (UNGA Resolution, 2008). The European Space Agency (ESA) also issued in 2002 its European Space Debris Safety and Mitigation Guidelines, respected by all its member states. (Kinkrad, 2004). China has reportedly worked in reducing and mitigating space debris since, in 2006, it released a white paper named "China's Space Activities" in which it showed how its policy focuses on mitigating space debris. In 2010 president Barack Obama went a step further and signed into law the Title 51 of the United States Code, entitled National and Commercial Space Programs. (Gabrynowicz, 2010). These regulations include some of the rules on space debris mitigation that were inspired by the UN COPUOS mitigation guidelines. These guidelines are highly influential and could serve as an example to other countries to pass laws concerning the limitation of space debris caused by the launch, deployment operation and disposal of satellites.

In 2015, the IADC[4] acknowledged the potential impact that large satellite constellations could have on the evolution of space debris and, consequently, initiated a series of investigations to identify potential mitigation actions which "could help inform the design and operation of such distributed architectures to ensure that they are developed in a predictable and sustainable manner." (IADC 15 03, 2017). In its Statement on Large Constellations of Satellites in Low Earth Orbit, the IADC offers a series of observations and guidance on the design and operation of such constellations systems that will ultimately be considered by operators and possibly implemented in their conceptual design. It has to be noted that the additional considerations embedded in the document does not intend to add nor to expand the IADC guidelines. Instead, it offers technical guidance such as the altitude separation, the spacecraft design or recommendations on collision avoidance, to best comply with the already mentioned guidelines.

Satellite operators willing to launch hundreds or thousands of satellites into space should bear in mind the following recommendations:

1. Operators should minimise potential break-ups by building each satellite in such manner that produces as few fragments as possible and that prevents accidental explosions and ruptures at end-of-mission.

2. Operators must minimise the potential for on-orbit break-ups during operational phases. When designing each satellite, it should be corroborated that failure leading to accidental break-ups is minimised.
3. Operators must avoid intentional destruction in outer space that may increase the probability of collision with other satellites. In case intentional destruction is needed, it shall be effectuated at the lowest altitude possible so that orbital fragments will easily enter back into the atmosphere and disintegrate.
4. Operators must prevent on-orbit collisions. In order to do so, operators shall conduct analysis and studies to estimate and limit the probability of collision with other objects that are known to be in the orbit where the satellite will be launched. The design of such the satellites launched into space should limit the aftermath of "collision with small debris which could cause a loss of control, thus preventing post-mission disposal."
5. Operators must make sure that all types of stored energy such as "residual propellants, batteries, high-pressure vessels, self-destructive devices, flywheels and momentum wheels", can be evacuated or released unless it poses a major risk for the payload. This guideline intends to prevent accidental break-ups caused by physical or chemical reactions. (IADC Space Debris Mitigation Guidelines, 2007).
6. Operators must de-orbit satellites that have ended their operational life and that may potentially interfere with the Low Earth Orbit (LEO). (Baker, 1989)Satellite operators should take all the precautionary methods in order to avoid environmental pollution, injury or damage to property. (IADC 15 03, 2017). This Post-Mission Disposal (PSD) aims to ensure that after its operational life, the satellite in question does not become space debris. Part of the PSD is the so-called "passivation", consisting of the elimination of all stored energy in order to minimise the chance of break-ups.
7. If operators decide to launch their satellites into the geostationary orbit, at the end of their operational life, satellites must be placed in graveyard orbits. (Larsen, 2017). This guideline can be complemented with the active role of the ITU that, for a long period of time, has recommended to place satellites, which are near to their end of operative life, in a disposal orbit at least 300 km away from the geostationary orbit. (Johnson & Loftus 1999).

Space debris is an issue for macro-constellations because of their quantity and the fact that they will need to be replaced at a relatively fast pace. In order to be compatible with the environment and somewhat viable it is absolutely crucial that the constellations deployed in LEO orbits will comply with the guidelines, in particular concerning its post-mission disposal.

CONCLUSION

The present chapter has illustrated the importance of the ITU procedures in order to understand what are the requirements that satellite operators need in order to put satellites in orbit and in the future what will they need to launch macro-constellations. A comprehensive understanding of the history, structure and legal framework of the ITU is essential to fully grasp the intricacies of the registration procedure.

This legal regime aims to manage and coordinate the use of frequencies and orbital slots that will play a fundamental role in the lifespan of the macro-constellation as a whole. Radiofrequencies and orbital slots are scarce resources that need to be preserved and regulated under both international and national regimes. As a matter of fact, the ITU recognises the paucity in these important resources in article 44 of its Constitution by stating that: "Member States shall bear in mind that radio frequencies and any associated orbits, including the geostationary-satellite orbit, are limited natural resources and that they must be used rationally, efficiently and economically". The phrasing of the article clearly gives away that frequencies and orbits are inherently finite. This situation generates competition between the satellite operators interested in using these limited resources. This competition is condemned to grow since macro-constellations will increase the demand. The growing demand and probabilities of harmful interferences is expected to drastically increase the cost of accessing these scarce resources. The ITU RR must be updated in order to face the reality of the satellite industry. Another discussed change would be the expansion of the authority of the ITU through a compulsory dispute settlement mechanism that would facilitate the resolution of disputes arising from the relevant activities. (Jakhu, 2007).

This chapter also addresses the issue of space debris in relation with macro-constellations. It is important to remark that macro-constellations themselves or the possible space debris it may generate will have an impact in other activities performed in outer space. Space activities are becoming more dangerous, for example, the International Space Station is placed at a LEO orbit, the same region proposed for large satellite constellations. It is important to take all the appropriate measures to mitigate the constant increase of space debris that will eventually have repercussion on other space activities such as those carried out on board of the ISS.

Although it is a very important starting point, we cannot only rely on the COPUOS space debris mitigation guidelines to put a solution to future problems. These guidelines need to be extended and complemented, e.g. through national legislation or *soft law* rules that could serve as a preparatory tool for the formation of future customary international law. Unfortunately, these guidelines are non-binding and, therefore a new way of implementing these guidelines needs to be explored.

As it has been explained in the chapter, the Radio Regulations of the ITU are binding upon its member states. The ITU counts with 193 member states, it is to say, 88 states more than states parties to the Outer Space Treaty. It seems that if we want to implement those guidelines the ITU regulations would be the perfect way to do it. Many of the guidelines focus on the end-of-life disposal of the satellites, and thus it could be seen as a legitimate competence of the Union. Even if debris mitigation is not part of the work of the ITU and is not mentioned in its main instruments, preserving the rights of other states to access radiofrequencies and orbital slots is one of its principal functions and it entails, in a way, the mitigation of space debris generated by non-functional satellites orbiting and occupying precious slots in such a limited resource. Unlike the Outer Space Treaty, the ITU regulations lack the uncertainty in the definition of a few concepts. Among the provisions of the Radio Regulations concepts such as "space object" cannot be found and more specific definitions of "satellite system" or "satellite network" are provided. (ITU Radio Regulations, 2015).

The ITU RR are periodically reviewed and, if necessary, revised at the world radiocommunication conference held every three to four years. This flexibility allows the ITU to be the perfect environment to deal with this subject. It is a matter of competence what makes it complicated to address these issues.

REFERENCES

Allison, A. L. (2014). The ITU and managing satellite orbital and spectrum resources in the 21st century. *SpringerBriefs in Space Development*.

Andrews, E. (1990). *Tiny Tonga seeks Satellite Empire in Space.* Retrieved from https://www.nytimes.com/1990/08/28/business/tiny-tonga-seeks-satellite-empire-in-space.html

Aoki, S. (2013). Efficient and equitable use of orbit by satellite systems: "Paper satellite" issues revisited. In *Proceedings of the International Astronautical Congress, IAC* (Vol. 14, pp. 11149-11155). IAF.

Baker, H. A. (1989). *Space debris: legal and policy implications*. Nijhoff.

Benko, M., & Schrogl, K. U. (2008). UN Committee on the Peaceful Uses of Outer Space: Adoption of the Resolution on Enhancing Registration Practice and of the UNCOPUOS Space Debris Mitigation Guidelines. *ZLW*, *57*, 335.

Codding, G. A., & Rutkowski, A. M. (1982). *The International Telecommunication Union in a changing world*. Artech House on Demand.

Constitution of the International Telecommunication Union (ITU Constitution), ATS (1994) 28; BTS 24 (1996)

Davenport, C. (2016). *Companies flood Earth's orbit with satellites, but no one's directing traffic*. Retrieved from: https://www.washingtonpost.com/business/economy/companies-flood-earths-orbit-with-satellites/2016/07/22/9db2e522-49ee-11e6-90a8-fb84201e0645_story.html?utm_term=.5066ccce7f07

De Man, P. (2013). The relevance of ITU regulations for clarifying the space debris concept and strengthening guidelines on the removal of satellites at the end of their functional life. *Annals of Air and Space Law*, *38*, 203-236.

De Selding, P. B. (2016). *Enough satellites to darken the skies*. Retrieved from: http://www.spacenewsmag.com/the-bottom-line/enough-satellites-to-darken-the-skies/

Delzeit, A. N., & Beal, R. F. (1996). The Vulnerability of the Pacific Rim Orbital Spectrum Under International Space Law. *NY Int'l L. Rev*., *9*, 69.

Detter, I. (1996), The effect of Resolutions of International Organizations. In Theory of International Law at the Threshold of the 21st Century– Essays in Honor of Krzysztof Skubiszewski. The Hague: Academic Press.

Frowein, J. A. (2004). The Internal and External Effects of Resolutions by International Organizations. *Völkerrecht—Menschenrechte—Verfassungsfragen Deutschlands und Europas*, 81-94.

Gabrynowicz, J. I. (2010). One half century and counting: The evolution of US national space law and three long-term emerging issues. *Harv. L. & Pol'y Rev*, *4*, 405.

Gorove, S. (1990). Environmental risks arising from space activities: focus on the Liability Convention. In *Environmental aspects of activities in outer space: state of the law and measures of protection*. Carl Heymanns.

Imburgia, J. S. (2011). Space debris and its threat to national security: A proposal for a binding international agreement to clean up the junk. *Vand. J. Transnat'l L.*, *44*, 589.

Jakhu, R. (2007). Legal issues of satellite telecommunications, the geostationary orbit, and space debris. *Astropolitics*, *5*(2), 173–208.

Jakhu, R. (2007). Legal issues of satellite telecommunications, the geostationary orbit, and space debris. *Astropolitics, 5*(2), 173-208.

Jakhu, R. S. (1983). The evolution of the ITU's regulatory regime governing space radio-communication services and the geostationary satellite orbit. *Annals of Air and Space Law*, *3*, 381–406.

Johnson, N. L., & Loftus, J. P. (1999). Reducing orbital debris- Standards and practices. *Launchspace*, *4*(2), 24–26.

Kerrest, A., & Smith, L. (2009). Article VII. In S. Hobe, B. Schmidt-Tedd, & K. U. Schrogl (Eds.), *Cologne Commentary on Space Law: Outer Space Treaty* (pp. 126–145). Carl Heymanns Verlag.

Kessler, D. J., & Cour-Palais, B. G. (1978). Collision frequency of artificial satellites: The creation of a debris belt. *Journal of Geophysical Research: Space Physics*, *83*(A6), 2637–2646. doi:10.1029/JA083iA06p02637

Klinkrad, H., Beltrami, P., Hauptmann, S., Martin, C., Sdunnus, H., Stokes, H., ... Wilkinson, J. (2004). The ESA space debris mitigation handbook 2002. *Advances in Space Research*, *34*(5), 1251–1259. doi:10.1016/j.asr.2003.01.018

Larsen, P. B. (2017). Small Satellite Legal Issues. *J. Air L. & Com*, *82*, 275.

Leinberg, G. (1989). Orbital space debris. *JL & Tech.*, *4*, 93.

Lyall, F. (1995). Posts and Telecommunications. In United Nations Legal Order (Vol. 2). Cambridge.

Lyall, F. (1996). Paralysis by phantom: problems of the ITU filling procedures. In *Proceedings of the Thirty-Ninth Colloquium on the Law of Outer Space* (p. 189). Academic Press.

Sands, P., Klein, P., & Bowett, D. W. (2009). *Bowett's law of international institutions*. Academic Press.

Savage, J. G. (1989). *The politics of international telecommunications regulation*. Westview Pr.

Standage, T. (1998). *The Victorian Internet: The remarkable story of the telegraph and the nineteenth century's online pioneers.* Phoenix.

Tchikaya, B. (1998). Le droit international des télécommunications. Presses universitaires de France.

The Convention on International Liability for Damage Caused by Space Objects, adopted by the General Assembly in its resolution 2777 (XXVI), 24 UST 2389, TIAS 7762, 961 UNTS 187.

This revised Radio Regulations, complementing the Constitution and the Convention of the International Telecommunication Union, incorporating the decisions of the World Radiocommunication Conferences of 1995 (WRC-95), 1997 (WRC-97), 2000 (WRC-2000), 2003 (WRC-03), 2007 (WRC-07), 2012 (WRC-12) and 2015 (WRC-15)

Thompson, J. C. (1996). Space for rent: the international telecommunications union, space law, and orbit/spectrum leasing. *J. Air L. & Com., 62*, 279.

Treaty on Principles Governing the Activities of States in the Exploration and Use of Outer Space, including the Moon and other Celestial. 18 UST 2410, TIAS 6347, 610 UNTS 205.

Vienna Convention on the Law of Treaties, 23 May 1969, United Nations, Treaty Series, vol. 1155, p. 331. (n.d.). Available at: https://www.refworld.org/docid/3ae6b3a10.html

Villa, M. (2016). *Builders of satellites large and small must work together to benefit science.* Retrieved from: http://www.spacenewsmag.com/commentary/builders-of-satellites-large-and-small%E2%80%A8-must-work-together-to-benefit-science/

Weeden, B. (2016). *Insight - Small Satellites and Space Situational Awareness.* Retrieved from: https://swfound.org/news/all-news/2016/09/insight-small-satellites-and-space-situational-awareness

White, R. L., White, H. M., & White, R. L. (1988). *The Law and Regulation of International Space Communication*. Artech.

ENDNOTES

1 Twenty countries formed the convention that established the International Telegraph Union: the Austrian empire; the Grand Duchy of Baden; the Kingdom of Bavaria; the Kingdom of Belgium; Kingdom of Denmark; Kingdom of Spain; French Empire; Kingdom of Greece; Free City of Hamburg; Kingdom of Hannover; Kingdom of Italy; Kingdom of the Netherlands; Kingdom of Portugal; Kingdom of Prussia; Russian Empire; Kingdom of Saxony; United Kingdoms of Sweden and Norway; Swiss Confederation; Ottoman Empire; and the Kingdom of Würtemberg.

2 WARC ORB-85 and 88 are specialised WARCs. Specialised WARCs generally examine issues relating to specific frequency bands or radio services. In this case, WARC ORB-85 and 88 deal with space services and orbital assignment.

3 To mention a few, the most important national agencies are the Bundesnetzagentur (Federal Network Agency), ANFR (Agence nationale des fréquences) and/or the Federal Communication Comission (FCC) in the United States.

4 The Inter-Agency Debris Coordination Committee, an inter-governmental forum established to coordinate efforts to deal with space debris, is formed by the Italian Space Agency (ASI), the Centre National d'Etudes Spatiales (CNES), China National Space Administration (CNSA), Canadian Space Agency (CSA), Deutsches Zentrum für Luft- und Raumfahrt e.V. (DLR), the European Space Agency (ESA), the Indian Space Research Organisation (ISRO), Japan Aerospace Exploration Agency (JAXA), the Korea Aerospace Research Institute (KARI), the National Aeronautics and Space Administration (NASA), the Russian State Space Corporation ROSCOSMOS (ROSCOSMOS), the State Space Agency of the Ukraine (SSAU), and the UK Space Agency (UKSA).

Chapter 10
On-Orbit Servicing and Active Debris Removal:
Technical Aspects

Martin J. Losekamm
Technical University of Munich, Germany

ABSTRACT

This chapter introduces the challenge of space debris and the concepts of on-orbit servicing and active debris removal. The evolution of the debris population is put into a historical perspective, observation and modelling methods are described, and internationally agreed mitigation strategies are briefly introduced. Proposed mission types and the required technologies for servicing and removal activities are detailed and their challenges explained. Where applicable, past, current, and proposed activities are summarized to illustrate the concepts.

INTRODUCTION

In 2009, the world witnessed the very first accidental collision involving an operational satellite in low Earth orbit. The US-built Iridium 33 commercial telecommunication and the Russian Kosmos-2251 military telecom satellites collided at a relative velocity of about 42,000 kilometres per hour. The breakup of the two spacecraft resulted in the creation of more than 1,000 pieces of debris larger than ten centimetres in size and a myriad of smaller ones.

In recent decades, the growing number of artificial satellites in orbit around Earth has led to a congestion problem in some heavily used orbital bands. Some regions in low Earth orbit, mostly used by Earth observation satellites, and in geostationary orbit, predominantly used for communication satellites, are at threat to become unusable in the foreseeable future. Mitigation measures aim to limit the number of new objects that unnecessarily remain in orbit after the end of their useful life. However, the number of dysfunctional spacecraft and debris objects in orbit today already presents a threat to the future usability of some orbital regions.

DOI: 10.4018/978-1-5225-7256-5.ch010

To overcome this challenge, several actions are needed: the limitation of debris objects created during launches, the extension of the useful life of spacecraft, and the deorbiting of unused assets. New concepts and technologies are needed to implement these objectives, going beyond the simple launch-and-discard practice that has been prevalent for many years.

This chapter explains the sources, evolution, observation, and modelling of space debris, as well as internationally-agreed mitigation guidelines to limit its population growth. The concepts of on-orbit servicing (OOS) and active debris removal (ADR) missions are introduced and the technological challenges are discussed. Current activities of the National Aeronautics and Space Administration (NASA), the European Space Agency (ESA), other space agencies, and commercial companies are briefly summarized. The chapter concludes with a brief economical assessment of OOS and ADR activities.

SPACE DEBRIS

Objects launched into orbits high enough that the drag of the residual atmosphere of Earth does not significantly impact their orbital motion require a very long time to fall back to Earth. Since the beginning of the Space Age, many objects were left in orbit, ranging from small ejectables[1] to defunct satellites and burned-out upper stages of rockets. While the latter mostly fall back relatively quickly due to the higher atmospheric drag, there still are a significant number of satellites in orbit launched during the first years of space exploration. A good example is the Vanguard 1 satellite, the oldest satellite still in orbit, which was launched into a 658-km by 3,840-km elliptical orbit in 1958 and has a projected lifetime of 240 years (Easton & Votaw, 1959; McLaughlin Green & Lomask, 1970). Spacecraft in higher, more circular orbits have orbital lifetimes measured in thousands of years. Satellites in the highest orbits used today stay there practically forever.

The continued launching of spacecraft into orbit leads to a congestion of the orbital bands most useful for the current and future space industry: the low Earth orbit (LEO) and the geostationary orbit (GEO). In GEO, the availability of orbital slots is inherently limited, constrained not only by the minimum safe distance between two objects, but also because of possible radio interference between satellites operating in the same communication bands. The allocation of slots is overseen by the International Telecommunications Union (ITU), but in some cases the inter-satellite spacing is already well below the recommended limits[2]. Although remarkable gains in efficiencies and performances of communication satellites have been achieved in the past decades, many expect that the demand for new platforms in GEO will continue to rise (ASD-Eurospace, 2012). This is not only due to the rising global demand of satellite communications, but also because of the increasing utilization of inter-spacecraft communication for both manned and unmanned systems. The rise to fame of small satellites in recent years may, however, diminish the demand for large GEO platforms.

Although the availability of orbits in LEO is not as limited as in GEO, many satellites have been placed in a small number of orbital bands of high operational value for Earth observation and imaging purposes. Recently, plans by companies such as SpaceX, OneWeb, Samsung, and others to launch so-called mega-constellations—consisting of hundreds or even thousands of data-relay satellites—have dramatically raised the demand for LEO orbits. Many of these satellites would be launched into long-lived orbits and potentially lack deorbiting systems (Chowdhury, 2018). Even at small failure rates, such large constellations will inevitably lead to a significant increase in the number of dead and non-manoeuvrable objects.

Sources and Classification of Space Debris

There are different sources and classes of space debris. In terms of mass and cross-sectional area, dysfunctional or abandoned spacecraft and rocket bodies contribute most to the debris population. During a rocket launch, the lower stages of the launch vehicle detach from the upper stages once their fuel tanks are empty. The first stage usually detaches at low altitudes and falls back to Earth's surface within minutes. Depending on the total number of stages, the second, third, or even fourth stages are set loose at much higher altitudes. Especially on missions to high orbits, these upper stages may remain in space for many years, often with leftover fuel in their tanks. Systems using hypergolic fuel[3] are of particular concern, since corrosion of their tanks can lead to spontaneous explosions even decades after launch. Nowadays, hypergolic fuel is mostly used by older Russian launchers, though the US and ESA used propellants of this class extensively in their Titan and Ariane launcher families, of which some still remain in orbit.

Satellites are launched into three principal categories of orbits. Earth observation satellites are placed in LEO up to altitudes of about 2,000 kilometres. Most communication satellites require constant positioning over a certain area and are thus placed in GEO at roughly 36,000 kilometres altitude. In between, navigation satellites—such as the US GPS constellation or the European Galileo satellites—circle the planet in medium Earth orbits (MEO) at around 20,000 to 25,000 kilometres. Many rocket stages remain in geostationary transfer orbit (GTO), which is used to ferry satellites into GEO and crosses the LEO and GEO regimes, thus endangering both commercially most interesting orbital regions. A small number of (mostly) scientific satellites are placed in highly elliptic orbits carrying them well beyond GEO.

The debris cloud around Earth is made up of a diverse mixture of objects, ranging from defunct satellites and spent upper stages to aluminium slag particles from solid rocket motor firings, paint flakes, and even droplets of liquid metal ejected from nuclear reactor cores. Large objects can be reliably tracked using optical and radar systems (Africano, Stansbery, & Kervin, 2004; Foster, Benbrook, & Stansbery, 2005; Lambour, Rajan, Morgan, Kupiec, & Stansbery, 2004; Schildknecht et al., 2004). The two largest systems relying on ground-based instruments are the Space Surveillance Network (SSN) operated by the US Strategic Command (Johnson, 1993) and the Space Surveillance System (SSS) operated by the Russian military (Dicky et al., 1993).

Both entities publish catalogues of tracked objects that characterize the debris population above ten centimetres fairly well. The SSN tracks about 17,000 objects with an aggregate mass of almost 7,000 metric tons, of which more than 75% are located in LEO (Klinkrad, 2015). Smaller objects can be tracked with newer, more powerful systems to a certain extent, albeit unreliably to date. The smallest particles can only be measured *in situ* using impact sampling methods, as has been done by several experiments in the past.

Several recorded satellite anomalies and failures (e.g. NOAA 7 in 1997, Cosmos-539 in 2002, and UARS in 2007) have been attributed to space debris based on the slight orbital change these satellites showed after the incidents (Krag, Flohrer, & Lemmens, 2013). In recent years, the topic of space debris has come to more widespread awareness, most notably through the test of a Chinese anti-satellite system targeting the non-operational Fengyun-1C satellite in 2007, and the 2009 collision of the American Iridium 33 and the Russian Cosmos 2251 satellites. These two incidents alone increased the population of trackable debris by 34% (3,400 fragments) and 22% (2,200 fragments), respectively (Klinkrad, 2015).

Figure 1 shows the historical evolution of the number of objects catalogued by the SSN from 1957 to 2018 (NASA Orbital Debris Program Office, 2018). The 2007 and 2009 fragmentation events can clearly be identified in the data. While the graph gives you an idea how the population has developed in

Figure 1. Historical evolution of the space debris population catalogued by the SSN as of April 2018. The graph was compiled from data published by NASA's Orbital Debris Program Office (2018)
Source: Author's elaboration

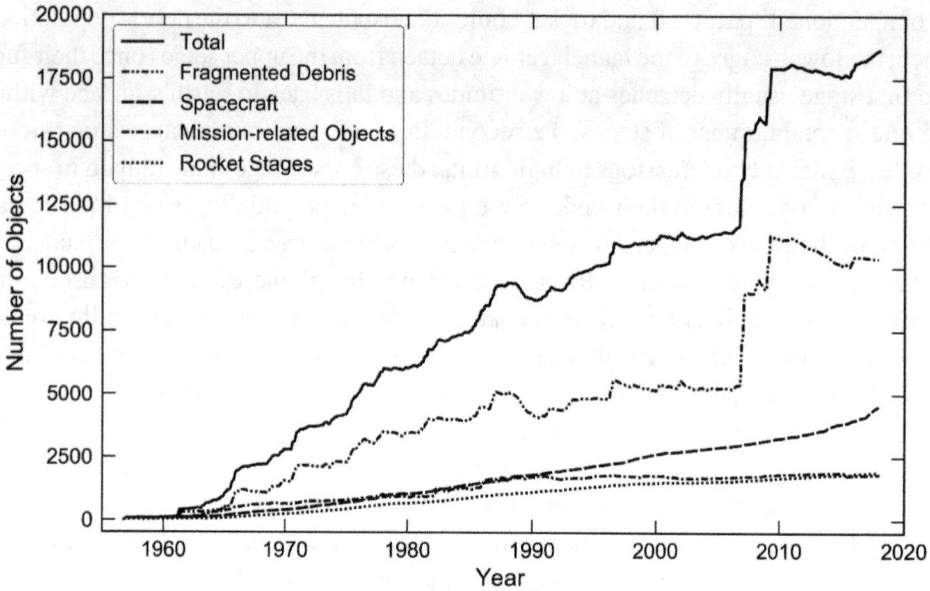

the past decades, it only includes a fraction of the debris objects in orbit. Millions of non-trackable fragments in the micrometre-to-centimetre regime are believed to exist, totalling several tens of tons in mass.

The likelihood that a debris object hitting a spacecraft critically endangers its mission rises with increasing impactor diameter and mass. Objects with an energy-to-mass ratio of more than 40 Joules per gram can cause a complete breakup of a spacecraft, possibly creating hundreds or even thousands of debris objects of various sizes (Krag et al., 2013). This energy threshold depends on the impact geometry (the angle and collision axis under which the objects hit each other) but can be roughly translated into size classes (Dolado-Perez, Pardini, & Anselmo, 2015; National Research Council, 1995). Objects under one millimetre in size do not present any critical danger to spacecraft but lead to surface degradation and possible damage to unprotected components. Particles larger than one millimetre might disable sensitive satellite subsystems at LEO collision velocities. Those larger than one centimetre are likely to render the complete spacecraft non-operational. Trackable objects larger than ten centimetres most likely lead to a catastrophic breakup.

In 1978, Donald Kessler and Burt Cour-Palais described the theoretical possibility that this creation of new debris could potentially lead to a runaway chain reaction in the most densely populated areas of space (Donald J. Kessler & Cour-Palais, 1978). Widely known as the *Kessler syndrome*, this phenomenon would render large portions of the currently populated orbital bands unusable once the rate of debris creation exceeds that of natural atmospheric cleansing. Many argue that the Kessler syndrome was effectively started by the high-profile fragmentation events mentioned above (Bonnal, Ruault, & Desjean, 2013), although its extent and time line, as well as a suitable strategy to prevent further population growth by actively removing debris objects from orbit, are still a field of active research (Anderson & Schaub, 2014; D. J. Kessler, Johnson, Liou, & Matney, 2010; Lewis, Newland, Swinerd, & Saunders, 2010; Rossi et al., 2016; White & Lewis, 2014a, 2014b).

Collision Avoidance

Even today, the risk of space debris to manned and unmanned systems in orbit is a real one. For assets exposed to high-flux environments, conjunction event analyses and collision avoidance manoeuvres are performed. According to researchers at the University of Southampton, there are more than 2,000 instances a day of tracked satellites passing each other within a range of five kilometres (Chowdhury, 2018). Between 2004 and 2012, more than 20 collision avoidance manoeuvres had to be performed for ESA satellites alone (Klinkrad, 2015).

In case of the International Space Station (ISS), a screening of all SSN catalogue objects for possible collisions is conducted at least three times a day for the next 72 hours (Klinkrad, 2015). If an acceptable risk threshold is exceeded, an avoidance manoeuvre is performed by boosting the station to a higher altitude. As of 2015, this has been done 25 times in the operational life of the ISS (NASA Orbital Debris Program Office, 2015). In case of insufficient reaction time, astronauts must retreat into their Soyuz capsules, of which one for every three astronauts is attached to the station as an emergency vehicle at all times.

For sub-catalogue debris, passive protection measures must be taken. Manned modules of the ISS, for example, are equipped with so-called Whipple shields, consisting of bumper and fabric layers made of Kevlar and Nextel and similar in construction to bulletproof vests (Christiansen, Crews, Williamsen, Robinson, & Nolen, 1995). They are designed to withstand impacts of particles up to about one centimetre in size traveling at velocities of ten kilometres per second. The energy imparted onto the shield during such an event is comparable to that released in the explosion of a hand grenade. Objects between one millimetres and one centimetre in size hit the ISS once every six months on average, debris larger than one centimetre at a rate of about one in 410 years (Klinkrad, 2015).

Observation and Modelling of the Debris Population

Over the past decades, space debris objects have been tracked by several tracking systems such as the previously mentioned SSN and its Russian pendant, the SSS. These systems were developed during the cold war to track hostile spacecraft and mostly use optical and radar systems to detect and track objects from the ground. This puts an inherent limit on the minimum size of objects that can be detected. Although some advances were made in the past few years, tracking still only works reliably for objects larger than about ten centimetres in size[4].

The smallest particles (micrometre-to-millimetres range) must be detected *in situ* using direct-impact sampling techniques. Special materials are launched into space aboard satellites (and previously aboard the Space Shuttle) and are subjected to the on-orbit environment. Particles hitting the plates create craters and holes of various sizes. After the plates are retrieved and returned to the ground, scientists count and characterize the holes and craters. Using these samples, they can derive the total population of particles using statistical methods.

With only limited tracking capabilities throughout much of its size range (in terms of population numbers, not mass), modelling and simulation are important tools to assess the current debris population and predict its future evolution. Three major models designed to give a realistic description of the present population have been developed (Yulin & Zhaokui, 2016): NASA's ORDEM model incorporates a large set of *in-situ* and ground-based observational data and describes the orbital debris environment for objects ranging from ten micrometres to ten meters in size (Flegel et al., 2010; Krisko, 2010; Liou et al.,

2002); ESA's MASTER model uses a detailed list of known historical events and validates the resulting population against historical telescope and radar data (Flegel et al., 2010; Krisko, Flegel, Matney, Jarkey, & Braun, 2015); the Russian SDPA-E is a semi-analytical stochastic model using size-dependent probability distributions to predict debris fluxes (Klinkrad, 2006; Nazarenko & Menshikov, 2001).

All three models are engineering tools created to support mission planning in LEO and GEO and have thus been designed to accurately simulate the current environment based on available observational data. Separate simulation tools have been developed for the prediction of the future evolution of debris, such as NASA's EVOLVE (Krisko, 2004) and LEGEND (Liou, Hall, Krisko, & Opiela, 2004), and the University of Southampton's DAMAGE (White & Lewis, 2014b) models, amongst others.

Mitigation Guidelines

No binding international agreement exists concerning the mitigation or active removal of space debris. In 1995, NASA was the world's first space agency to issue a comprehensive set of orbital debris mitigation guidelines (NASA Office of Safety and Mission Assurance, 2017). After a multi-year effort, the Inter-Agency Space Debris Coordination Committee (IADC) published its own set of guidelines in 2002, which was also presented to the United Nations (UN) Committee for Peaceful Use of Outer Space (COPUOS). A Code of Conduct was signed by all major European space agencies in 2006 and translated into mandatory requirements by ESA, to be applied to all its missions procured in or after 2008. The UN mitigation and space sustainability guidelines—based on the work done by the IADC—were endorsed by the UN General Assembly in the same year (United Nations Office for Outer Space Affairs, 2010) and extended in 2018 (Weeden & Samson, 2018). Several national guidelines and requirements are based on these international efforts.

All documents share the same foundation of basic requirements a space mission must fulfil. The following points are a summary of the guidelines published by the UN:

1. Limit debris released during normal operations.
2. Minimize the potential for break-ups of spacecraft.
3. Minimize the probability of accidental collisions.
4. Avoid intentional destruction.
5. Minimize the potential for breakups resulting from stored energy (e.g. in the form of fuel or batteries).
6. Limit the long-term presence of spacecraft and launch vehicles in LEO.
7. Limit the long-term interference of spacecraft and launch vehicles with the GEO region.

Some documents are more specific concerning the meaning of the expression *long-term* and provide exact numbers. The European Code of Conduct for example states that spacecraft must re-enter the atmosphere within 25 years of the end of their useful life. Compliance with these guidelines can be achieved through the choice of a suitable orbit (natural decay) or the use of deorbiting devices.

Studies predict that even total compliance with these guidelines will most probably not be sufficient to stop the growth of the debris population. Liou and Johnson (2008) performed simulations of the future evolution of the trackable population using the LEGEND model. They assumed that no new launches happened after 2005 and tried to analyse the natural evolution due to breakups, collisions, and atmospheric cleansing. Their simulations show that even if we had stopped launching new satellites in 2005, the population would still increase by itself through naturally-occurring fragmentation events.

Active Measures: ADR and OOS

If even total compliance with the guidelines described above may not result in a reduction of the debris environment, active measures to limit the number of spacecraft or to remove objects from orbit must be taken. If it were possible to remove objects that are most likely to add significantly to the population, the probability for a runaway effect to occur could be reduced. Simulation tools like EVOLVE, LEGEND, and DAMAGE are being used to identify such objects and calculate the probability for them to break up in catastrophic collisions.

In addition to actively removing objects, reducing the number of satellites launched into orbit by increasing their useful lifetime is another possibility to consider. Depending on the application and orbit altitude, satellite lifetimes are either limited by fuel capacity or by aging instrumentation and faulty control electronics. In many cases, topping up fuel tanks or swapping out a few systems can prolong a satellite's lifespan quite significantly. Even though such operations are technically challenging, they may even hold an economic advantage for satellite operators.

The rest of this chapter introduces these two approaches in depth and also discusses the impact and economic aspects of such missions.

ON-ORBIT SERVICING

Although it has only recently come to more widespread attention, on-orbit servicing has been practiced since the early days of human space flight. The United States' first space station, Skylab, was launched in 1973 on a Saturn V booster and experienced severe technical difficulties immediately after lift-off. After a brief but intensive training period for the astronauts involved, NASA used the first manned Skylab mission (SL-2) to repair the damage that was inflicted to the station when its meteoroid shield inadvertently deployed early during ascent and was subsequently torn away by atmospheric drag. SL-2 saved the eight-month mission of Skylab and demonstrated the effectiveness of crew members performing complex and unplanned repair tasks (Waltz, 1993).

The Space Transportation System (STS), more commonly known under the name Space Shuttle, for the first time allowed to capture spacecraft in orbit for service operations. The five missions to the Hubble Space Telescope are some of the most famous servicing missions conducted using the STS (Treat, Scalamogna, & Conant, 2015), although others have been performed as well. Hubble also was the first telescope specifically designed to be serviced by astronauts. The recovery of the Palapa B2 and Westar 6 satellites during STS-51A in 1984 marked the first time that artificial objects were actively removed from orbit and, in this case, brought back to Earth (*On-Orbit Satellite Servicing Study*, 2010). Other examples of successful missions include the on-orbit repair of the Solar Maximum Mission (SMM) during STS-41C in 1984, as well as the repair of the SYNCOM-IV and Intelsat 6 satellites (Saleh, Lamassoure, Hastings, & Newman, 2003).

One of the most extensive historical studies of OOS is the Space Assembly, Maintenance and Servicing (SAMS) Study, a joint effort of the US Department of the Air Force, the US Strategic Defense Initiative Office, NASA, and the Lockheed Missiles & Space Company. Its final report was published in 1986 and includes very detailed descriptions of the SAMS architecture (*Space Assembly, Maintenance And Servicing (SAMS) Study, Final Report*, 1986). The study foresaw several of the enabling technologies that were at the time thought to be key components of OOS: remotely piloted orbital transfer and manoeuvring vehicles, a facility for the on-orbit handling and storage of cryogenic propellants, a propellant transfer system, a teleoperated satellite servicer system, and a crewed orbital transfer module that could carry astronauts from a servicing facility at the space station—back then it was stilled called Freedom and was solely an American project that only later evolved into ISS program—to remote servicing locations (Saleh et al., 2003). The program was terminated when the study failed to sufficiently demonstrate the cost effectiveness of OOS, mainly due to the ambitiously grand scale of the planned activities.

More recently, the assembly of the ISS would not have been possible without the extensive involvement of astronauts and robotic manipulators. Although humans played a vital role in many of the assembly operations performed, it has become increasingly clear that robotic systems—such as the Canadarm2, built by MDA Space Missions and officially known as the Space Station Remote Manipulator System (SSRMS)—are very reliable and versatile tools that can be extensively used for on-orbit servicing and assembly purposes. NASA has also been performing a technology demonstration operation for robotic refuelling aboard the ISS since 2011. During phase I and phase II of the Robotic Refueling Mission (RRM), the Canadarm2 and its Special Purpose Dexterous Manipulator (SPDM or Dextre) were successfully used to perform a series of refuelling tests on hardware that had not been designed for such manipulations (Leone, 2014). NASA plans to continue the tests with new experiment hardware.

Concept of On-Orbit Servicing

OOS refers to the performance of activities to maintain, repair, upgrade, refuel, or deorbit a spacecraft while it is in or near its operational orbit (Krolikowski & David, 2013). All these activities require the servicer spacecraft to approach, rendezvous and interact with the asset to be serviced. The interaction can take on different forms, depending on the objective of the servicing mission.

Plans for OOS build on the fundamental assumption that spacecraft can be built to be serviced, meaning that critical components and systems prone to failure or aging can be repaired or replaced. The rationale behind extending a satellite's operational life is that the cost for servicing is assumed to be less than the cost for building, qualifying, and launching a new satellite. Of particular interest are large and expensive geostationary communication platforms. These satellites have typical lifetimes of about 15 years (Anderson & Schaub, 2014; Saleh & Castet, 2011), at the end of which most of them run out of fuel, become non-manoeuvrable, and as a consequence cannot keep their position within their assigned orbital slots. Because of the threat these uncontrollable assets pose to other satellites, they need to be moved out of GEO into higher or lower *graveyard* orbits. If, however, satellite operators were able to refuel their satellite's propulsion systems, they may operate them for another couple of years before the need for a replacement arises.

Other concepts foresee more challenging tasks. Certain systems in satellites are more likely to fail than others, potentially rendering it inoperable well before its intended end of life. On Earth, you can bring your car to a service station for repairs if it breaks down. In space, there is no service station for damaged satellites—at least not yet. Bringing it back to Earth is too expensive and, after the retirement of the Space Shuttle, not even possible at the moment. Overcoming this profound constraint, the basic principle of OOS is to bring the service station to the damaged satellite instead of vice versa—eliminating the need for an expensive "towing vehicle" like the Shuttle.

Technical Aspects

OOS requires several key technologies and concepts that are not yet fully developed and are unique in their level of complexity. Current plans exclusively rely on automated or teleoperated[5] systems, without any astronauts required. Figure 2 shows the mission elements that are generally required for both OOS and ADR missions. The *target* is the object to be serviced or deorbited; the *chaser* could be a servicing satellite for OOS or a deorbiting satellite in case of an ADR mission. The chaser is controlled by an operator on the ground through radio-communication links, either directly or via data-relay satellites in higher (mostly geostationary) orbits, such as the US Tracking and Data Relay Satellites (TDRS). In some cases, the operator may also have limited control over the target. The Chaser and the target may also exchange information directly to enable (semi-) autonomous operations.

During the Space Shuttle missions, astronauts could adapt their strategy and tools when they encountered problems during operations. A servicer spacecraft cannot be adapted or changed once in orbit and must thus be able to cope with a broad range of tasks and challenges, some of which may not have been anticipated before launch. Until recently, complex automated or teleoperated tasks have hardly been conducted on Earth, let alone in space. Advances in the fields of robotics, system automation, sensor systems, and real-time data visualization are thus key to successful servicing missions. While some of these advances are needed for future ground-based robotic systems as well, others are unique to the space industry and thus require dedicated development efforts.

Figure 2. OSS and ADR mission elements. Communication paths are depicted as arrows. Dashed arrows indicate an optional path. A chaser can be a servicing satellite for OOS or a deorbiting satellite in case of an ADR mission.
Source: Author's elaboration

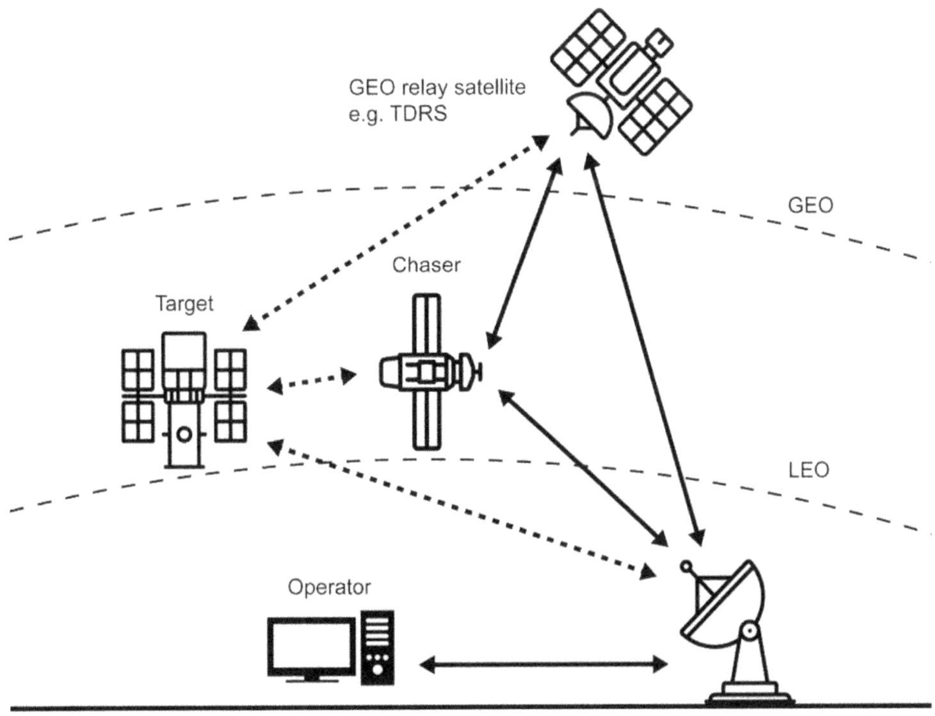

Mission Design

Scientists and engineers envision two different classes of servicing missions. One concept foresees to build a single satellite dedicated to servicing one specific target. This strategy makes sense whenever very complicated tasks specific to only one target must be performed. The servicer can be built to perfectly execute the required operations, even if special tooling and equipment are required. The drawback is that delivering a dedicated satellite to orbit and scrapping it after just one use is very expensive. However, this approach might be worth the (financial) effort, especially in the case of large scientific or government satellites for which no replacements are available.

The second concept builds on adaptable multi-purpose servicer satellites that can be used to repair or refuel several targets. Consequently, the tools and spare parts aboard these servicers need to be more generic, probably requiring a standardized, easily swappable set of systems. This approach is likely to be more cost-effective, even though it requires a more complicated servicer build-up. Another aspect to consider are the orbital manoeuvres needed to steer a servicer to different targets. This capability requires better propulsion systems and more fuel, adding significantly to the cost of launching such a satellite. Yet, this concept may be very well suited for operators of large fleets of satellites, such as communication service providers like Intelsat and SES.

Spacecraft Design

Servicer spacecraft must have several design features not commonly found on other satellites. First, they need very accurate position and attitude control systems to be able to approach their targets with sufficient precision to grab them. This in itself is not entirely unique to OOS satellites but is typically not required for most spacecraft. While it can be assumed that many targets will still be under control by ground operators at the time they are being serviced, sometimes this may not be the case. Out-of-control satellites can spin and tumble, complicating the capture process. The servicer's propulsion system must be able to match the spin rate of the target if a reliable connection between the satellites shall be made. After the connection is made, the system must be able to de-spin and stabilize both satellites.

Second, the servicer must be equipped with an actuator that can rigidly attach it to the target. While a rigid connection is not necessarily required for ADR, it is essential for OOS operations. Third, the servicer needs to have a set of manipulators and tools that can be used to perform repairs. These range from relatively simple mechanical manipulators, to automatic screw drivers and cutters, to electrical connectors that can connect to computer interfaces. Fourth, a set of spare parts is required to swap with damaged or inoperable ones on the target satellite. These might be single electrical or mechanical components, or complete systems that can be swapped. In-space additive manufacturing could also be used to manufacture required tools and devices directly in orbit ("Technology Quarterly: Space," 2016).

Another important part of the servicer satellite is its sensor suite. It is required to provide the operator on the ground as much information as possible about the target and the current state of the tools. Cameras play a very important role in providing optical information, but they are lacking haptic feedback that is needed for some tasks. Research is currently underway in the field of torque-controlled robotic arms and similar technologies, which can give an operator an indication of how much force is required to perform a certain movement. Also, a single camera feed does not offer any sense of depth perception. Several solutions to this problem are being investigated; stereo cameras, radar, lidar[6] are the most notable ones. Efforts are underway to use virtual reality as well.

One of the main challenges for teleoperation is the time delay introduced by the distance between satellite and ground station. Appropriate training is necessary to be able to cope with this delay between action (for example the movement of a joystick to fire a thruster) and the reaction of the spacecraft (i.e. the movement in the direction of the thruster firing). If the time delay is too large, the human brain tends to think that the command was not accepted and will redo the action. This can lead to drastically wrong decisions being made if the operator is not properly trained.

Capture and Docking Mechanism

In case of OOS operations, the servicer and target must be rigidly connected. Otherwise the physical law of action and reaction (Newton's third law of motion) will lead to the rotation or movement of the two spacecraft relative to each other once the servicer's actuators (tools and manipulators) apply a force. Future satellites can be equipped with docking ports that allow a direct docking between spacecraft, similar to the docking of the Space Shuttle, the Russian Soyuz capsules, or the European Automated Transfer Vehicle (ATV) to the ISS.

Most satellites already in orbit today are not equipped with such docking ports and thus must be captured using a different mechanism. Several approaches are being studied, some of which use the nozzles of propulsion or attitude control systems as grappling ports (de Peuter et al., 1994; Kaiser, Sjöberg, Delcura, & Eilertsen, 2008). Most concepts use one or more robotic arms to grab a satellite and pull it close to the servicer (Biesbroeck, Soares, Hüsing, & Innocenti, 2013; Metz, 2012). This approach is similar to the grappling and berthing of the Cygnus and Dragon cargo craft to the ISS using the station's robotic arm.

Servicing Operations

The actual task of servicing a satellite can take on many forms. One of the easiest operations is the refuelling of propellant tanks—at least when neglecting the fact that older tanks were not designed to be refuelled and lack the required nozzles and valves. As mentioned earlier, NASA successfully performed several tests on such systems during phase I and II of its RRM. Future satellite generations can be equipped with the required interfaces to allow refuelling operations. A single refuelling satellite might deliver propellants to as many as ten or twenty satellites, depending on the maximum initial launch mass available.

With the advent of new heavy launch vehicles like the Falcon Heavy, heavier satellites carrying more propellant may be launched. However, one point to consider is that larger servicer satellites may not always be economically favourable. The more fuel a satellite carries, the more targets it must visit before it becomes profitable. This is not much of a problem if it is released into an orbit with many targets grouped closely together. If, however, the targets are situated in orbits with largely different altitudes or inclinations, a lot of fuel must be used to move between these orbits. This sets a limit on the amount of fuel a single servicer should carry for a given deployment scenario.

Systems that can be repaired or swapped are batteries, solar panels, computer systems, bearings, pumps, motors, and attitude and orbit determination and control system (AOCS) components such as gyroscopes, reaction wheels, or propulsion units. On most of today's satellites, these systems are not accessible and are thus hard or impossible to service. Working on these systems requires cutting open the outer insulation of the satellite, removing covers and housing, and then take apart systems that were not meant to be taken apart. Electronic boards for instance are often secured by glue, making their swapping all but impossible. Other systems are easier to exchange. Solar panels, for example, are attached to the outside of a spacecraft, with only few cables connected to its inner systems. The constant bombardment of the solar cells with cosmic and solar radiation, sunlight, and micrometeorites leads to a degradation of their surface and thus to a reduction of energy output. Exchanging them brings the satellite's power supply system back to its initial performance and thus allows longer operations.

According to Graham and Kingston (2015), the commercially most interesting mission types can be categorized as follows:

1. **On-Orbit Refuelling:** The servicer rigidly attaches to the satellite and transfers fuel to extend its operational life. The servicer is a multi-use vehicle capable of refuelling several satellites.
2. **AOCS Life Extension:** The servicer rigidly attaches to the satellite and extends its operational lifetime by carrying out station-keeping and attitude control manoeuvres.
3. **Deorbit:** The servicer attaches to the satellite and removes it from its operational orbit by either deorbiting it (LEO) or moving it to a graveyard orbit (GEO). The servicer may either be a multi-use or purpose-built vehicle.
4. **Array Operations:** The servicer rigidly attaches to the satellite and connects to its electrical sub-system. It then uses its own power generation capabilities (i.e. solar arrays) to provide electrical power to the satellite and thus extend its operational lifetime.
5. **Mechanical Intervention:** The servicer uses a robotic arm to assist with the deployment of equipment that has not deployed correctly after launch.

The spacecraft to be serviced may or may not be cooperative. If it is under control, it might provide some feedback or active control to facilitate rendezvous and docking. Most spacecraft in orbit today are not designed to be serviced, and as such have no docking ports or access panels. Attaching to the satellite thus is one of the major challenges of many proposed OOS missions. Accessing the subsystems to be serviced is another. However, it is to be expected that more and more satellites launched in the future will be designed to be serviceable.

Design for Servicing

Future generations of satellites may rely on standardized interfaces, easing the swapping of systems. Instead of exchanging single components, a complete subsystem can be swapped without the need for special tooling. As interfaces will be standardized, systems can be exchanged between different platforms and capabilities can be easily upgraded by installing a more powerful version of a particular unit. This modular approach is already being tested and used by some researchers and companies.

Active and Proposed Missions

The Canadian satellite manufacturer MacDonald, Dettweiler and Associates (MDA Corp.) and the communication service provider Intelsat announced a partnership to develop a servicing capability for commercial applications in 2011. The services were to be performed by a vehicle called Spacecraft Infrastructure Servicing (SIS), which would have been able to refuel up to 1000 kilograms of fuel to as many as ten satellites in GEO. Intelsat had committed to an agreement that would have seen five of its own satellites be refuelled, while MDA planned to sell the remaining five services to at least one more client. However, in February 2012 MDA and Intelsat announced that they had ended their agreement and "indicated that a governmental customer base was necessary to sustain their business beyond its initial servicing missions" (Krolikowski & David, 2013).

After acquiring the California-based satellite manufacturer Space Systems Loral (SSL) in 2012, MDA began cooperating with NASA and DARPA on the Restore-L and Robotic Servicing of Geosynchronous Satellites (RSGS) programs, respectively (Henry, 2017a). Restore-L shall for the first time demonstrate the refuelling and relocation of a satellite not designed to be serviced, a likely target being the government-owned Landsat 7 (Alessandro, 2016). The mission successfully passed its preliminary design review in December 2017 (Russel, 2017). RSGS aims at demonstrating the refuelling, relocation, repair, and assembly of satellites in GEO using two robotic arms and may even be used to attach new payloads to operating satellites (Roesler, Jaffe, & Henshaw, 2017). After RSGS' main objectives have been achieved, MDA will be allowed to use the spacecraft for the remainder of its life. In 2017, the company merged with SSL and DigitalGlobe into Maxar Technologies, Ltd., and announced that it will use the RSGS spacecraft to re-birth SIS. Luxembourg-based satellite operator SES agreed to be the first customer for SIS (Henry, 2017a).

Before the RSGS program was initiated, DARPA's Phoenix program was envisioned to develop satellite scavenger vehicles that would be able to perform on-orbit refurbishment of defunct space systems. Existing hardware from several decommissioned satellites would be used to assemble a new and functional spacecraft in orbit. DARPA scrapped the program in favour of RSGS after concluding that servicing operational satellites likely results in a higher payoff than refurbishing defunct ones (Roesler et al., 2017).

Orbital ATK provides life-extension services to GEO satellites using its so-called Mission Extension Vehicle (MEV), which attaches itself to the target satellite und provide supplemental capabilities such as attitude control and orbital manoeuvring. As of April 2018, Intelsat has ordered MEVs for two of its satellites, with the first one likely to launch in 2019 (Erwin & Henry, 2018). Orbital recently announced the development of a Mission Robotic Vehicle, capable of carrying up to twelve Mission Extension Pods that can be attached to satellites individually (Foust, 2018b). The pods are based on the MEV but only provide station-keeping capabilities, providing a life extension of up to five years.

In Europe, Effective Space announced in January 2018 that it signed a first contract for its Space Drone life-extension vehicle, which uses electric propulsion elements to manoeuvre a target satellite in GEO (Foust, 2018a). Airbus Defence and Space announced in 2017 that it is developing an autonomous servicing vehicle called Space Tug, capable of refuelling, repairing, and monitoring the health of satellites (Henry, 2017b). No costumer for the Space Tug has been announced as of April 2018.

ACTIVE DEBRIS REMOVAL

The conceptual difference between OOS and ADR is that plans for servicing assume that there is a sufficiently large number of satellites whose life could be extended by refuelling or rather simple repairs. Obviously, some system failures may just be too complicated to repair, and some satellites really have reached the end of their useful life and are no longer needed. As per international debris mitigation guidelines, those systems should in the future use technologies to deorbit themselves. In cases of a complete spacecraft failure, however, this may not always be possible, even if the satellite is equipped with the appropriate technologies. In that case, the active removal of the object from orbit is necessary.

This also applies to the already existing population of space debris, of which most satellites and other objects have no means to actively deorbit.

Some authors regard deorbiting as a one of many capabilities that flexible OOS vehicles must offer. Others count ADR satellites as a different class of spacecraft, mostly due to the relaxed capture requirements described in the next section. Either way, an ADR mission will perform one of two actions, depending on the orbit the target is in: deorbit the object such that it burns up in Earth's atmosphere or move it to a graveyard orbit. The former method is the best option for objects in LEO, where slight changes in orbit are sufficient to steer an object onto a controlled re-entry path. In GEO, a re-entry is not generally a viable option. Instead, unused assets are moved to an orbit a few hundred kilometres above or below GEO where they remain practically forever.

Technical Aspects

With the increasing importance of ADR missions, a capture strategy different from the two employed for OOS (see Capture and Docking Mechanism) has become the focus of many studies. Termed as *soft capture*, it builds on two assumptions: (1) damage to debris objects is not critical if it creates no additional fragments; (2) a rigid connection between target and chaser is not an essential requirement for the deorbit process. These assumptions allow the use of less complex systems that avoid the use of highly accurate and expensive robotic systems. Most proposed concepts use nets (Retat & Others, 2012; Zhai, Qiu, Liang, & Li, 2009), harpoons (Nishida, Kawamoto, Okawa, Terui, & Kitamura, 2009; Reed, Busquets, & White, 2012), or even biologically-inspired tentacles (Innocenti, Soares, & Delaval, 2013) to create a flexible, (in many cases) tethered connection between the two satellites.

Such a flexible attachment makes orbital manoeuvring of the aggregated system more complex. Groups working on these concepts hope that these difficulties can be overcome and are already testing some components and strategies. One example is the free-flying SPHERES robots aboard the ISS, which are being used to study the behaviour of rigidly-attached and tethered systems (Mantellato et al., 2017; Sternberg et al., 2016; Zachrau, 2018). Other groups are investigating approaches that lie somewhere in between hard and soft capture, making use of biologically-inspired adhesives to create a semi-rigid connection (Trentlage, Mindermann, Ben Larbi, & Stoll, 2016).

Deorbiting can be achieved using passive or active devices. The former is mainly used by smaller spacecraft not capable of carrying active devices and include drag sails (Visagie, Lappas, & Erb, 2015) and electrodynamic tethers (Forward, Hoyt, & Uphoff, 2000). Active deorbiting is done using the chaser's main propulsion system. In most scenarios, the chaser will stay attached to the target and deorbit with it, but there exist concepts that reuse a chaser several times.

Yet another possibility is the of use lasers (Ebisuzaki et al., 2015; Phipps et al., 2012) or ion beams (Bombardelli, 2012; Kitamura, Hayakawa, & Kawamoto, 2014; Schaub & Sternovsky, 2014) that ablate material from debris objects. The ablated material would act as a primitive thruster and could be used to change the orbital parameters of the object such that it re-enters Earth's atmosphere by itself. The main advantage of this scenario is that the chaser spacecraft need not be attached to the object and can thus stay at a safe distance from it.

Impact of Removal Missions

The impact of ADR missions on the evolution of the debris population and the resulting probability to stop the chain reaction of the Kessler syndrome is a field of active research. Most IADC member agencies use their evolutionary simulation tools to assess the hypothetical impact of ADR missions (Anderson & Schaub, 2014; Lewis et al., 2010; Liou, 2011; Liou et al., 2004; White & Lewis, 2014b). The results are not completely consistent, but the selection of several high-priority targets that are most likely going to add significantly to the population is a common trend (Bonnal et al., 2013; Braun et al., 2013; Lewis et al., 2010; van der Pas, Lousada, Terhes, Bernabeu, & Bauer, 2014; White & Lewis, 2014a). Simulations using the CASACDE and DAMAGE models, for example, generally show similar trends but still produce significantly different results (White & Lewis, 2014a). Most predictions do however agree that a business-as-usual strategy will result in a more or less significant increase of the debris population.

Active and Proposed Missions

A high-profile, recently planned debris removal mission was the Deutsche Orbitale Servicing Mission (DEOS). It aimed at developing a vehicle that can capture and dock with a tumbling and uncooperative target in LEO and deorbit the mated pair in a controlled manner. The project was conducted as a cooperation between major German and European companies and the German space agency and was intended to serve as a demonstrator for future missions. It was cancelled in 2014 due to cost overruns.

A similar concept was developed during the e.Deorbit Concurrent Design Facility (CDF) study as part of ESA's Clean Space initiative (Biesbroeck et al., 2013). The e.Deorbit program was initially conceived to bring ESA's defunct Envisat satellite back into the atmosphere by 2023. The mission was to use nets, harpoons, or a robotic arm to grab the eight-ton satellite and actively deorbit it. As of March 2018, ESA has shifted the program's focus to study synergies between the original mission and OOS activities in the areas of satellite navigation, robotics, and operational procedures (Werner, 2018).

In April 2018, the RemoveDEBRIS ADR demonstrator, built by a consortium of European and South African universities and companies, was launched to the ISS (Russel, 2018). The platform will be released from the station and test several ADR technologies: the capture of debris objects using nets and harpoons, vision-based navigation using LIDAR and optical cameras, and deorbiting using a drag sail (Forshaw et al., 2016).

Outside of Europe, the Japanese company Astroscale has been working on the development of deorbiting and ADR technologies since 2013. Their IDEA-OSG 1 satellite, intended to perform *in-situ* analyses of the debris environment, was destroyed during the launch failure of a Soyuz rocket in November 2017. The company plans to launch ELSA-d, a satellite that shall demonstrate end-of-life services, in early 2020 (Blackerby, 2018).

COMMERCIAL VIABILITY OF OOS AND ADR

In general, from a commercial perspective, not all missions that are technically feasible pay off financially and can thus considered to be viable. While most studies focus on technical feasibility, many disregard the fact that OOS may not be inherently beneficial to the industry as a whole. This is to be expected in the early stages of technology development, when researchers and companies are first and foremost interested in creating a technology baseline.

Depending on the size of their business, operators respond quite differently to satellite failures. While most larger operators can rely on existing free capacities of other platforms, and may even relocate satellites to provide better coverage, smaller companies usually do not have this option. They will go to considerable lengths to extend the life of existing platforms in case a replacement is delayed or destroyed upon launch (Graham & Kingston, 2015). According to Lamassoure (2001), the following classification of on-orbit servicing should be made, as seen from the customer's perspective: (1) life extension, (2) upgrade, and (3) modification. Life extension includes all activities that prolong the operational life of a satellite in its initial design, while measures (2) and (3) imply that modifications are made to the spacecraft. Upgrading a satellite includes any measures that seek to add functionality to better meet the system's mission objectives, while modifications are made to meet new mission goals.

Many currently performed servicing studies identify the fleet of communications satellites in GEO as the most likely target of first commercial activities, due to several reasons. First, these satellites are mostly high-performance platforms designed to operate for as many as ten to 15 years. Saleh and Castet (2011) performed a reliability analysis of geosynchronous communications satellites and established that over 90% do not fail during a 12-year operational period. They rather run out of fuel and must therefore be moved out of GEO. Second, a servicer spacecraft needs less fuel to move between targets in GEO than in LEO, although the fuel required to get there in the first place is considerably higher. Third, the high cost of such communications satellites may render their servicing economically more viable, given that a service mission comes at a sufficiently low cost.

Graham and Kingston (2015) performed a systematic assessment of the commercial viability of OOS activities from the customer's perspective. Assuming a zero-profit return—the minimum requirement for financial viability—they showed that a broad range of servicing missions seem to be viable, including refuelling, life extension of the AOCS, and array operations. Introducing a 13% profit return for a mature servicing industry, only refuelling and life extension measures that increase the satellites capacity from 0% to 100% (i.e. the repair of failed satellites) remain viable, plus a small number of other measures performed on the highest earning satellites in orbit. For a nascent industry with 22.65% profit element, only the refuelling and AOCS missions were found to be viable. According to the assessment, the timing of servicing missions is crucial: missions "should only be carried out when required, e.g. refuelling should be carried out when the spacecraft is about to run out of fuel and not on arrival" (Graham & Kingston, 2015) in orbit.

CONCLUSION

OOS and ADR require much of the same technologies. Progress in the fields of robotics, autonomous systems, teleoperations, and sensor systems must be achieved to make them feasible. Many milestones have been reached during the past two decades and unmanned servicing missions have been getting more and more attractive recently. Considering that many of the required technologies are also important for planned exploration and mining missions, it seems likely that it will not take long until the first true OOS mission will take to the skies.

Therefore, it currently seems that refuelling missions to completely inoperable satellites in GEO are the best option for establishing an OOS industry. The creation of additional demand by governmental agencies could potentially have a beneficial impact, without the requirement that profit return is achieved. Several commercial companies are already developing OOS capabilities and the first contracts with customers, governmental and commercial alike, have been signed.

Even more so than servicing, ADR missions are seriously being considered as a measure for reducing the likelihood of a catastrophic worsening of the on-orbit debris situation. Space agencies, the UN, and universities alike are conducting research on mission concepts, removal strategies, and the required technologies. It does not seem unlikely that a first true ADR mission will be conducted sooner than the first fully-fledged OOS operation. There are indications that several countries may already have some limited capabilities in orbit, but certainly no fully capable spacecraft exists to date.

Yet, states, international organizations, and private actors considering to engage in OOS and ADR activities not only have to overcome technical challenges but must also implement appropriate control measures to ensure that such missions are fully consistent with the existing norms of national and international law. Property rights and damage liability are just two of the challenging aspects that have not been fully incorporated into the international legal framework yet. Progress is thus not only required in the realm of the engineers and researchers, but also in those of the lawyers and politicians.

REFERENCES

Africano, J. L., Stansbery, E. G., & Kervin, P. W. (2004). The optical orbital debris measurement program at NASA and AMOS. *Advances in Space Research*, *34*(5), 892–900. doi:10.1016/j.asr.2003.02.022

Alessandro, A. (2016). *NASA's Restore-L Mission to Refuel Landsat 7*. Demonstrate Crosscutting Technologies. Retrieved April 30, 2018, from https://www.nasa.gov/feature/nasa-s-restore-l-mission-to-refuel-landsat-7-demonstrate-crosscutting-technologies

Anderson, P. V., & Schaub, H. (2014). Local debris congestion in the geosynchronous environment with population augmentation. *Acta Astronautica*, *94*(2), 619–628. doi:10.1016/j.actaastro.2013.08.023

ASD-Eurospace. (2012). *Space Trends: Global Space Activity Overview 1986-2011* (1st ed.). Paris, France: Eurospace.

Biesbroeck, R., Soares, T., Hüsing, J., & Innocenti, L. (2013). The e.Deorbit CDF Study: A Design Study for the Safe Removal of a Large Space Debris. In L. Ouwehand (Ed.), *Proceedings of the 6th European Conference on Space Debris*. Darmstadt, Germany: European Space Agency.

Blackerby, C., Okamoto, A., Fujimoto, K., Okada, N., Forshaw, J. L., & Auburn, J. (2018, September 14). ELSA-d: An In-Orbit End-of-Life Demonstration Mission. In *Proceedings of the 69th International Astronautical Congress*. Bremen, Germany: International Astronautical Federation.

Bombardelli, C. (2012). A Plan to De-orbit Envisat. In *Proceedings of the 2nd European Workshop on Active Debris Removal*. Paris, France: Centre national d'études spatiales.

Bonnal, C., Ruault, J.-M., & Desjean, M.-C. (2013). Active debris removal: Recent progress and current trends. *Acta Astronautica, 85*, 51–60. doi:10.1016/j.actaastro.2012.11.009

Braun, V., Lüpken, A., Flegel, S., Gelhaus, J., Möckel, M., Kebschull, C., ... Vörsmann, P. (2013). Active debris removal of multiple priority targets. *Advances in Space Research, 51*(9), 1638–1648. doi:10.1016/j.asr.2012.12.003

Chowdhury, H. (2018, March 29). Mega-constellations of satellites increase space junk risk. *Financial Times*. Retrieved from http://www.ft.com

Christiansen, E. L., Crews, J. L., Williamsen, J. E., Robinson, J. H., & Nolen, A. M. (1995). Enhanced meteoroid and orbital debris shielding. *International Journal of Impact Engineering, 17*(1–3), 217–228. doi:10.1016/0734-743X(95)99848-L

de Peuter, W., Visentin, G., Fehse, W., Elfving, A., Brown, D., & Ashford, E. (1994). Satellite Servicing in GEO by Robotic Servicing Vehicle. *ESA Bulletin*, (78): 33–39.

Dicky, V., Khutorovsky, Z., Kuricshah, A., Menshicov, A., Nazirov, R., Pitsyck, A., ... Yurasov, V. (1993). The Russian space surveillance system and some aspects of spaceflight safety. *Advances in Space Research, 13*(8), 21–31. doi:10.1016/0273-1177(93)90564-R

Dolado-Perez, J. C., Pardini, C., & Anselmo, L. (2015). Review of uncertainty sources affecting the long-term predictions of space debris evolutionary models. *Acta Astronautica, 113*, 51–65. doi:10.1016/j.actaastro.2015.03.033

Easton, R. L., & Votaw, M. J. (1959). Vanguard I IGY Satellite (1958 Beta). *The Review of Scientific Instruments, 30*(2), 70–75. doi:10.1063/1.1716492

Ebisuzaki, T., Quinn, M. N., Wada, S., Piotrowski, L. W., Takizawa, Y., Casolino, M., ... Mourou, G. (2015). Demonstration designs for the remediation of space debris from the International Space Station. *Acta Astronautica, 112*, 102–113. doi:10.1016/j.actaastro.2015.03.004

Erwin, S., & Henry, C. (2018, January 4). Orbital ATK lands second Intelsat satellite servicing deal. *SpaceNews*. Retrieved from http://www.spacenews.com

Flegel, S., Krisko, P., Gelhaus, J., Wiedemann, C., Möckel, M., Vörsmann, P., ... Matney, M. (2010). Modeling the Space Debris Environment with MASTER2009 and ORDEM2010. In *Proceedings of the 38th COSPAR Sceintific Assembly*. Bremen, Germany: COSPAR.

Forshaw, J. L., Aglietti, G. S., Navarathinam, N., Kadhem, H., Salmon, T., Pisseloup, A., ... Steyn, W. H. (2016). RemoveDEBRIS: An in-orbit active debris removal demonstration mission. *Acta Astronautica, 127*, 448–463. doi:10.1016/j.actaastro.2016.06.018

Forward, R. L., Hoyt, R. P., & Uphoff, C. W. (2000). Terminator Tether™: A Spacecraft Deorbit Device. *Journal of Spacecraft and Rockets, 37*(2), 187–196. doi:10.2514/2.3565

Foster, J. L., Benbrook, J. R., & Stansbery, E. G. (2005). Detection of small radar cross-section orbital debris with the Haystack radar. *Advances in Space Research, 35*(7), 1210–1213. doi:10.1016/j.asr.2005.05.041

Foust, J. (2018a, January 17). Effective Space signs first contract for satellite life extension services. *SpaceNews*. Retrieved from http://www.spacenews.com

Foust, J. (2018b, March 14). Orbital ATK unveils new version of satellite servicing vehicle. *SpaceNews*. Retrieved from http://www.spacenews.com

Graham, A. R., & Kingston, J. (2015). Assessment of the commercial viability of selected options for on-orbit servicing (OOS). *Acta Astronautica*, *117*, 38–48. doi:10.1016/j.actaastro.2015.07.023

Henry, C. (2017a, June 29). MDA restarts satellite servicing business with SES as first customer. *SpaceNews*. Retrieved from http://www.spacenews.com

Henry, C. (2017b, September 28). Airbus to challenge SSL, Orbital ATK with new space tug business. *SpaceNews*. Retrieved from http://www.spacenews.com

Innocenti, L., Soares, T., & Delaval, J. (2013). ESA Clean Space Initiative. In L. Ouwehand (Ed.), *Proceedings of the 6th European Conference on Space Debris*. Nordwijk, The Netherlands: European Space Agency.

Johnson, N. L. (1993). US space surveillance. *Advances in Space Research*, *13*(8), 5–20. doi:10.1016/0273-1177(93)90563-Q

Kaiser, C., Sjöberg, F., Delcura, J. M., & Eilertsen, B. (2008). SMART-OLEV—An orbital life extension vehicle for servicing commercial spacecrafts in GEO. *Acta Astronautica*, *63*(1–4), 400–410. doi:10.1016/j.actaastro.2007.12.053

Kessler, D. J., & Cour-Palais, B. G. (1978). Collision frequency of artificial satellites: The creation of a debris belt. *Journal of Geophysical Research*, *83*(A6), 2637. doi:10.1029/JA083iA06p02637

Kessler, D. J., Johnson, N. L., Liou, J.-C., & Matney, M. (2010). The Kessler Syndrome: Implications to Future Space Operations. In S. C. McQuerry (Ed.), *Proceedings of the 33rd Annual AAS Rocky Mountain Guidance and Control Conference*. Breckenridge, CO: Univelt.

Kitamura, S., Hayakawa, Y., & Kawamoto, S. (2014). A reorbiter for large GEO debris objects using ion beam irradiation. *Acta Astronautica*, *94*(2), 725–735. doi:10.1016/j.actaastro.2013.07.037

Klinkrad, H. (2006). *Space Debris: Models and Risk Analysis*. Berlin, Germany: Springer.

Klinkrad, H. (2015). Hazards of Orbital Debris. In J. N. Pelton & F. Allahdadi (Eds.), *Handbook of Cosmic Hazards and Planetary Defense*. Cham, Switzerland: Springer.

Krag, H., Flohrer, T., & Lemmens, S. (2013). Consideration of Space Debris Mitigation Requirements in the Operation of LEO Missions. In *Space Operations: Experience, Mission Systems, and Advanced Concepts* (pp. 413–429). Reston, VA: American Institute of Aeronautics and Astronautics. doi:10.2514/5.9781624102080.0413.0430

Krisko, P. H. (2004). EVOLVE historical and projected orbital debris test environments. *Advances in Space Research*, *34*(5), 975–980. doi:10.1016/j.asr.2003.02.025

Krisko, P. H. (2010). NASA's new orbital debris engineering model, ORDEM2010. In *Proceedings of the Fourth IAASS Conference*. Noordwijk, The Netherlands: European Space Agency.

Krisko, P. H., Flegel, S., Matney, M. J., Jarkey, D. R., & Braun, V. (2015). ORDEM 3.0 and MASTER-2009 modeled debris population comparison. *Acta Astronautica, 113*, 204–211.

Krolikowski, A., & David, E. (2013). Commercial On-Orbit Satellite Servicing: National and International Policy Considerations Raised by Industry Proposals. *New Space, 1*(1), 29–41. doi:10.1089pace.2013.0002

Lamassoure, E. S. (2001). *A Framework to Account for Flexibility in Modeling the Value of On-Orbit Servicing for Space Systems*. Cambridge, MA: Massachusetts Institute of Technology.

Lambour, R., Rajan, N., Morgan, T., Kupiec, I., & Stansbery, E. (2004). Assessment of orbital debris size estimation from radar cross-section measurements. *Advances in Space Research, 34*(5), 1013–1020. doi:10.1016/j.asr.2003.02.043

Leone, D. (2014, August 11). NASA Ready to Put Robotic Refueling to the Test. *SpaceNews*. Retrieved from http://www.spacenews.com

Lewis, H. G., Newland, R. J., Swinerd, G. G., & Saunders, A. (2010). A new analysis of debris mitigation and removal using networks. *Acta Astronautica, 66*(1–2), 257–268. doi:10.1016/j.actaastro.2009.05.010

Liou, J.-C. (2011). An active debris removal parametric study for LEO environment remediation. *Advances in Space Research, 47*(11), 1865–1876. doi:10.1016/j.asr.2011.02.003

Liou, J.-C., Hall, D., Krisko, P., & Opiela, J. (2004). LEGEND – a three-dimensional LEO-to-GEO debris evolutionary model. *Advances in Space Research, 34*(5), 981–986. doi:10.1016/j.asr.2003.02.027

Liou, J.-C., & Johnson, N. L. (2008). Instability of the present LEO satellite populations. *Advances in Space Research, 41*(7), 1046–1053. doi:10.1016/j.asr.2007.04.081

Liou, J.-C., Matney, M. J., Anz-Meador, P. D., Kessler, D. J., Jansen, M., & Theall, J. R. (2002). *The New NASA Orbital Debris Engineering Model ORDEM 3.0 (NASA/TP-2002-210780)*. Houston, TX: National Aeronautics and Space Administration.

Mantellato, R., Lorenzini, E. C., Sternberg, D., Roascio, D., Saenz-Otero, A., & Zachrau, H. J. (2017). Simulation of a tethered microgravity robot pair and validation on a planar air bearing. *Acta Astronautica, 138*, 579–589. doi:10.1016/j.actaastro.2016.12.029

McLaughlin Green, C., & Lomask, M. (1970). *Vanguard - A History (SP-4202)*. Washington, DC: National Aeronautics and Space Administration.

Metz, M. (2012). DLR Perspective on Sustainable Use of Space. In *Proceedings of the CleanSpace Workshop on Active Debris Removal*. Darmstadt, Germany: European Space Agency.

NASA Office of Safety and Mission Assurance. (2017). *NASA Procedural Requirements for Limiting Orbital Debris (NPR 8715.6B)*. Washington, DC: National Aeronautics and Space Administration.

NASA Orbital Debris Program Office. (2015). Two More Collision Avoidance Maneuvers for the International Space Station. *Orbital Debris Quarterly News, 19*(4).

NASA Orbital Debris Program Office. (2018). Monthly Number of Objects in Earth Orbit by Object Type. *Orbital Debris Quarterly News, 22*(1).

National Research Council. (1995). *Orbital Debris: A Technical Assessment*. Washington, DC: National Academies Press.

Nazarenko, A. I., & Menshikov, I. L. (2001). Engineering model of space debris environment. In H. Sawaya-Lacoste (Ed.), *Proceedings of the Third European Conference on Space Debris*. Noordwijk, The Netherlands: European Space Agency.

Nishida, S.-I., Kawamoto, S., Okawa, Y., Terui, F., & Kitamura, S. (2009). Space debris removal system using a small satellite. *Acta Astronautica*, *65*(1–2), 95–102. doi:10.1016/j.actaastro.2009.01.041

On-Orbit Satellite Servicing Study: Project Report (NP-2010-08-162-GSFC). (2010). Greenbelt, MD: National Aeronautics and Space Administration.

Phipps, C. R., Baker, K. L., Libby, S. B., Liedahl, D. A., Olivier, S. S., Pleasance, L. D., ... Valley, M. T. (2012). Removing orbital debris with lasers. *Advances in Space Research*, *49*(9), 1283–1300. doi:10.1016/j.asr.2012.02.003

Reed, J., Busquets, J., & White, C. (2012). Grappling System for Capturing Heavy Space Debris. In *Proceedings of the 2nd European Workshop on Active Debris Removal*. Paris, France: Centre national d'études spatiales.

Retat, I. (2012). Net capture system: a potential orbital space debris removal system. In *Proceedings of the 2nd European Workshop on Active Debris Removal*. Paris, France: Centre national d'études spatiales.

Roesler, G., Jaffe, P., & Henshaw, G. (2017, March 8). Inside DARPA's Mission to Send a Repair Robot to Geosynchronous Orbit. *IEEE Spectrum*. Retrieved from https://www.spectrum.ieee.org

Rossi, A., Lewis, H., White, A., Anselmo, L., Pardini, C., Krag, H., & Bastida Virgili, B. (2016). Analysis of the consequences of fragmentations in low and geostationary orbits. *Advances in Space Research*, *57*(8), 1652–1663. doi:10.1016/j.asr.2015.05.035

Russel, K. (2017). *Restore-L Mission Passes Preliminary Design Review*. Retrieved April 30, 2018, from https://www.satellitetoday.com/government-military/2017/12/26/restore-l-mission-passes-preliminary-design-review/

Russel, K. (2018, April 4). SpaceX Lofted an Experimental Debris Removal Satellite for a UK University. *Via Satellite*. Retrieved from https://www.satellitetoday.com

Saleh, J. H., & Castet, J.-F. (2011). Geosynchronous communication satellites: system reliability and failures. In *Spacecraft Reliability and Multi-State Failures: A Statistical Approach*. West Sussex, UK: Wiley. doi:10.1002/9781119994077.app1

Saleh, J. H., Lamassoure, E. S., Hastings, D. E., & Newman, D. J. (2003). Flexibility and the Value of On-Orbit Servicing: New Customer-Centric Perspective. *Journal of Spacecraft and Rockets*, *40*(2), 279–291. doi:10.2514/2.3944

Schaub, H., & Sternovsky, Z. (2014). Active space debris charging for contactless electrostatic disposal maneuvers. *Advances in Space Research, 53*(1), 110–118. doi:10.1016/j.asr.2013.10.003

Schildknecht, T., Musci, R., Ploner, M., Beutler, G., Flury, W., Kuusela, J., ... de Fatima Dominguez Palmero, L. (2004). Optical observations of space debris in GEO and in highly-eccentric orbits. *Advances in Space Research, 34*(5), 901–911. doi:10.1016/j.asr.2003.01.009

Space Assembly, Maintenance And Servicing (SAMS) Study, Final Report. (1986). Sunnyvale, CA: Lockheed Missiles & Space Company, Inc.

Sternberg, D., Hilton, A., Miller, D., McCarthy, B., Jewison, C., & Roascio, D., … Saenz-Otero, A. (2016). Reconfigurable ground and flight testing facility for robotic servicing, capture, and assembly. In *Proceedings of the 2016 IEEE Aerospace Conference*. Big Sky, MT: IEEE. 10.1109/AERO.2016.7500526

Technology Quarterly. Space. (2016, August 27). *The Economist, 420*(9004).

Treat, J., Scalamogna, A., & Conant, E. (2015). The Secret to Hubble's Success. *National Geographic*. Retrieved April 30, 2018, from http://www.nationalgeographic.com/hubble-timeline/

Trentlage, C., Mindermann, P., Ben Larbi, M. K., & Stoll, E. (2016). Development and Test of an Adaptable Docking Mechanism Based on Mushroom-Shaped Adhesive Microstructures. In *Proceedings of AIAA SPACE 2016*. Reston, VA: American Institute of Aeronautics and Astronautics. 10.2514/6.2016-5486

United Nations Office for Outer Space Affairs. (2010). *Space Debris Mitigation Guidelines of the Committee on the Peaceful Uses of Outer Space*. Vienna, Austria: United Nations.

van der Pas, N., Lousada, J., Terhes, C., Bernabeu, M., & Bauer, W. (2014). Target selection and comparison of mission design for space debris removal by DLR's advanced study group. *Acta Astronautica, 102*, 241–248. doi:10.1016/j.actaastro.2014.06.020

Visagie, L., Lappas, V., & Erb, S. (2015). Drag sails for space debris mitigation. *Acta Astronautica, 109*, 65–75. doi:10.1016/j.actaastro.2014.12.013

Waltz, D. M. (1993). *On-Orbit Servicing of Space Systems* (1st ed.). Malabar, FL: Krieger.

Weeden, B., & Samson, V. (2018, April 4). New UN Guidelines For Space Sustainability Are A Big Deal. *Breaking Defense*. Retrieved from https://www.breakingdefense.com

Werner, D. (2018, March 21). ESA to investigate links between debris removal and satellite servicing. *SpaceNews*. Retrieved from http://www.spacenews.com

White, A. E., & Lewis, H. G. (2014a). An adaptive strategy for active debris removal. *Advances in Space Research, 53*(8), 1195–1206. doi:10.1016/j.asr.2014.01.021

White, A. E., & Lewis, H. G. (2014b). The many futures of active debris removal. *Acta Astronautica, 95*, 189–197. doi:10.1016/j.actaastro.2013.11.009

Yulin, Z., & Zhaokui, W. (2016). Space Traffic Safety Management and Control. *IEEE Transactions on Intelligent Transportation Systems, 17*(4), 1189–1192. doi:10.1109/TITS.2015.2494686

Zachrau, H. J. (2018, March 14). *SPHERES Tether Slosh*. Retrieved April 30, 2018, from https://www.nasa.gov/mission_pages/station/research/experiments/2553.html

Zhai, G., Qiu, Y., Liang, B., & Li, C. (2009). System Dynamics and Feedforward Control for Tether-Net Space Robot System. *International Journal of Advanced Robotic Systems*, *6*(2), 10. doi:10.5772/6791

ADDITIONAL READING

Fehse, W. (2003). *Automated Rendezvous and Docking of Spacecraft*. Cambridge, UK: Cambridge University Press. doi:10.1017/CBO9780511543388

Johnson, N. L., Stansbery, E., Whitlock, D. O., Abercromby, K. J., & Shoots, D. (2008). History of On-Orbit Satellite Fragmentations (14th ed., NASA/TM-2008-214779). Houston, TX: National Aeronautics and Space Administration.

Klinkrad, H. (2006). *Space Debris: Models and Risk Analysis*. Berlin, Germany: Springer.

Krolikowski, A., & David, E. (2013). Commercial On-Orbit Satellite Servicing: National and International Policy Considerations Raised by Industry Proposals. *New Space*, *1*(1), 29–41. doi:10.1089pace.2013.0002

National Research Council. (1995). *Orbital Debris: A Technical Assessment*. Washington, D.C.: National Academies Press.

On-Orbit Satellite Servicing Study: Project Report (NP-2010-08-162-GSFC). (2010). Greenbelt, MD: National Aeronautics and Space Administration.

Pelton, J. N., & Allahdadi, F. (Eds.). (2015). *Handbook of Cosmic Hazards and Planetary Defense*. Cham, Switzerland: Springer.

Waltz, D. M. (1993). *On-Orbit Servicing of Space Systems* (1st ed.). Malabar, FL: Krieger.

KEY TERMS AND DEFINITIONS

Active Debris Removal: The removal of a space debris object from orbit such that it either burns up in Earth's atmosphere or is placed in an orbit of little operational value.

Chaser: The spacecraft "chasing" after a damaged satellite or space debris object that it shall repair or deorbit.

Deorbiting Device: A device used to remove a spacecraft or debris object from orbit, which may be either active (propulsion unit) or passive (drag sail, electrodynamic tether).

Geostationary Orbit: A spacecraft in geostationary orbit (GEO) orbits Earth at an altitude of 35,786 kilometers above the equator and appears at a fixed position in the sky. Its orbital period is equal to Earth's rotational period.

Low Earth Orbit: A low Earth orbit is an orbit around Earth with an altitude of less than 2,000 kilometers.

On-Orbit Servicing: A spacecraft is used to repair, upgrade, refuel, or deorbit another spacecraft in orbit.

Satellite: An artificial satellite is a man-made object intentionally placed into orbit. Satellites are oftentimes manned or unmanned spacecraft.

Space Debris: Space debris is a term for the mass of defunct, artificially created objects in Earth orbit.

Target: The target of an on-orbit servicing or active debris removal mission can be either another spacecraft or a space debris object.

ENDNOTES

1 Ejectables are objects ejected from a satellite during its operational life, for example fragments of explosive bolts used to release solar panels or antennas.

2 Currently, the slot size over North America is defined as two degrees of longitude in size, which corresponds to roughly 1,470 kilometers at GEO altitude. Within each slot, a satellite operator usually keeps a satellite within a box of typically 0.1 degrees of longitude (70 kilometers) in length. Operators may form clusters of satellites by placing more than one satellite into one slot.

3 Hypergolic fuel spontaneously ignites by itself once its components mix. Thus, combustion does not need to be initiated but starts by itself, making hypergolic propellants a good choice if high reliability of (rarely used) rocket motors is required. The main disadvantage is the corrosive nature of its components.

4 Optical tracking has been successfully demonstrated down to sizes of three centimeters and less. However, the telescopes used do not offer sufficient coverage to add reliable data to catalogues.

5 Teleoperated activities are conducted by a remote operator via radio link to the servicer satellite. Camera and sensor data substitute for direct optical or haptic feedback, albeit with a time delay that is proportional to the distance between the satellite and the ground station.

6 Lidar uses an operating principle similar to radar. Instead of radio waves, laser light is shone onto an object, reflected by it, and subsequently detected by a light sensor.

Chapter 11
On-Orbit Servicing and Active Debris Removal:
Legal Aspects

P. J. Blount
University of Luxembourg, Luxembourg

ABSTRACT

This chapter examines the legal issues associated with the activities of on-orbit servicing and active debris removal. It addresses these legal issues from the perspective of military, civil, and commercial space activities. It argues that the major international legal challenges for these technologies are avoiding their weaponization in order to use them to provide for the sustainability of the space environment. Further, it argues that the development of domestic authorization regimes will be necessary precursor to the deployment of OOS and ADR.

INTRODUCTION

Innovative space technologies have both the ability to advance human use and exploration of space and to disrupt the status quo that governs the space domain. Both On-orbit Servicing (OOS) and Active Debris Removal (ADR) technologies serve as excellent examples of the promises and dangers of innovative technologies. As noted in the technical chapter on these technologies, active debris removal and on-orbit servicing both contain a shared critical function: they both are premised on the technical capability to manoeuvre to, rendezvous or come into close contact with, and manipulate an object in space. It is important to note that 'object in space,' as used here, is not meant to be an equivalent to the legal term "space object," which "is arguably the most important term in all the corpus of space law because the entire legal regime for space is centred on what is launched into and used in outer space" (Hearsey, 2015, p. 191). Here it simply means a mass in space, regardless of whether that mass constitutes a space object in the legal sense.

DOI: 10.4018/978-1-5225-7256-5.ch011

This common capability of rendezvous and manipulation means that these technologies can be analysed along similar lines due to their overlapping capability. Therefore, this legal analysis will first examine these technologies based on the common rendezvous and manipulate functionality through the lens of three general bodies of space law: the international law applying to military uses of space, the international law applying to civil uses of space; and the domestic law applying to commercial uses of space. This three-pronged general analysis will then be followed by two sections that investigate legal issues raised by the divergent applications of these two technologies.

ON-ORBIT SERVICING AND ACTIVE DEBRIS REMOVAL

As emerging innovative technologies, OOS and ADR are not dealt with explicitly by space law, and neither the space treaties nor any known national law directly addresses these two capabilities. At the international level, this is because the space treaties, and in particular the Treaty on Principles Governing the Activities of States in the Exploration and Use of Outer Space, Including the Moon and Other Celestial Bodies (hereinafter the Outer Space Treaty), were negotiated to allow for new technologies and innovations to emerge in light of the unknown course for the development of technology (Blount, 2018). The result is that international space law is built on principles that apply to space activities comprising a broad range of technologies. At the national or domestic level, the lack of direct regulation can be attributed to a general lag in the development of the law. These technologies are still nascent in development and, as such, have not yet had sufficient time to warrant specific domestic regulatory actions.

Since one of the key applications of both OOS and ADR is the mitigation and reduction of space debris, it should be noted, in particular, that no binding international agreement exists concerning the mitigation or active removal of space debris. After a multi-year effort, the Inter-Agency Space Debris Coordination Committee (IADC) published a set of technical mitigation guidelines in 2002. These guidelines were presented to the United Nations (UN) Committee for Peaceful Use of Outer Space (COPUOS), which in turn adapted the guidelines and adopted its own set of debris mitigation measures. While both the IADC and UN COPUOS guidelines represent positive steps forward, neither set of rules are legally binding on states. Despite the fact that these guidelines are not legally binding, numerous states including the US, France, Russia, and China as well as the European Space Agency have adopted some form of debris mitigation guidelines. These national frameworks, though, generally do not speak to the specifics of OOS or ADR.

In light of these legal lacunae, the following legal analysis will attempt to explain how these new two categories of future space missions may fit into the current legal framework and will emphasize critical issues that result from the various gaps in the law (Losekamm et al., 2015, p. 1-2; Masson-Zwaan & Richards, 2015).

Military Uses of Outer Space

One of the threshold questions of legality in space law is whether a space activity is intended for peaceful purposes. While on their face OOS and ADR are not military space activities, the core capability of manoeuvre, rendezvous, and manipulation is of a dual use nature, allowing also military capabilities. In other words, this type of technology has the capability for being used to interfere with another states space activities in a non-peaceful manner. As a result, this analysis will begin by evaluating the legality of these technologies in light of their dual use potential.

From the very beginning of the Space Age, space was considered to be a strategic asset by states engaging in space activities (Bourbonnière, 2005). Recognizing the unique attributes of space, states negotiated the Outer Space Treaty and the four treaties considered to be its progeny as a way of stabilizing their relations in the new domain of outer space (on the Outer Space Treaty generally, *see* Dembling and Arons, 1967). The legal regime that was created centred on the principle of making space a realm to be used by states for peaceful purposes. While 'peaceful purposes' is a threshold for a legal space activity, the legal regime does not completely exclude military activities. This is because the space treaties are, in essence, a series of security agreements that are specifically intended to address questions of international peace and security in outer space. The idea of international peace and security in international law has never meant exclusively non-militarized or even non-weaponized. As a result, there is a great deal of legal ambiguity as to the exact limits of what is and is not a peaceful purpose (Blount, 2009; Bourbonnière and Lee, 2007, p. 877; Bourbonnière, 2005, p. 15-18). Specific to this discussion, the rendezvous and manipulation capabilities of OOS and ADR raise questions related the weaponization of space and the extent to which these technologies can be perceived as peaceful.

Weaponization of outer space has long been a contentious issue and has a colourful history including technologies such as fractional orbital bombs (FOBS), the Strategic Defense Initiative (SDI), and most recently, anti-satellites (ASAT) activities by the United States and China. The capability to interfere with another state's space asset is one of great strategic importance, and the use of rendezvous and manipulation has certainly been contemplated for its military uses. For example, the *Space Shuttle* was an early entrant into OOS technology as it was used to service and repair on orbit assets. However, when the United States was beginning the *Shuttle* program, the USSR characterized the *Shuttle* as an ASAT weapon (Grego, 2012). Despite the use of the *Space Shuttle* for peaceful servicing missions of civil satellites, rendezvous missions that are funded by the military, such as the Defense Advanced Research Projects Agency's (DARPA) Phoenix Project, raise specific questions as to whether this leads to the weaponization of space and if so what laws there are to limit such weaponization (DARPA, 2016).

Weaponization is a grey legal area in the law of outer space. While states, when negotiating the space treaties, were eager to limit the potential conflicts in space, they were not correspondingly keen on limiting their strategic uses of outer space, especially in light of the dawning age of intercontinental ballistic missiles. As a result, the only article in the Outer Space Treaty that directly addresses weapons and military uses of outer space is Article IV. This article precludes states from placing *any* weapons on the Moon or other celestial bodies, but it only precludes states from placing *some* weapons, namely weapons of mass destruction (WMD), into Earth orbit. This means that conventional weapons are not explicitly legally precluded from being placed into Earth orbit or the general void of space (Bourbonnière, 2005, p. 7).

Attempts have been made to establish legal limitations on the placement of weapons in orbit. These efforts have focused on evolving policy regimes known as the Prevention of an Arms Race in Outer Space (PAROS) at the international level (*see generally* Rathberger, Remuss, and Schrogl, 2009, p. 34). The attempted development of PAROS can be observed primarily in two international organizations. First, the UN General Assembly passes a resolution on PAROS each year, but it should be noted that the United States routinely votes against or abstains from this resolution (Federation of American Scientists, n.d.). Second, the Conference on Disarmament has PAROS on its list of competencies, but the body has been deadlocked since 1995 when China "insisted" on linking PAROS to the fissile material cut-off treaty

(Rathberger, Remuss, and Schrogl, 2009, p. 34). These two organizations clearly show a divergence in international agreement on the weaponization of space. While PAROS is overwhelmingly passed by the UNGA as a non-binding resolution annually, it became a sticking point as a legal limitation at the deadlocked CD, "which has served to underscore the incompleteness of the existing legal framework based on the peaceful purposes orientation of state action in outer space" (Meyer, 2014, p. 2). In short, PAROS can be said to be an international policy, but a policy with no legal force.

Much of what makes legal obligations linked to PAROS so difficult to achieve is linked directly to a debate over what constitutes a weapon, which has been characterized as a "matter of degree" (Mineiro, 2008, p. 446; see also Moltz, 2008, p. 42-43). Specifically, the problem is that "many space systems designed for peaceful purposes have the *capacity* to destroy or interfere" with other space objects (Mineiro, 2008, p. 447). This quandary as seen in the *Space Shuttle* example from above is not new, since potentially everything in space has some potential to be used as a weapon (Wright, Grego, and Grunlund, 2005; Moltz, 2008, p. 42-43). The pertinent question, then, is whether OOS or ADR technologies might violate the principle of peaceful purposes in light of the weaponization debate. Since these technologies are specifically designed to interact with and in some case deorbit objects in space, they have more than a latent capability in interfere with other spacecraft.

The answer to the question of whether or not these technologies constitute weapons is most likely 'no.' Dual use technologies have been a feature of space exploration ever since states initiated their respective space activities (Dembling and Arons, 1967, p. 435). The successful use of the *Shuttle* to rendezvous with and repair satellites illustrates through state practice that there is no legal prohibition that disallows the deployment and use of these technologies for peaceful purposes. However, if these technologies were used for ASAT activities they may violate Art. 2(4) of the UN Charter. This clause prohibits states from using aggressive force against another state, thus while the law does not prohibit OOS and ADR technology de facto, it does prohibit certain uses of that technology that impinge on the rights of other states by interfering with those state's space activities. This accords with space law since, as stated, a central goal of the regime is the maintenance of international peace and security.

The lack of legal prohibition is not the end of the story, however, and simply saying that these technologies are commercial in nature should not cut the conversation short. "[T]he constant march of technology" is making weaponizable dual use technologies easier to build (Smith, 2011, p. 42). These technologies have increasing strategic value, and even if commercially developed, they have the potential to lead to heightened security concerns and could possibly strike the opening bell of an arms race in outer space (Lewis, 2013, p. 73). If these technologies develop in the commercial sector, home states can and likely will gain access to these technologies. In such scenarios, states will need to paint clear policy redlines about the acceptable uses of these technologies in order to retain the strategic peacefulness of outer space through clear signalling to other states. This is because a deployment or perceived deployment of weapons in in space could constitute a "fundamental change in circumstances" under the Vienna Convention on the Law of Treaties that could result in defections from the Outer Space Treaty regime (Bourbonnière and Lee, 2007, p. 876; *but see* Tate, 2006).

From a space security perspective then, these technologies have the potential to be very destabilizing, because of their inherent potential to be used in non-peaceful ways. As a result, states must take measures to be clear about their intent when engaging in these activities regardless of whether the primary purpose is commercial, civil, or military. Such signalling and best practices have the ability to set normative behaviour for future space activities and assure other states of the regularity of space operations in order to avoid conflict.

Civil Space Activities

As already noted, the concept of the "peaceful uses" of outer space is a dominant theme in the development of space technologies. Though a security treaty, the bulk of the Outer Space Treaty addresses the peaceful use and exploration of outer space. As a result, states have traditionally emphasized their peaceful civil space programs as they develop space technology, much of which is dual use in nature. For instance, this was a central factor during the establishment of NASA, which by law is "devoted to peaceful purposes for the benefit of all mankind" (National Aeronautics and Space Act, Sec. 102; Logsdon, 1998, p. 2, 7-8, 35). While military capability is important to most nations, states use civil space agencies to represent their peaceful interest in space and to coordinate international cooperation among space actors. This section will explore the international legal regime as it applies to these civil space agencies in their development of OOS and ADR technologies. Specifically, it will address the concepts of free access to space in conjunction with the principle of due regard to other space actors, liability for space activities, and jurisdiction over space objects.

Article I of the Outer Space Treaty serves as a linchpin in ensuring the peaceful uses of space exploration (Dembling and Arons, 1967, p. 431). This article states that outer space "shall be free for exploration and use by all states without discrimination of any kind." This is a significant statement as it gives states a great deal of leeway in their outer space activities, especially in light of the above noted ambiguity in the legal content of "peaceful purposes." For instance, North Korea has used arguments based on Article I to justify an Inter-Continental Ballistic Missile (ICBM) program veiled as a space program (Elleman, 2013; Blount, 2011). While the principle of free access is legally very broad, it is not without limitation. A basic principle of physics (at least non-quantum physics) is that two objects cannot be in the same place at the same time. This physical restriction means that states do not have a complete freedom of access to space, but instead must coordinate their activities in order to avoid "harmful interference" with the activities of others (Outer Space Treaty, Art. IX; *see* also International Telecommunication Union, 2016, Art. 1.169). As a result, Article IX of the Outer Space Treaty requires that states act "with due regard to the corresponding interests of other State Parties to the Treaty" and to engage in "international consultations" when their space activities could cause "potentially harmful interference."

The specific requirements of Article IX are ambiguous at best, but it has been argued that they set a *de minimis* standard for information sharing that can facilitate the coordination of outer space activities among states (Marder, 2008; Blount, 2011). Despite this ambiguity, Article IX does set a threshold for understanding activities such as those contemplated by OOS and ADR. States pursuing such technologies have a legal obligation to use these technologies in such a way that they do not interfere with other states space activities. This would include making sure that the spacecraft employed in such activities are not involved in collisions with the spacecraft of other states, but Article IX likely has a number of heightened requirements in relation to OOS and ADR. Since these operations require that spacecraft come into close proximity to other space objects, states will need to make sure that they are not a source of more space debris through collisions with their intended targets. Additionally, OOS and ADR operators will need to ensure that once they have engaged with a space object that they manoeuvre it in such a way that it does not harmfully interfere with other space objects.

While the Outer Space Treaty does not attempt to establish a regime in which states must coordinate their space activities, neither does it contemplate a seamless system free of risk. Indeed, the treaty consistently treats space as a realm of high risk, thus Article VII of the Outer Space Treaty makes states liable for accidents they cause in space. Article VII was expanded on in the 1972 Liability Convention, which established an international liability regime for states engaging in outer space activities. At its core, the Liability Convention holds a launching state liable for damage caused by its space objects on a fault-based standard for damage caused in space and on an absolute or strict liability standard for damage caused on the surface of the earth or to aircraft in flight (Arts. II and III). States engaged in OOS and ADR will be liable for damage that they caused under the terms of the treaty, but the Liability Convention adds a bit of complexity. It is quite possible that either an OOS or and ADR mission might be targeting a space object that does not belong to the state engaging in the mission. Under the Liability Convention the "launching state" is liable for damage caused "by its space object" (Arts. II and III). "Launching state" is defined as a state which launches a space object, procures the launch of the space object, and from whose territory or facility a space object is launched (Art. I). This means that liability may be apportioned over a number of states. OOS and ADR create questions of liability since between the targeted object and the targeting object there may be multiple launching states. Questions of causation will likely arise in a variety of scenarios involved with these technologies. For example, if the rendezvous inadvertently bumps the object out of its orbit and it collides with a third party's spacecraft, then a question of which space object caused the damage will be raised. Other scenarios flow from this. What if the collision happens during the engagement? What if the engagement is flawless yet the OOS created a flaw in the serviced spacecraft that results in damage to a third-party craft? The legal rabbit hole is vast, though Article V does contemplate scenarios through which states can apportion fault through bilateral agreements.

While the risk of collision is one that states must take seriously, it should not overshadow the fact that states have a duty under the Outer Space Treaty to avoid the "harmful contamination" of outer space (Outer Space Treaty, Art. IX). While there has been little elaboration as to what exactly this phrase means, states have engaged in a cooperative regime, though not legally binding, to mitigate the creation of space debris through mechanisms such as the IADC Debris Mitigation Guidelines. These efforts are being further elaborated in the Long Term Sustainability Guidelines emerging from the UN Committee on the Peaceful Uses of Outer Space (UNCOPUOS), which requests that states "Investigate and consider new measures to manage the space debris population in the long term" (Guideline 28). So, while there is risk involved with OOS and ADR, these technologies have an important role to play in the overall management of space in the long term and to help states avoid future liability based on previously created debris.

Finally, much of this discussion has pivoted on an assumption that a state engaging in OOS or ADR has some sort of right to engage with the target of these activities. It has also been acknowledged, though, that these activities may not be limited to space objects owned by the states engaging in these activities. States will need to be wary of issues of ownership and jurisdiction when engaging with space objects. Article VIII of the Outer Space Treaty grants the state, upon whose registry a space object is entered, jurisdiction and control over that object, and it further states that the presence of an object in space does not affect the ownership of that space object. This means that states engaged in OOS and ADR will need to be wary of a potential bevy of states, including the launching state or states, the state of registration, and the state of ownership, which for obvious reasons raise numerous legal questions about which state or states must be approached before engaging in a rendezvous procedure. Presumably, the registration state, which has jurisdiction, would be sufficient, but the potential liability issues raised for launching states, may give rise to some right of objection. Further, the concept of salvage found in maritime law, which allows abandoned craft on the seas to become to property of the individuals that salvage it, has not been adopted into outer space, meaning that states must have affirmative consent from the owner and states with jurisdiction before engaging with that object. Such consent would be the result of the "consultations" between states envisioned in Article IX of the Outer Space Treaty. Without such consent, space objects remain under the jurisdiction and control of the states that have registered them, and interference could be treated as a hostile act. Arguably, this includes debris that results from these objects, especially since liability for damage caused by debris attaches to the original launching state (US Department of Defense, 2008). As a result, even though debris creates risk to space activities and reduces the sustainability of space as a whole, States are not at liberty to actively engage with debris without appropriate consultations under Article IX.

While states may seek to employ OOS and ADR technologies to reduce the amount of debris in space in order to promote the safety, security, and sustainability of outer space, there are significant legal hurdles to the widespread employment of these technologies. States are free to clean up their own debris, but they are not free to clean up the debris of others without negotiations on a bilateral basis, even if such debris poses a risk to active spacecraft. If future OOS and ADR operations are going to be a salve to the space debris problem, then states will need to negotiate multilateral norms governing this type of activity.

Commercial Space Activities

While the military and civil aspects of OOS and ADR are important, it is the commercial sector that seems to be poised to develop these capabilities. In particular, OOS holds great potential for commercial actors as it could be the gateway to extending the life of space assets thereby lowering costs and increasing profits over the course of a satellite's operational lifetime. This section seeks to highlight the core legal issues associated with the commercial deployment of OOS and ADR.

The Outer Space Treaty was adopted among and between states, and as a result, it only provides a legal framework governing state activities. While the obligations in the Outer Space Treaty are owed by states to other states, international obligations often come in forms that require a state to govern its domestic actors in such a way as to ensure compliance. The Outer Space Treaty does not expressly address commercial activities in space (Losekamm, et al., 2015, p. 5). It does not, however, leave a void. Instead, through Article VI, it gives states the affirmative duty to directly oversee nongovernmental actors and the obligation of responsibility for the actions of these same nongovernmental actors. The term nongovernmental actors include within its ambit commercial actors. First, Article VI gives States "international responsibility for national activities in outer space . . . whether such activities are carried on by governmental agencies or by non-governmental entities." In other words, the state is on the hook for the harm its nongovernmental actors engaged in space activities do to other states and their citizens. This clause is unique in international law - similar rules do not apply to other global commons such as the sea, and it represents the high level of care that the drafters wanted to establish in space. Second, the clause states that these entities "shall require authorization and continuing supervision by the appropriate State Party to the Treaty." The clause obligates states to develop some sort of domestic framework for governing non-state actors, and it pushes regulation of specific capabilities down to the domestic level. Some commentators have argued that domestic regulation, rather than international agreement, is the best way to address legal issues associated with innovative emerging technologies. For example, Mirmina argues that domestic regulatory systems are better suited to addressing space debris than international law (2004).

The obligations under Article VI have been traditionally implemented at the domestic level through licensing regimes that grant government authorities oversight of private space activities. These frameworks leave a patchwork of legal questions as many of them are incomplete or nascent. This is because the obligation in the Outer Space Treaty gives states wide latitude in how they adopt their respective regimes. Even established regimes, such as that of the United States have areas of fragmentation in the state's supervision of space actors. Such fragmentation is best seen in the partial on-orbit jurisdiction deployed by the US regime, which leaves a variety of space activities external to any one administrative authority's jurisdiction to authorize and supervise. This is the result of how the US allocates licensing jurisdiction across different government agencies based on the nature of the space activity. Specifically, the Federal Aviation Administration (FAA) has jurisdiction over launch and re-entry; the National Atmospheric and Oceanic Administration (NOAA) has jurisdiction over remote sensing; and the Federal Communications Commission (FCC) has jurisdiction over telecommunication activities from space. (Subcommittee on Space, 2014). It should be noted that ADR and OOS are not addressed directly by US regulations. In the United States, the current state of the law is such that the most that a potential operator seeking authorization to engage in this sort of activity might hope for is a favourable payload review from the FAA, which would only address the safety of operations. So, even a favourable payload review would leave open the question of whether the activity had actually been authorized and possibly create a situation in which there was no continuing supervision outside of the FCC authority over the electromagnetic spectrum used by such a craft.

The US system differs from many other domestic legal systems that treat space activities as a single category and set up a centralized licensing authority. For instance, in the United Kingdom, space activities are governed by the Outer Space Act 1986 and the more recent UK Space Industry Act of 2018. Under UK law, all space activities require licensing, and a "space activity" includes "(a) launching or procuring the launch or the return to earth of a space object or of an aircraft carrying a space object, (b) operating a space object, or (c) any activity in outer space" (UK Space Industry Act, 2018, Sec. 1). Under this regime and those similar to it OOS and ADR will be evaluated like other space activities to determine whether the state is comfortable licensing them. The disparity in licensing regimes, though, points to the fact that there is a still a lack of clarity on how these activities should be evaluated with regards to safety, security and liability.

The gaps in authorization and continuing supervision at the domestic level create a legal vacuum in which two forces work to cancel each other out. Without regulatory certainty investors are unwilling to invest in expensive ventures, and without significant investment from industry, regulators are able to punt on regulation until there is enough investment for there to be a serious undertaking. As noted above, there are numerous questions attached to the commercial viability of these technologies, and regulatory uncertainty certainly fuels this fire. This is not to say that states should leap to develop regulations for authorization and continuing supervision of activities that they are unsure how to regulate. The state's exposure to risk for space activities of non-governmental actors, through Article VI of the Outer Space Treaty, incentivizes states to move cautiously in the development of new licensing regimes. In particular, states need to be sure before they authorize that they have the capability to continually supervise the non-governmental actor. Regulations, must also balance the exposure of the state with the burdens that it enacts in order to ensure that regulatory requirements linked to authorization and supervision do not destroy the economic feasibility of the activities. Though international law is silent on OOS and ADR, it establishes a framework through which states have the primary responsibility to ensure that activities fall within the realm of peaceful uses and do not harmfully interfere with other state's activities. States should approach the drafting of such laws with caution in order to maintain proper control without unduly inhibiting the commercial industry.

This is not to paint a completely bleak picture for the future of these technologies. The legal issues raised above do not necessarily represent insurmountable obstacles. Commercial actors engaged in high stakes ventures will institute private legal parameters through contractual agreements. So, while authorization and supervision will likely require regulatory change in most legal systems, private contracts are capable of addressing gaps in the law, including a number of the issues identified above. Issues such as permission to engage with a space object, the allocation of risk and liability of the parties involved, and mission parameters would all be ripe for private agreement. Indeed, the nature of these contracts and their impact on state interests would make them an important part of a submission to a licensing authority, meaning that the state could have a role in defining the minimum requirements within such contracts. Similar targeted arrangements, in the form of international agreements, could be used among states to allocate jurisdiction as contemplated in the Liability Convention, which allows several launching states to apportion liability among themselves by agreement (Losekamm et al., 2015, p. 5). Contractual law and specific international arrangements present paths to overcome a number of legal gaps, but may not be enough to overcome problems caused by the effects of regulatory uncertainty on private capital investment. This is because investors are reluctant to invest in ventures that are merely "not illegal" through silence in the law, which creates investment risk in the uncertain political realm.

Contract law also serves another function. Contractual elements that become widely used can over time become standardized as industry best practices (Losekamm, 2015, p. 8). An accumulation of such best practices can result in normative behaviour similar to a *lex mercatoria* of space (Peculijic, 2016). A *lex mercatoria*, or law among merchants, is a system of regulation through which merchants establish regularity through norms in their interactions. These norms can then have bearing in judicial determinations on where fault lies in business deals gone bad or where rights reside. Contract clauses can help shape both domestic and international law by developing a set of standards in the form of best practices that are easily adoptable because of their proven effectiveness. As a result, private contracts have a great deal of potential for ensuring that operators and states are covered across regulatory gaps that exist in the initial stages of the deployment of technology.

To this point, this chapter has covered general legal issues that are applicable to both OOS and ADR based on their overlapping core functionality. This chapter will now turn to discussions of issues that are particular to OOS and ADR respectively.

On Orbit Servicing: Specific Issues

Two quasi-legal issues present themselves specific to on-orbit servicing. The first is the question of risk and insurance. Such activities, outside of the sparse examples noted above, have few antecedents. The nature of the activity, namely approaching and rendezvousing on-orbit is much more laden with risk than routine satellite orbit maintenance. This means that insurance markets will initially have a difficult time calculating risk in order to determine the coverage and the premiums due (on this problem, generally, see Ordyna, 2010). This is exacerbated by the lack of clarity in relation to applicable tort law (for example, *see* Kaplan, 1988). A further question lies in whether the state regulator needs to impose insurance minimums to ensure that an accident in space does not overly expose the state within the international liability regime. While insurance is generally a matter for private markets, the risk involved with space activities has challenged insurance markets' ability to provide coverage. This has in the past led to domestic legal mechanisms that relieve all or part of this burden. The insurance issue is ripe for ADR as well, however, since ADR may be similar to the act of controlled re-entry, insurers will have more data to work with in assessing coverage and premiums.

The second issue is that of standardization. As noted above, in order for satellite operators to routinely avail themselves of OOS services, some sort of standard interface will need to exist. The question is who should develop such standards. International standards can be developed in a variety of ways. Standards can come from industry associations, international organizations, domestic regulation, or simply through adoption by the market (Jayakar, 1998). The adoption of standards in and of itself is not novel, but in the context of outer space and its emphasis on international cooperation and safety and security of operations, the question arises as to the extent to which the international community should be involved in the adoption of a standardized interface. This question becomes particularly salient, in light of the increased safety such operations would result in, which, for instance, is one of the purposes behind the standardized docking mechanism on spacecraft (International Docking Standard, n.d.). Additionally, from a commercial perspective a market-driven, proprietary standard may be untenable as it would trap operators into using a particular company for servicing, thereby reducing the effects of competition in the market.

Active Debris Removal: Specific Issues

The section of this chapter above that discussed due regard under Article IX of the Outer Space Treaty, presents the problem as a negative obligation, in other words an obligation that is framed as 'thou shall not do space activities that cause interference.' However, the Article IX obligation arguably states due regard as a positive and active obligation inasmuch as it requires that states "shall conduct all their activities in outer space" with due regard. This is important because it raises the possibility that the development of debris removal technologies, could mean that states, in the future, could become obligated to actively remove debris in order to fulfil the Article IX obligations to not harmfully contaminate the space environment and to not cause harmful interference with the space activities of others. Space debris is a problem that affects all space actors, and as such the creation of space debris affects the "corresponding interests" of all states (Outer Space Treaty, 1967, Art. IX). So, while there is no legal provision banning space debris, there is increasing consensus that states are obligated to implement debris mitigation measures in order to limit the creation of new debris. The development of ADR technology coupled with this growing consensus could result in a norm according to which states begin to actively remove any new (and possibly) old debris for which they are responsible in order to address the corresponding interests of all states. While the development of such a norm would certainly be distant, as the technology would have to develop to a point that it was widespread and reliable, such a positive obligation would raise questions about which situations a state would be required to act in due regard. As the law and policy surrounding space debris develops the actual development of cost-effective debris removal technologies, could change the nature of state obligations towards mitigating debris.

The flip side of this positive obligation, though, surrounds situations in which an entity can use active debris removal to protect its space object by removing a space object of another state or private actor without express permission. States are obligated to "avoid . . . harmful contamination," which might certainly encompass the prevention of debris creation through on orbit collision (Outer Space Treaty, 1967, Art. IX). Such action clearly interferes with the "jurisdiction and [the legal fiction of] control" deployed in Article VIII of the Outer Space Treaty, and possibly the UN Charter's Art. 2(4) prohibition on the use of force against another state, both addressed above. States and commercial actors may currently avail themselves of a number of untested legal doctrines such as necessity, implied consent, self-defence, force majeure, or distress (Bourbonniere, 2005, p. 22-23). The developing notion of space traffic management, which would provide a regime of coordination among space actors to avoid on orbit incidents, could potentially contain protocols for best practices in such situations, but the adoption of an international space traffic management regime seems unlikely in the near term (see International Academy of Astronautics, 2006; Smitham, 2012; and Rendleman, 2012).

CONCLUSION

Since the inception of OOS there has been little to no development of the legal regime governing it. This is largely the result of the fact that OOS has only happened on a limited scale and ADR has yet to be effectively achieved. As a result, these activities are governed by broad treaty principles at the international level, and domestically to the extent that states have expanded their jurisdiction over these activities. The new commercial interest in these technologies is creating a need for legal answers so that the economic potential of these technologies can be unlocked. The emergence of regulation is likely to occur at the national level initially, and domestic regulation could play a large role in expanding the international regime to more fully cover these technologies. Currently, a number of states are liberalizing their domestic regimes in order to accommodate innovative technologies, but these regimes usually lack regulations specific to particular technologies, so it will be a matter of time before legal clarity is attained as to the various aspects of OOS and ADR.

OOS and ADR both present an array of opportunities for the development of a safe, secure, and sustainable space environment. But as "disruptive technologies," both must bridge a number of technical and legal gaps (not to mention economic and political gaps) (Losekamm, 2015, p. 1). As this chapter has shown, though, these are not unattainable solutions. They will, however, require close work between the legal and technical community at both the domestic and international levels in order to develop the technology as one that holds more promise for the space environment than it does threat.

REFERENCES

Blount, P. J. (2009). Limits on Space Weapons: Incorporating the Law of War Into the Corpus Juris Spatialis. *Proceedings of the 51st Colloquium on the Law of Outer Space*. Available at http://papers.ssrn.com/abstract=1393321

Blount, P.J. (2011). Developments in Space Security and Their Legal Implications. *Law/Technology, 44*(2), 18-39.

Blount, P. J. (2018). Innovating the Law: Fifty Years of the Outer Space Treaty. In M. Hofmann & P.J. Blount, (Eds.), Innovation in Outer Space: International and African Legal Perspectives (pp. 31-52). Oxford, UK: Nomos.

Bourbonniére, M. (2005). National-Security Law in Outer Space: The Interface of Exploration and Security. *Journal of Air Law and Commerce, 70*, 3–62.

Bourbonnière, M., & Lee, R. J. (2007). Legality of the Deployment of Conventional Weapons in Earth Orbit: Balancing Space Law and the Law of Armed Conflict. *European Journal of International Law*, *18*(5), 873–901. doi:10.1093/ejil/chm051

Charter of the United Nations. June 26, 1945.

Convention on International Liability for Damage Caused by Space Objects, 1972, 961 UN.T.S. 187.

DARPA. (2016, Mar. 25). *Program Aims to Facilitate Robotic Servicing of Geosynchronous Satellites.* Retrieved from: http://www.darpa.mil/news-events/2016-03-25

Dembling, P., & Arons, D. (1967). The Evolution of the Outer Space Treaty. *Journal of Air Law and Commerce*, *33*, 419–456.

Department of Defense. (2008, Feb. 14). Transcript: Department of Defense News Briefing with Deputy National Security Advisor Jeffrey, Gen. Cartwright and NASA Administrator Griffin. In P. J. Blount & J. I. Gabrynowicz (Eds.), *USA-193: Selected Documents (51). University*. National Center for Remote Sensing, Air, and Space Law.

Elleman, M. (2013). Prelude to an ICBM? Putting North Korea's Unha-3 Launch into Context. *Arms Control Today*, *43*(2), 8–13.

Federation of American Scientists. (n.d.). *Prevention of an Arms Race in Outer Space*. Retrieved from https://fas.org/programs/ssp/nukes/ArmsControl_NEW/nonproliferation/NFZ/NP-NFZ-PAROS.html

Grego, L. (2012). *A History of Anti-satellite Programs, 4*. Retrieved from: http://www.ucsusa.org/sites/default/files/legacy/assets/documents/nwgs/a-history-of-ASAT-programs_lo-res.pdf

Hearsey, C. M. (2015). *The Evolution of Outer Space Law: An Economic Analysis of Rule Formation* (Masters Thesis). Available from ProQuest Dissertations and Theses database. (UMI No. 10003525)

Inter-Agency Space Debris Coordination Committee. (2007). *IADC Debris Mitigation Guidelines*. Retrieved from: http://www.unoosa.org/documents/pdf/spacelaw/sd/IADC-2002-01-IADC-Space_Debris-Guidelines-Revision1.pdf

International Academy of Astronautics. (2006). *Cosmic Study on Space Traffic Management*. IAA.

International Docking Standard. (n.d.). Retrieved from: http://www.internationaldockingstandard.com/

International Telecommunication Union. (2016). *Radio Regulations*. ITU.

Jayakar, K. (1998). Globalization and the Legitimacy of International Telecommunications Standard-Setting Organizations. *Indiana Journal of Global Legal Studies*, *5*, 711–738.

Kaplan, L. S. (1988). Space-Specific Remedies for Torts in Outer Space: What Path Will US Law Follow. *International Law*, *22*, 1145–1156.

Lewis, J. A. (2013). Reconsidering Deterrence for Space and Cyberspace. In M. Krepon & J. Thompson (Eds.), Anti-Satellite Weapons, Deterrence and Sino-American Space Relations (pp. 61–80). Washington, DC: Stimson Center.

Logsdon, J. M. (Ed.). (1998). *Legislative Origins of the National Aeronautics and Space Act of 1958 : Proceedings of an Oral History Workshop : Conducted April 3, 1992.* Washington, DC: Government Printing Office.

Losekamm, M. J., Hacker, J., Sardesi, N., Peculjic, A. N., & Vigneron, A. (2015). Legal and Political Implications of Future On-Orbit Servicing Missions. Paper presented at the International Astronautical Congress, Jerusalem, Israel.

Marder, E. (2008). *CPR for the OST: How China's Anti-Satellite Weapons Test Can Breathe New Life into Article IX of the Outer Space Treaty.* Center for Defense Information.

Masson-Zwaan, T., & Richards, B. (2015, Dec. 8). International Perspectives on Space Resource Rights. *Space News.* Retrieved from: http://spacenews.com/op-ed-international-perspectives-on-space-resource-rights/

Meyer, P. (2014). *The Diplomacy of Space Security: Whither the International Code of Conduct?* Simons Papers in Security and Development. Retrieved from http://summit.sfu.ca/item/14921

Mineiro, M. C. (2008). The United States and the Legality of Outer Space Weaponization: A Proposal for Greater Transparency and a Dispute Resolution Mechanism. *Annals of Air and Space Law, 2008,* 441–466.

Mirmina, S. A. (2004). The Regulation of Orbital Debris through National Measures. *Air and Space Law, 29*(2), 137–146.

Moltz, J. C. (2008). *The Politics of Space Security: Strategic Restraint and the Pursuit of National Interests.* Stanford, CA: Stanford University Press.

National Aeronautics and Space Act of 1958, As Amended. (2008). Retrieved from: https://history.nasa.gov/spaceact-legishistory.pdf

Ordyna, P. (2010). Insuring Human Space Flight: An Underwriter's Dilemma. *Journal of Space Law, 36,* 231–251.

Outer Space Act. 1986. United Kingdom. 1986 c. 38.

Peculjic, A. N. (2016). Identifying Elements of *Lex Mercatoria* in the Space Domain. In R. Moro-Aguilar, P.J. Blount, & T. Masson-Zwaan (Eds.), *Proceedings of the International Institute of Space Law 2015* (pp. 143-154). Hague. Eleven.

Rathgeber, W., Remuss, N. K., & Schrogl, K. U. (2009). Space Security and the European Code of Conduct for Outer Space Activities. *Disarmament Forum, 4,* 33–41.

Rendleman, J. (2012). *Space Traffic Management - Private Regulation?* Paper presented at AIAA SPACE 2012 Conference & Exposition.

Smith, M. V. (2011). Spacepower and Warfare. *Joint Force Quarterly, 60,* 42.

Smitham, M. C. (2012). *The Need for a Global Space-Traffic-Control Service: An Opportunity for Us Leadership. Maxwell Papers.* Air War College.

Space Industry Act, United Kingdom. 2018 c. 5.

Subcommittee on Space. (2014). *Space Traffic Management: How to Prevent A Real Life 'Gravity,' Hearing May 9, 2014*. Retrieved from https://www.gpo.gov/fdsys/pkg/CHRG-113hhrg88144/pdf/CHRG-113hhrg88144.pdf

Tate, L. (2006). The Status of the Outer Space Treaty at International Law during War and Those Measures Short of War. *Journal of Space Law*, *32*, 177–202.

Treaty on Principles Governing the Activities of States in the Exploration and Use of Outer Space, Including the Moon and Other Celestial Bodies, October 10, 1967, 610 U.N.T.S. 205

United Nations Committee on the Peaceful Uses of Outer Space. (2016). Guidelines for the long-term sustainability of outer space activities: First set, UN Doc. A/71/20.

Wright, D., Grego, L., & Gronlund, L. (2005). *The Physics of Space Security: A Reference Manual*. Washington, DC: American Academy of Arts and Sciences. Retrieved from http://www.amacad.org/publications/Physics_of_space_security.pdf

Chapter 12
In-Situ Resource Utilization:
Technical Aspects

Claas Tido Olthoff
Technical University of Munich, Germany

Philipp Reiss
Technical University of Munich, Germany

ABSTRACT

Human spaceflight is an expensive endeavor. Every kilogram that needs to be transported to low Earth orbit or beyond costs tens of thousands of dollars, with the cost increasing exponentially the farther humanity extends its reach into the solar system and beyond. It is therefore prudent, if not necessary, to consider the use of resources that are available at the destination of a given exploration mission. This concept is called in-situ resource utilization (ISRU). The processes that are required to extract useful materials from the local environment can not only be used to support a human crew, but also to obtain resources that are of value on Earth and can thus be returned there for commercial gain. This chapter provides background information on ISRU in general and discusses the most important technologies and processes that are currently employed or under development.

INTRODUCTION

In the year 2012, the U.S.-based private company Planetary Resources announced its ambitious plans to mine asteroids for their resources. Yet, it is not the only "new space" company turning its interest towards this "galactic gold rush": several other tech billionaires also entered the competition in this new chapter of space activities. Furthermore, the United States adopted a statement in their 2016 national legislation that explicitly supports private companies in acquiring property rights over resources extracted from asteroids and other celestial bodies. In Europe, Luxembourg is following this path and even going a step further, announcing that they will develop a legal and regulatory framework "confirming certainty about the future ownership of minerals extracted in space from Near Earth Objects such as asteroids" (Tynan). It is evident that outer space in-situ resource utilization and exploitation is increasingly seen as a lucrative endeavour, and a probable driver for future space activities.

DOI: 10.4018/978-1-5225-7256-5.ch012

Paraphrasing the historical speech of J.F. Kennedy, Naveen Jain, the founder of Moon Express, said: "We choose to go to the Moon not because it is easy, but because it is a great business" (Scoles). Moon Express was the first private company to receive approval for a commercial space mission beyond Earth orbit by the Federal Aviation Administration. As a matter of fact, as of 2018 all concepts for space mining missions are still in the developing phase. And even though there is a significant interest and momentum among private companies, and a number of space faring nations are trying to foster the environment for the required legal developments, we are still years apart from real space mining. Therefore, this chapter will depict how far we have come with in-situ resource utilization and asteroid mining technologies and relevant governing legal regime until now.

BACKGROUND

Asteroid mining describes the exploitation of resources that are typically rare or depleted through consumption on Earth, mainly minerals and metals. Such elements could either be mined from asteroids and Near-Earth Objects (NEO) to restock terrestrial supplies, or to be utilized in-situ for space exploration – an approach known as In-Situ Resource Utilization (ISRU). In this regard, asteroid mining also addresses the extraction of resources that are required for life support and propellant production, such as water, oxygen or hydrogen. Due to Earth's strong gravity well, mining extra-terrestrial resources during a long-term mission is more efficient than launching new resupplies from Earth. Asteroid mining can be the initial step for ISRU as it provides the raw materials that can further be processed according to the respective needs of a mission.

The classical strategy in space exploration is to equip a mission with all resources that are required, such as consumables for life support or propellant for spacecraft, over the full duration of the mission. ISRU revolutionizes this approach, since a majority of the required resources are collected or generated in-situ during the mission. This enables significant savings in mass and thus in cost, since only an initial supply of resources and the basic tools to obtain new resources in-situ needs to be launched into space. The dedicated purpose of ISRU is to "harness and utilize space resources to create products and services which enable and significantly reduce the mass, cost, and risk of near-term and long-term space exploration" (Sanders et al., 2005). ISRU also offers more flexibility and risk reduction, because changing needs during a mission can be addressed and satisfied faster by using locally available resources instead of launching resupply missions from Earth.

DEFINITION OF ISRU

The term ISRU covers different sub-technologies, among them In-Situ Consumable Production (ISCP), In-Situ Propellant Production (ISPP), and In-Situ Fabrication and Repair (ISFR). ISCP includes the extraction of oxygen for life support systems. ISPP describes the production of rocket fuel by using extracted hydrogen, methane or oxygen. ISFR covers technologies that produce tools, construction material or spare parts.

Potential resources vary between the different locations. The most abundant resource on planetary bodies is the soil and rock material that covers the bedrock, the so-called regolith. Either the regolith itself can be considered as a resource, or the elements that are bound to or implanted in the regolith.

On larger planetary bodies such as Mars, the atmosphere can be used as a resource. Other secondary resources include sunlight for power generation, or temperature and pressure gradients to drive machines or generators.

ISRU is not a stand-alone science, since it has multiple interconnections between mission phases and technologies. In the mission planning, ISRU needs to be included at early stages with the goal to identify the mission supply requirements, prospect for available resources at the destination, select exploitable resources and processes, develop processing technologies, and plan the required infrastructure. An infrastructure that allows ISRU on a large scale, for instance to supply planetary settlements, typically includes the following elements:

- **Mining:** The initial step for the utilization of resources is mining of the raw material for later processing. The establishment of a mine requires careful sampling and evaluation, considering factors such as mineral abundances, accessibility, transportation need, energy supply, human supervision, and crew time.
- **Transport:** The transport connects the mining site with the beneficiation site, since both might not be at the same location. The logistical factors to be considered are the same as for the mining.
- **Beneficiation:** The raw material is prepared in a suitable format for the processing and to improve the yield. This covers a variety of processes, such as sorting minerals according to their type, quality, or grain size. Relevant methods include sieving and crushing, as well as electrostatic, centrifugal or pneumatic separation.
- **Processing:** The raw or beneficiated feedstock is processed by adequate chemical processes with the goal of yielding useable resources, such as gases (oxygen, hydrogen, noble gases), metals (iron, titanium, aluminium, noble metals), or ceramics (glass and silicates).
- **Production:** The production aims to refine the resources and generate consumables, equipment, and purge gases for life support systems, as well as semi-finished products (e.g. casting or sintering of plates, beams, bricks, formed goods, and tubes) as construction material for habitats, landing pads, or roads. This element also includes the storage of the produced goods.
- **Disposal:** Waste disposal is the inversion of the initial transportation step. Used minerals and by-products are either returned to the mining site or to a deposit. The same logistical considerations as for mining and transport apply for the disposal.

NEEDS AND USES

Resources for Consumption

In the scope of this section, consumables include everything that both the crew and the exploration system technologies consume. These resources are the result of the ISCP process. The required consumables can take various shapes and forms, but on the most basic level, they can be broken down into the three groups: oxygen, water, and fuel for the spacecraft. One notable exception here is food, which is one major consumable for the crew. However, even looking ahead many decades into humanities path of space exploration, mankind will not be traveling to a location that provides food directly to the crew in the form of local fauna and flora. For this reason, food is excluded in the following discussion.

Oxygen

Oxygen is the most immediate consumable for the crew, because the human body can only survive a few minutes without it. An average person needs about 0.816 kg of oxygen per day (National Aeronautics and Space Administration [NASA], 2015).

Oxygen also has other uses beyond direct crew consumption. In fuel cells it can be combined with hydrogen to generate electricity and as a by-product water (Vielstich, Lamm, Gasteiger, & Yokokawa, 2010). Finally, it can be used as an oxidizer in rocket engines.

Water

Potable water is the second most important consumable for the crew. Without water, the human body can survive for a few days, at most. An average person needs about 2.5 kg of water per day for direct consumption (National Aeronautics and Space Administration [NASA], 2015). The actual requirement for water per crew member is significantly higher than that, because other processes related to sustaining a human crew on a long duration exploration mission also need water. The crew hygiene functions like laundry and body cleansing are the largest contributors (National Aeronautics and Space Administration [NASA], 2015).

Through electrolysis, water, hydrogen, and oxygen are closely coupled as consumables. Water can be electrolyzed to produce breathing oxygen for the crew. Hydrogen is needed for a variety of CO_2 reduction technologies detailed in the section "CO_2 Capture and Reduction". If hydrogen and oxygen are cooled to cryogenic temperatures, they become liquids and can be used as fuel and oxidizer in rocket engines.

Fuel

The combination of liquid hydrogen and liquid oxygen is the fuel of choice for space exploration. Both components can be produced from water through electrolysis, but require cryogenic cooling for storage, a technology that may be prohibitive in terms of mass and volume and therefore cost. Due to the abundance of CO_2 in the Martian atmosphere, researchers are pursuing ways to use the byproduct of the Sabatier process, gaseous methane, as fuel for rocket engines (Sullivan, Linne, Bryant, & Kennedy, 1995). The only inputs that would be needed for this process of ISPP are hydrogen and energy. On very cold planetary surfaces like Titan or Pluto, hydrocarbons like methane and ethane already exist in liquid form (Stofan et al., 2007) and could be directly processed.

Miscellaneous

There are a few other substances that can be considered consumables, but cannot be as clearly defined as the three mentioned above. One is make-up gas to maintain the pressure in a crewed habitat at a certain level or to purge processing systems and reactors. There is no perfectly sealed container, so every spacecraft or habitat will have a certain leakage rate. Especially with atmospheres at sea-level pressure, like the one on the International Space Station (ISS) (Reuter & Reysa, 1997), it is not possible to simply add more oxygen, because this would quickly violate the strict upper limits for oxygen content (max. 24.1%) that are in place for fire safety (Weirauch, Andrew, F., 2011). The ISS uses nitrogen to replenish the atmosphere (Reuter & Reysa, 1997), but other inert gases can be used as well.

Strictly speaking, (electrical) energy is also a consumable resource. However, the destinations of most exploration missions currently under consideration are close enough to the Sun that photovoltaic systems can be used to generate electricity. Farther away from the Sun, the mass penalty for huge solar arrays becomes too large and alternative energy sources need to be found.

Resources for Construction

Especially for planetary missions, the use of local resources for the construction of habitats, launch and landing pads or other infrastructure is envisioned by many studies (Fateri & Gebhardt, 2015; Montes et al., 2015; Werkheiser, Fiske, Edmunson, & Khoshnevis). The simplest form of this utilization is covering the habitat with a layer of regolith (i.e. lunar or Martian soil) for radiation protection. Studies suggest (Miller et al., 2009; Newly, John, E., Wilson, & Townsend, Lawrence, W., 1988; Simonsen & Nealy, 1991) that a regolith layer of 0.5-1 m is sufficient to protect the crew even from the most severe space weather.

With more effort, regolith can be sintered or molten to provide rigid building blocks for habitat structures or even roads (Roberson, Hintze, & O'Connor). Processes like 3D printing are also being investigated.

Aside from these uses in civil construction, processed regolith and other mineral resources can be used as raw materials for producing spare parts. The resupply with spare parts is a critical aspect of long term or even permanently crewed outposts (Do, Owens, Ho, Schreiner, & Weck, 2016).

Resources for Commercial Use

Due to the large launch mass penalty that entering deep gravitational wells like the Moon and Mars entails, commercial use of resources found in space is currently only viable in the context of asteroid mining. Analyses of the composition of several asteroids that have orbits similar to Earth's have shown that some contain large quantities of fairly pure rare earth elements (REE) like yttrium and neodymium (Badescu, 2013). Enhanced levels of REE might also be found on the Moon in so-called KREEP-rich areas (Crawford, 2015). Despite their name, REE are quite abundant on Earth, but they naturally occur in very dispersed form and are therefore expensive to purify (McLeod & Krekeler, 2017). REE are an increasingly critical materials group for terrestrial applications, and considered critical to the growing global green economy (U.S. Department of Energy, 2011). This provides a commercial opportunity, if the process of mining these minerals from space and returning them to Earth can be made cheaper than processing the deposits on Earth. There are a few start-up companies like Planetary Resources and Deep Space Industries who are pursuing commercial asteroid mining. The near-term focus of both companies are technology demonstration missions using CubeSats. Since no one has actually performed a successful and financially viable asteroid mining operation, it is impossible to estimate the cost of such an undertaking.

Another substance that is often mentioned in the context of commercial space activities is the mining of helium-3 (^3He) on the lunar surface. This isotope of helium is a preferred fuel for nuclear fusion reactors since it does not produce high energy neutrons that would irradiate the reactor. ^3He is implanted into the lunar regolith by the solar wind to a concentration between 1 and 50 ppb (parts per billion, 1 ppb = 0,0000001%) (Cocks, 2010). The following is a thought experiment assuming an attempt to cover the world's electricity consumption entirely by ^3He-based fusion. The world electricity consumption in 2012 was about 18,9 petawatt hours (PWh) (International Energy Agency, 2014). In the given reference, the energy consumption is actually given in a unit called million tons of oil equivalent (Mtoe). The number cited is 1626 Mtoe and 1000 Mtoe are equal to 11.63 petawatt hours. Assuming one metric ton of ^3He can produce a constant output of 10 GW for one year (Santarius, 2004), 216 metric tons of ^3He would be required *annually* to sustain humanity's electricity needs. Given the low concentration of ^3He in the lunar regolith, for *each ton* of ^3He, 150 million tons of regolith need to be mined and processed (Sviatoslavsky, 1993). To cover the world's electricity demand *for one year* over 32 billion tons of lunar regolith would have to go through the ISRU process chain described in the section "Regolith Processing" and *then* be transported back to Earth, which adds even more cost to the endeavour. It therefore seems that ^3He mining on the lunar surface will not be an economically viable means of supplying humanity with energy in the near to midterm future.

TECHNOLOGIES

Oxygen Production

Electrolysis

Electrolysis is a very mature and proven technology in human spaceflight. First used on Soviet space stations, an electrolyzer has been utilized on the ISS since the year 2000 (Schaezler, Ryan, N. & Cook, Anthony, J., 2015; Wieland, Paul, O., 1998). In the context of ISRU, once the water has been extracted from the local environment, electric energy is used to split water into hydrogen and oxygen. This process is shown in the following reaction.

$$2H_2O + energy \rightarrow 2H_2 + O_2$$

The hydrogen can be used directly as rocket fuel or as an input to other processes described later in this section. The oxygen can also be used as rocket fuel or for crew consumption.

Chemical Reduction of Regolith

Various approaches exist to extract oxygen from the mineral phase of the regolith, whereas one distinguishes these methods by the chemical process that is required for the extraction. The main categories include: chemical reduction through solid/gas interaction, electrochemical reduction via silicate/oxide melt, pyrolysis, and aqueous solutions. A thorough evaluation of the major technologies to extract oxygen from lunar regolith is provided by Taylor and Carrier (1993) and Schwandt, Hamilton, Fray, and Crawford (2012a). Amongst them, the most promising regarding cost, schedule, technical considerations, and currently most developed technologies are hydrogen reduction, carbothermal reduction, molten silicate electrolysis, and vapour phase reduction by pyrolysis (Horneck et al., 2003; Monchieri et al., 2010). A more recent development is the electrochemical reduction of lunar regolith to extract oxygen, a process derived from metal purification (Schwandt, Hamilton, Fray, & Crawford, 2012b).

Hydrogen Reduction

Hydrogen reduction is ideally applied to regolith with high iron oxide content, such as found in ilmenite ($FeTiO_3$). Other sources of oxygen for the reduction with hydrogen are titanium dioxide (TiO_2) and silicon dioxide (SiO_2). The process requires temperatures between 700 °C and 1000 °C (Gibson & Knudsen, 1985; Schwandt et al., 2012a) with an optimum at 1000 °C (Yoshida et al., 2000), just below the sintering temperature of the regolith. In the presence of hydrogen, the oxygen is removed from the mineral phase and forms water:

$$FeTiO_3 + H_2 = Fe + TiO_2 + H_2O$$

The water generated in this reaction can further be processed via electrolysis to produce hydrogen and oxygen. Since hydrogen is required as a reactant in the first step, the hydrogen produced by the electrolysis can be recovered and fed back into the first reaction step. Oxygen yield from lunar soils strongly depends on the initial iron content of the sample, but generally lies in the range of 1% to 5% by mass for lunar regolith (Alan, 1997). Allen, Morris, and McKay (1996) found that the susceptibility to reduction of Fe^{2+} in mineral phases can be arranged as follows: ilmenite > glass > olivine > pyroxene.

In 1993, Taylor et al. (1993) built a demonstrator for the hydrogen reduction of ilmenite. A sample of 20 mg was heated to 1000 °C while hydrogen was blown over it and kept circulating. The produced water was condensed or separated using a drying agent. Within 1.5 h more than 95% of the ilmenite were reduced, which represented an oxygen yield of 10% by mass.

Two recent NASA research projects implemented the hydrogen reduction process to produce oxygen at a targeted rate of up to 1000 kg per year on a possible future lunar outpost (Larson, 2012). During the developments of NASA's *ROxygen* project, 1% to 2% by mass of oxygen release could be realized in a small reactor, whereas in a representative system with hydrogen feedback and contaminant scrubber only 0.2 to 0.5% by mass oxygen could be achieved (Lee, Oryshchyn, Paz, Reddington, & Simon, 2013). Laboratory tests of the PILOT facility yielded 1% by mass oxygen for a 11.6 kg batch of lunar regolith simulant (Clark, Keller, & Kirkland, 2009).

Carbothermal Reduction

Instead of using hydrogen for reduction, methane can be used. This allows a larger variety of feedstock, since both silicate and oxide minerals can be processed. The reaction temperature above 1600 °C however is significantly higher than for the hydrogen reduction process (Rosenberg, Beegle, Robert L., Jr., Guter, Miller, & Rothenberg, 1992; Rosenberg, Musbah, & Rice, 1996), and implies sintering and melting of the regolith. The latter can be avoided by reducing the reaction temperature to around 1000 °C at the cost of a reduced oxygen yield. Another drawback is the significant amount of reactants that need to be resupplied. Despite these challenges, the technology was successfully evaluated in a laboratory environment (Sanders & Duke, 2005) and in field tests. The following three process steps (reduction, methanation, and electrolysis) are required to produce up to 50% oxygen from ilmenite (Schwandt et al., 2012a):

$$FeTiO_3 + CH_4 = Fe + TiO_2 + CO + 2H_2$$

$$2CO + 6H_2 = 2CH_4 + 2H_2O$$

$$H_2O = H_2 + \frac{1}{2}O_2$$

Molten Silicate Electrolysis

Molten silicate electrolysis has the advantage that oxygen is readily released without further necessary reduction steps such as water electrolysis. Silicate minerals such as anorthite, olivine, and pyroxene are melted in a furnace equipped with two electrodes. The metal part is collected at the cathode, whereas oxygen is released at the anode. The biggest challenge with this process is to provide the required power to heat the regolith to temperatures between 1300 °C and 1700 °C and the material stress that comes with such high temperatures (e.g. increased wear of the electrodes is to be taken into account). Also the energy demand strongly depends on the amount of iron oxide, in a way that efficient oxygen production can only be realized with less than 5% by mass iron oxide in the melt (Taylor & Carrier, 1993).

Recent developments aim at improving the process efficiency and investigate material issues for the reactor design. A laboratory test setup to study molten silicate electrolysis at a scaled-up level has been developed at the Massachusetts Institute of Technology together with NASA (Sibille et al.).

Vapour Phase Reduction

Pyrolysis processes aim for a decomposition of the regolith under high temperatures well above 2000 °C. For the vapour phase reduction, the soil is evaporated to decompose iron oxide into iron and oxygen in gaseous form:

$$FeTiO_3 = Fe + TiO_2 + \frac{1}{2}O_2$$

Considering the additional release of oxides from magnesium, aluminium, and calcium, theoretical oxygen yields up to 50% are possible (Schwandt et al., 2012a). Once the oxygen is released, it needs to be quickly cooled down, condensed, and collected, before recombination with the metal occurs. Besides the recuperation of the oxygen, producing the elevated temperature required for the vaporization is a major challenge. One possible approach is to use a solar furnace where concentrated solar light is focused on the regolith (Senior, 1992).

Electrochemical Reduction

Lunar regolith can be almost complete reduced to pure metal (>99%) through electrochemical reduction. A process formerly known as FFC-Cambridge process (Schwandt, Doughty, & Fray, 2010), originally derived from industrial purification of metal oxides, has recently been applied to lunar regolith (Ellery et al., 2017; Schwandt et al., 2012b) with the goal to extract oxygen. The process employs an electrolytic cell with a cathode made of metal oxide, a carbon-based anode, and molten calcium chloride ($CaCl_2$) as electrolyte. For a lunar ISRU application the cathode would be made of sintered porous regolith. At 900

°C reaction temperature, the cathode decomposes and liberates oxygen ions (O^{2-}), which are transported through the electrolyte to the anode. If the anode is carbon-based, the ions discharge and form carbon oxides (CO and CO_2), which are subsequently released as gas. Using different anode materials, such as tin oxide (SnO_2) or calcium titanate and calcium ruthenate ($CaTi_xRu_{1-x}O_3$), the oxygen ions can be discharged and released as molecular oxygen (O_2).

The main advantage of this technology over the previously discussed reduction processes is the extremely high theoretical oxygen yield at moderate temperatures. An additional benefit is that pure metal is produced at the cathode as a by-product of the reduction process, which could subsequently be used for construction and fabrication purposes. A cathode made of sintered lunar anorthite ($CaAl_2Si_2O_8$) for example is reduced to Al, Ca, and Si after the oxygen is extracted between 900 °C and 1100 °C. Applying acid leaching by HCl, anorthite could even be reduced to SiO_2, which also yields $CaCl_2$ that can be used to replenish the electrolyte (Ellery et al., 2017). Theoretically there is no consumption of the electrolyte for the actual reduction process, but in practice a certain amount of electrolyte gets lost through the removal of the porous metal cathode after reduction.

CO_2 Capture and Reduction

Another means of producing oxygen from local resources is capturing carbon dioxide and then extracting the oxygen via chemical processes. To capture CO_2 on spacecraft, adsorber or absorber beds are usually used. These beds contain metal oxides (Butler, Faszcza, & Thomas, 1997), zeolites (Wieland, Paul, O., 1998) or amines (Chullen, Campbell, Papale, Hawes, & Wichowski, 2015) that release the captured CO_2 when subjected to vacuum and/or high temperatures, after which they can then be reused. In spacecraft atmospheres the concentration of CO_2 is mainly influenced by the number of crew members and their activity level. With respect to ISRU, the most likely use case will be processing of the atmosphere of Mars. It contains 96% CO_2 (Mahaffy et al., 2013) so capturing it only requires a compressor and molecular filters to remove the other constituents.

Once the CO_2 is isolated, it can be further processed to separate the carbon from the oxygen. There are several processes to do this, three of them will be presented in the following sections.

Sabatier Reaction

The most mature technology is the Sabatier process. Developed in the early 20th century (Sabatier, 1913) it is described by the following chemical reaction:

$$CO_2 + 4H_2 \rightarrow CH_4 + 2H_2O$$

The reaction requires elevated temperatures between 200 °C and 400 °C (Bockstahler et al., 2015) and is more effective when used with a catalyst.

On the product side methane and water are found. The water is electrolyzed to produce oxygen and hydrogen. The water can be used directly for crew consumption. As it can be seen from the chemical reaction, the process requires hydrogen. Half of that hydrogen can be recovered after the electrolysis, the other half is bound in the methane. The methane can be used as rocket fuel (Zubrin, Muscatello, & Berggren, 2013). For this use, the methane can either be used directly in its gaseous form, or after liquefaction.

A Sabatier reactor is currently in use on board the ISS (Samplatsky, Grohs, Edeen, Crusan, & Burkey, 2011) coupled with the CO_2 capture system to regenerate oxygen from crew exhaled CO_2 and thereby reducing the required resupply mass. On the ISS the methane is dumped overboard and hydrogen is resupplied from the ground.

Bosch Reaction

If the use of methane as rocket fuel is precluded, as it is on the ISS, the Bosch reaction offers the possibility of producing water and pure carbon instead of methane as is shown in the chemical reaction below (Abney, Morgan, B. et al., 2014).

$$CO_2(g) + 2H_2(g) \rightarrow C(s) + 2H_2O(g)$$

As it can be seen, this reaction alleviates the need for hydrogen resupply because all of the required hydrogen can be recovered by electrolyzing the product water. The Bosch process is currently under development at NASA (Abney, Morgan, B. et al., 2014), no unit has been flown in space however.

Solid Oxide Electrolysis

Solid Oxide Electrolysis (SOE) uses temperatures around 1000 °C to create carbon monoxide and molecular oxygen. Due to the toxic nature of CO a second process needs to be employed to create different compounds. An example for such a process could be the Boudouard reaction (input pending...) (equation 10) (Ebbesen & Mogensen, 2009).

$$2CO_2 + energy \rightarrow 2CO + O_2$$

$$2CO \rightarrow C + CO_2$$

This process is the least developed of the three presented here, but a prototype SOE system is scheduled to fly to Mars onboard NASA's Mars 2020 rover (Hoffman, Rapp, & Hecht).

Water Processing

Water Extraction From Regolith

Water can be produced from regolith via reduction processes, such as the ones that have been described in previous sections of this chapter. However, water is also readily available in some cold regions such as the poles on Moon and Mars or on comets, meteorites, and asteroids. Extraction of the water can be achieved by heating the hydrated soil. The temperature hereby varies depending on the form of the water deposit. To liberate highly volatile water that is loosely bound (through physisorption and chemisorption) on the surface of soil particles in lunar cold traps, temperatures around -150 °C to -90 °C are sufficient (Hibbitts et al., 2011). Stronger bonds through dissociative chemisorption are resolved at temperatures around 130 °C to 230 °C.

On the Moon, water ice can be present as small-grained water frost, mixed ice with regolith grains, or subsurface ice layers (Hayne et al., 2015). Recent remote sensing data confirmed that water ice is present in permanently shadowed areas at the lunar poles, indicating water abundances of <0.5% by mass (Sunshine et al., 2009), 0.1% to 2.0% by mass in cold traps with temperatures below -160 °C (Hayne et al., 2015) and up to 5.6 ± 2.9% by mass (Colaprete et al., 2010) in the lunar south pole crater Cabeus.

Water on Mars is highly abundant and available mainly in the form of water ice. It exists as subsurface ice deposits, surface ice caps at the polar and mid-latitude regions, and as ice patches in craters. Although these reservoirs might be hard to access, recent studies have also found locations on Mars with more easily accessible shallow water ice deposits that exist in only 1-2 m depth (Dundas et al., 2018). The total water ice reservoir on Mars was estimated to be $>5 \cdot 10^6$ km^3, which corresponds to an approximately 35 m thick layer over the entire planet (Christensen, 2006). Water-equivalent hydrogen in the range of 20% to 100% was detected at latitudes above ±50°, with a global average of at least 14% (Feldman, 2004).

Potable Water Processes

The water that is extracted from lunar or Martian soil will not immediately be potable. It will contain trace amounts of contaminants, some of which might be harmful to the crew. Water purification systems have been in use on Earth for many decades, and on the ISS since 2008 (Williams, Dake, & Gentry, 2012). The international community has been gaining operational experience with this system, which has the added challenge of needing to operate in a microgravity environment. Systems employed on the surface of the Moon or Mars may have different challenges due to the gravitational conditions there. The water recovery system on the ISS consists of a Water Processing Assembly (WPA) that processes so called “gray water” and a Urine Processing Assembly (UPA) that converts urine to gray water. Technology derived from the WPA could be used to process water from ISRU sources, its multiple filtration beds and catalytic reactors (Carrasquillo, Cloud, & Kundrotas, 2004) can remove many substances from the input water. However, since the exact composition of trace contaminants in the water gained from an ISRU system is unknown until samples can be taken, any future system will have to be constructed in a flexible manner to enable the broadest possible spectrum of contaminant removal.

Once the water is processed, it needs to be made biologically inert to prevent microbial growth in water lines and storage tanks. For this either silver or iodine can be used, both are currently being employed on the ISS, as is irradiation using ultra violet light (Bobe et al., 2005; Carter, Tobias, & Orozco, 2012).

Regolith Processing

Regolith is a valuable raw material for the production of basic construction material, semi-finished products, tools and spare parts. It can be utilized to produce bricks and blocks to build habitats, thermal insulation, or radiation shields. The glass and ceramic content of the regolith can be used to create semiconductors, optical wires, or as high-performance construction material. The metal content on the other hand can serve as basis for structural parts, pressure vessels, pipes, and a multitude of common tools (Sastri, Duke, & Haskin, 1992). Whereas the extraction of glass and metal from regolith requires energy-extensive beneficiation processes, the utilization of the crude regolith material is less energy demanding and hence most promising.

Regolith can directly be processed by combining it with a binder material to form a cement-like compound. This process is very similar to what is used on Earth for construction. It can also be processed by sintering and melting through solar (Meurisse, Makaya, Willsch, & Sperl, 2018), laser (Fateri & Gebhardt, 2015), or microwave (Taylor & Meek, 2005) heating. Such techniques have a particularly high potential when being combined with additive manufacturing, or 3D printing, technologies (Lim & Anand, 2015). 3D printing of regolith can be applied in various ways to enable in-situ fabrication of habitats and infrastructures for future planetary settlements. One example is the combination of inflatable structures with 3D-printed regolith. Here an inflatable crew module is deployed on the lunar surface and covered with printed regolith to provide thermal insulation and protection against radiation and micrometeorites. First technological demonstrations of printing such larger scale structures already exist (Cesaretti, Dini, Kestelier, Colla, & Pambaguian, 2014), but to this date still rely on terrestrial binders. Technologies that use lunar soil only as a raw material are therefore being studied intensively to enable more sustainable in-situ fabrication.

CONCLUSION

Space agencies have been pursuing the idea of ISRU since the 1970's (Sanders, Larson, Interbartolo, Mueller, & Muscatello, 2012), resulting in a variety of studies and system architectures (Drake, 2009; Monchieri et al., 2010; National Aeronautics and Space Administration [NASA], 2005). The initial technology demonstrations will be carried out by robotic precursor missions in the coming decade. The proposed Mars 2020 rover for instance includes the MOXIE payload to demonstrate the extraction of CO_2 from the Martian atmosphere through solid oxide electrolysis. For lunar ISRU NASA has been developing the payloads RESOLVE (Kleinhenz, Sacksteder, & Nayagam, 2008; Quinn et al., 2015) and VAPoR (Glavin et al., 2012) to test ISRU extraction processes on the Moon. ESA is developing the PROSPECT (Carpenter & Fisackerly, 2017) instrument package for the Russian Luna-27 lander, which is to detect and characterize lunar volatiles and also includes a first technology demonstration for lunar ISRU processes.

Overall, ISRU is a very dynamic technical field with many ideas and concepts being discussed in the scientific and engineering communities. It is obvious, that exploration missions would become much cheaper endeavours, if not all required resources have to be brought along from Earth. However, the technical challenges that have to be overcome and the unknowns that have to be resolved are so numerous, that there are no immediate ISRU applications in sight for planned human exploration missions.

Yet, overcoming technical challenges is not the only barrier in order to start conducting ISRU missions. Existing legal constrains as well as normative grey areas also need to be resolved before states and private actors engage in these activities. The following chapter will give an overview of the legal challenges that this new class of missions are facing.

REFERENCES

Abney, M. B., Mansell, J., Rabenberg, E., Stanley, M., Edmunson, J., Alleman, E., . . . Dumez, S. (2014). Series-Bosch Technology For Oxygen Recovery During Lunar or Martian Surface Missions. *Proceedings of the 44th International Conference on Environmental Systems.*

Alan, C. C. (1997). *Oxygen Extraction from Lunar Samples*. Retrieved from https://curator.jsc.nasa.gov/lunar/lnews/lnmar97/oxygen.htm

Allen, C. C., Morris, R. V., & McKay, D. S. (1996). Oxygen extraction from lunar soils and pyroclastic glass. *Journal of Geophysical Research, 101*(E11), 26085–26095. doi:10.1029/96JE02726

Badescu, V. (Ed.). (2013). *Asteroids: Prospective energy and material resources*. Heidelberg, Germany: Springer. doi:10.1007/978-3-642-39244-3

Bobe, L. S., Samsonov, N. M., Soloukhin, V. A., Farafonov, N. S., Novikov, V. M., Andreichuk, P. O., ... Skuratov, V. M. (2005). Water Supply of the Crew of a Space Station Through Water Recovery and Water Delivery: SRV-K and SPK-U System Operation on ISS. *Proceedings of the 35th International Conference on Environmental Systems.* 10.4271/2005-01-2806

Bockstahler, K., Matthias, C., Lucas, J., Witt, J., Hovland, S., & Laurini, D. (2015). Development Status of the Advanced Closed Loop System ACLS. *Proceedings of the 45th International Conference on Environmental Systems.*

Butler, L. B., Faszcza, J. J., & Thomas, G. A. (1997). Development Status of the Metal Oxide Regenerable CO2 Removal System for the NASA EMU. *Proceedings of the 27th International Conference on Environmental Systems.* 10.4271/972505

Carpenter, J., & Fisackerly, R. (2017). PROSPECT: ESA's package for resource observation and in situ prospecting for exploration, commercial exploitation and transportation. *Proceedings of the 48th Lunar and Planetary Science Conference.*

Carrasquillo, R. L., Cloud, D., & Kundrotas, R. (2004). Status of the Node 3 Regenerative ECLSS Water Recovery and Oxygen Generation Systems. *Proceedings of the 34th International Conference on Environmental Systems.* 10.4271/2004-01-2384

Carter, L., Tobias, B., & Orozco, N. (2012). Status of ISS Water Management and Recovery. *Proceedings of the 42nd International Conference on Environmental Systems.* 10.2514/6.2012-3594

Cesaretti, G., Dini, E., de Kestelier, X., Colla, V., & Pambaguian, L. (2014). Building components for an outpost on the Lunar soil by means of a novel 3D printing technology. *Acta Astronautica*, *93*, 430–450. doi:10.1016/j.actaastro.2013.07.034

Christensen, P. R. (2006). Water at the Poles and in Permafrost Regions of Mars. *Elements*, *2*(3), 151–155. doi:10.2113/gselements.2.3.151

Chullen, C., Campbell, C., Papale, W. G., Hawes, K., & Wichowski, R. (2015). Rapid Cycle Amine 3.0 System Development. *Proceedings of the 45th International Conference on Environmental Systems.*

Clark, D. L., Keller, B. W., & Kirkland, J. A. (2009). Field Test Results of the PILOT Hydrogen Reduction Reactor. *AIAA SPACE Conference & Exposition.*

Cocks, F. H. (2010). 3He in permanently shadowed lunar polar surfaces. *Icarus*, *206*(2), 778–779. doi:10.1016/j.icarus.2009.12.032

Colaprete, A., Schultz, P., Heldmann, J., Wooden, D., Shirley, M., Ennico, K., ... Sollitt, L. (2010). Detection of water in the LCROSS ejecta plume. *Science, 330*(6003), 463–468. doi:10.1126cience.1186986 PMID:20966242

Crawford, I. A. (2015). Lunar resources. *Progress in Physical Geography, 39*(2), 137–167. doi:10.1177/0309133314567585

Do, S., Owens, A., Ho, K., Schreiner, S., & de Weck, O. (2016). An independent assessment of the technical feasibility of the Mars One mission plan – Updated analysis. *Acta Astronautica, 120*, 192–228. doi:10.1016/j.actaastro.2015.11.025

Drake, B. G. (2009). *Human Exploration of Mars Design Reference Architecture 5.0* (No. NASA/SP-2009.566).

Dundas, C. M., Bramson, A. M., Ojha, L., Wray, J. J., Mellon, M. T., Byrne, S., ... Holt, J. W. (2018). Exposed subsurface ice sheets in the Martian mid-latitudes. *Science, 359*(6372), 199–201. doi:10.1126cience.aao1619 PMID:29326269

Ebbesen, S. D., & Mogensen, M. (2009). Electrolysis of carbon dioxide in Solid Oxide Electrolysis Cells. *Journal of Power Sources, 193*(1), 349–358. doi:10.1016/j.jpowsour.2009.02.093

Ellery, A., Lowing, P., Wanjara, P., Kirby, M., Mellor, I., & Doughty, G. R. (2017). FCC Cambridge Process with Metal 3D Printing as Universal In-Situ Resource Utilisation. *Proceedings of the 14th Symposium on Advanced Space Technologies in Robotics and Automation (ASTRA).*

Fateri, M., & Gebhardt, A. (2015). Process Parameters Development of Selective Laser Melting of Lunar Regolith for On-Site Manufacturing Applications. *International Journal of Applied Ceramic Technology, 12*(1), 46–52. doi:10.1111/ijac.12326

Feldman, W. C. (2004). Global distribution of near-surface hydrogen on Mars. *Journal of Geophysical Research, 109*(E9), E09006. doi:10.1029/2003JE002160

Gibson, M. A., & Knudsen, C. W. (1985). Lunar Oxygen Production from Ilmenite. In W. W. Mendel (Ed.), Lunar Bases and Space Activities of the 21st Century (pp. 543–550). Academic Press.

Glavin, D. P., Malespin, C., ten Kate, I. L., Getty, S. A., Holmes, V. E., Mumm, E., . . . Mahaffy, P. R. (2012). Volatile Analysis by Pyrolysis of Regolith for planetary resource exploration. In *2012 IEEE Aerospace Conference: Big Sky, Montana*, March 3 - 10, 2012 (pp. 1–11). Piscataway, NJ: IEEE. 10.1109/AERO.2012.6187065

Hayne, P. O., Hendrix, A., Sefton-Nash, E., Siegler, M., Lucey, P. G., Retherford, K. D., ... Paige, D. (2015). Evidence for exposed water ice in the Moon's south polar regions from Lunar Reconnaissance Orbiter ultraviolet albedo and temperature measurements. *Icarus, 255*, 58–69. doi:10.1016/j.icarus.2015.03.032

Hibbitts, C., Grieves, G., Poston, M. J., Dyar, M. D., Alexandrov, B., Johnson, M., & Orlando, T. M. (2011). Thermal stability of water and hydroxyl on the surface of the Moon from temperature-programmed desorption measurements of lunar analog materials. *Icarus, 213*(1), 64–72. doi:10.1016/j.icarus.2011.02.015

Hoffman, J. A., Rapp, D., & Hecht, M. The Mars Oxygen ISRU Experiment (MOXIE) on the Mars 2020 Rover. AIAA SPACE 2014 Conference and Exposition. doi:10.2514/6.2015-4561

Horneck, G., Facius, R., Reichert, M., Rettberg, P., Seboldt, W., Manzey, D., ... Gerzer, R. (2003). Humex, a study on the survivability and adaptation of humans to long-duration exploratory missions, part I: Lunar missions. *Advances in Space Research, 31*(11), 2389–2401. doi:10.1016/S0273-1177(03)00568-4 PMID:14696589

International Energy Agency. (2014). *Key World Energy Statistics 2014.* IEA.

Kleinhenz, J., Sacksteder, K., & Nayagam, V. (2008). Lunar Resource Utilization: Development of a Reactor for Volatile Extraction from Regolith. In *46th AIAA Aerospace Sciences Meeting and Exhibit.* Reston, VA: American Institute of Aeronautics and Astronautics. 10.2514/6.2008-1415

Larson, W. (2012). *Hydrogen Reduction of Regolith.* Retrieved from http://isru.nasa.gov/Hydrogen-Reduction-of-Regolith.html

Lee, K. A., Oryshchyn, L., Paz, A., Reddington, M., & Simon, T. M. (2013). The ROxygen Project: Outpost-Scale Lunar Oxygen Production System Development at Johnson Space Center. *Journal of Aerospace Engineering, 26*(1), 67–73. doi:10.1061/(ASCE)AS.1943-5525.0000230

Lim, S., & Anand, M. (2015). In-Situ Resource Utilisation (ISRU) derived extraterrestria construction processes using sintering-based additive manufacturing techniques – focusing on a lunar sur- face environment. *European Lunar Symposium.*

Mahaffy, P. R., Webster, C. R., Atreya, S. K., Franz, H., Wong, M., Conrad, P. G., ... Trainer, M. (2013). Abundance and isotopic composition of gases in the martian atmosphere from the Curiosity rover. *Science, 341*(6143), 263–266. doi:10.1126cience.1237966 PMID:23869014

McLeod, C., & Krekeler, M. (2017). Sources of Extraterrestrial Rare Earth Elements: To the Moon and Beyond. *Resources, 6*(3), 40. doi:10.3390/resources6030040

Meurisse, A., Makaya, A., Willsch, C., & Sperl, M. (2018). Solar 3D printing of lunar regolith. *Acta Astronautica, 152*, 800–810. doi:10.1016/j.actaastro.2018.06.063

Miller, J., Taylor, L., Zeitlin, C., Heilbronn, L., Guetersloh, S., DiGiuseppe, M., ... Murakami, T. (2009). Lunar soil as shielding against space radiation. *Radiation Measurements, 44*(2), 163–167. doi:10.1016/j.radmeas.2009.01.010

Monchieri, E., Hovland, S., Masi, M., Sliepcevich, A., Lavagna, M., Hoppenbrouwers, T., & Venditti, F. (2010). ESA Lunar In-Situ Resource Utilisation (ISRU) Concept Design and Breadboarding Activities. *61st International Astronautical Congress.*

Montes, C., Broussard, K., Gongre, M., Simicevic, N., Mejia, J., Tham, J., ... Davis, G. (2015). Evaluation of lunar regolith geopolymer binder as a radioactive shielding material for space exploration applications. *Advances in Space Research, 56*(6), 1212–1221. doi:10.1016/j.asr.2015.05.044

National Aeronautics and Space Administration (NASA). (2005). *NASA's Exploration Systems Architecture Study* (NASA TM No. NASA-TM-2005-214062). NASA.

National Aeronautics and Space Administration (NASA). (2015). *Life Support Baseline Values and Assumptions Document.* Houston, TX: NASA.

Newly, J. E., Wilson, J. W., & Townsend, L. W. (1988). Solar-Flare Shielding With Regolith at a Lunar-Base Site. Langley, VA: Academic Press.

Quinn, J., Smith, J. J. C., Paz, A., Colaprete, A., Elphic, R., & Zacny, K. (2015). Resource Prospector. *The RESOLVE Payload*, *1863*, 2046.

Reuter, J. L., & Reysa, R. P. (1997). International Space Station Environmental Control and Life Support System Design Overview Update. *Proceedings of the 27th International Conference on Environmental Systems*. 10.4271/972333

Roberson, L., Hintze, P., & O'Connor, G. A Conceptual Study for the Autonomous Direct Forming of Lunar Regolith into Flexlock Geomats for Lunar Surface Operations. *47th AIAA Aerospace Sciences Meeting including The New Horizons Forum and Aerospace Exposition*. 10.2514/6.2009-1013

Rosenberg, S. D. Beegle, R. L., Jr., Guter, G. A., Miller, F. E., & Rothenberg, M. (1992). The onsite manufacture of propellant oxygen from lunar resources. In M. F. McKay, D. S. McKay, & M. B. Duke (Eds.), *Space Resources Materials* (3rd ed.; pp. 162–185). Academic Press. Retrieved from http://ntrs.nasa.gov/search.jsp?R=19930007699

Rosenberg, S. D., Musbah, O., & Rice, E. E. (1996). Carbothermal Reduction of Lunar Materials for Oxygen Production on the Moon: Reduction of Lunar Simulants with Carbon. In *Lunar and Planetary Science Conference XXVII* (pp. 1103–1104). Academic Press.

Sabatier, P. (1913). *La Catalyse en chimie organique*. Academic Press.

Samplatsky, D., Grohs, K., Edeen, M., Crusan, J., & Burkey, R. (2011). Development and Integration of the Flight Sabatier Assembly on the ISS. In *41st International Conference on Environmental Systems*. Reston, VA: American Institute of Aeronautics and Astronautics. 10.2514/6.2011-5151

Sanders, G. B., & Duke, M. (2005). *In-Situ Resource Utilization (ISRU)*. Capability Roadmap Progress Review.

Sanders, G. B., Larson, W. E., Interbartolo, M. A., Mueller, R. P., & Muscatello, A. C. (2012). *Early In-Situ Resource Utilization (ISRU) Leading to Robust Sample Return and Human Exploration Missions*. Academic Press.

Sanders, G. B., Romig, K. A., Larson, W. E., Johnson, R., Rapp, D., Johnson, K. R., . . . Zubrin, R. (2005). *Results from the NASA Capability Roadmap Team for In-Situ Resource Utilization (ISRU)*. Retrieved from http://ntrs.nasa.gov/search.jsp?R=20110024178

Santarius, J. F. (2004, September). *Lunar 3He and Fusion Power*. IEEE Rock River Valley Section, Rockford, IL. Retrieved from http://fti.neep.wisc.edu/presentations/jfs_ieee0904.pdf

Sastri, S., Duke, M. B., & Haskin, L. A. (1992). PART 3—Manufacturing and Fabrication. In M. F. McKay, D. S. McKay, & M. B. Duke (Eds.), Space Resources Materials (pp. 270–306). Academic Press.

Schaezler, R. N., & Cook, A. J. (2015). Report on ISS O2 Production, Gas Supply & Partial Pressure Management. *Proceedings of the 45th International Conference on Environmental Systems*.

Schwandt, C., Doughty, G. R., & Fray, D. J. (2010). The FFC-Cambridge Process for Titanium Metal Winning. *Key Engineering Materials, 436*, 13–25. . doi:10.4028/www.scientific.net/KEM.436.13

Schwandt, C., Hamilton, J. A., Fray, D. J., & Crawford, I. A. (2012a). Oxygen from Lunar Regolith. In V. Badescu (Ed.), *Moon* (pp. 165–187). Springer. doi:10.1007/978-3-642-27969-0_7

Schwandt, C., Hamilton, J. A., Fray, D. J., & Crawford, I. A. (2012b). The production of oxygen and metal from lunar regolith. *Planetary and Space Science, 74*(1), 49–56. doi:10.1016/j.pss.2012.06.011

Scoles, S. (n.d.). *Luxembourg's Bid to Become the Silicon Valley of Space Mining*. Retrieved from https://www.wired.com/2017/01/luxembourg-setting-silicon-valley-space-mining/

Senior, C. L. (1992). Lunar oxygen production by pyrolysis. In *Space Programs and Technologies Conference*. Reston, VA: American Institute of Aeronautics and Astronautics. 10.2514/6.1992-1663

Sibille, L., Sadoway, D., Sirk, A., Tripathy, P., Melendez, O., Standish, E., & … . Poizeau, S. (n.d.). Recent Advances in Scale-Up Development of Molten Regolith Electrolysis for Oxygen Production in Support of a Lunar Base. *47th AIAA Aerospace Sciences Meeting including The New Horizons Forum and Aerospace Exposition*. 10.2514/6.2009-659

Simonsen, L. C., & Nealy, J. E. (1991). *Radiation Protection for Human Missions to the Moon and Mars* (NASA Technical Paper No. 3079). Hampton, VA: NASA.

Stofan, E. R., Elachi, C., Lunine, J. I., Lorenz, R. D., Stiles, B., Mitchell, K. L., … West, R. (2007). The lakes of Titan. *Nature, 445*(7123), 61–64. doi:10.1038/nature05438 PMID:17203056

Sullivan, T. A., Linne, D., Bryant, L., & Kennedy, K. (1995). In-situ-produced methane and methane/carbon monoxide mixtures for return propulsion from Mars. *Journal of Propulsion and Power, 11*(5), 1056–1062. doi:10.2514/3.23935

Sunshine, J. M., Farnham, T. L., Feaga, L. M., Groussin, O., Merlin, F., Milliken, R. E., & A'Hearn, M. F. (2009). Temporal and spatial variability of lunar hydration as observed by the Deep Impact spacecraft. *Science, 326*(5952), 565–568. doi:10.1126cience.1179788 PMID:19779149

Sviatoslavsky, I. N. (1993). *The Challenge of Mining He-3 on the Lunar Surface: How All the Parts Fit Together: Prepared for Space 94, The 4th International Conference and Exposition on Engineering, Construction and Operations in Space, and The Conference and Exposition/Demonstration on Robotics for Challenging Environments, February 26 - March 3, 1994, Albuquerque NM*. Madison, WI: Academic Press.

Taylor, L. A., & Carrier, W. D. (1993). Oxygen Production on the Moon: an Overview and Evaluation. In J. Lewis, M. S. Matthews, & M. L. Guerrieri (Eds.), Resources of Near Earth Space (pp. 69–108). The University of Arizona Press.

Taylor, L. A., Jerde, E. A., McKay, D. S., Gibson, M. A., Knudsen, C. W., & Kanamori, H. (1993). Production of O2 on the Moon: A Lap-Top Demonstration of Ilmenite Reduction with Hydrogen. *Proceedings of the 24th Lunar and Planetary Science Conference*.

Taylor, L. A., & Meek, T. T. (2005). Microwave Sintering of Lunar Soil: Properties, Theory, and Practice. *Journal of Aerospace Engineering, 18*(3), 188–196. doi:10.1061/(ASCE)0893-1321(2005)18:3(188)

Tynan, D. (n.d.). *Galactic gold rush: the tech companies aiming to make space mining a reality*. Retrieved from https://www.theguardian.com/science/2016/dec/06/space-mining-moon-asteroids-tech-companies

U.S. Department of Energy. (2011). *Critical Materials Strategy*. Academic Press.

Vielstich, W., Lamm, A., Gasteiger, H. A., & Yokokawa, H. (2010). *Handbook of Fuel Cells*. Chichester, UK: John Wiley & Sons, Ltd. doi:10.1002/9780470974001

Weirauch, A. (2011). *F*. Houston, TX: ISS Pressurized Volume Hardware Common Interface Requirements Document.

Werkheiser, M. J., Fiske, M., Edmunson, J., & Khoshnevis, B. (2015). On The Development of Additive Construction Technologies for Application to Development of Lunar/Martian Surface Structures Using In-Situ Materials. In *AIAA SPACE 2015*. Conference and Exposition; doi:10.2514/6.2015-4451

Wieland, P. O. (1998). Living Together in Space: The Design and Operation of the Life Support Systems on the International Space Station (NASA TM No. 1998-206956). Huntsville, AL: NASA.

Williams, D., Dake, J., & Gentry, G. (2012). International Space Station Environmental Control and Life Support System Status for the Prior Year: 2010 - 2011. *Proceedings of the 42nd International Conference on Environmental Systems*. 10.2514/6.2012-3612

Yoshida, H., Watanabe, T., Kanamori, H., Yoshida, T., Ogiwara, S., & Eguchi, K. (2000). Experimental Study on Water Production by Hydrogen Reduction of Lunar Soil Simulant in a Fixed Bed Reactor. *Second Space Resources Roundtable*.

Zubrin, R. M., Muscatello, A. C., & Berggren, M. (2013). Integrated Mars In Situ Propellant Production System. *Journal of Aerospace Engineering*, *26*(1), 43–56. doi:10.1061/(ASCE)AS.1943-5525.0000201

Chapter 13
In-Situ Resource Utilization:
Legal Aspects

Philip De Man
University of Leuven, Belgium

ABSTRACT

This chapter discusses whether and how the existing rules of international space law can be applied to regulate the activity of space resource exploitation, with a focus on in-situ resource utilization. As such, the chapter aims to establish how natural resources should be defined in the context of space law, how the activity of exploitation compares to other types of activities in space, in particular the prohibited activity of national appropriation, and the extent to which the type and supply of space resource may impact on its regulation. The chapter concludes with a brief overview of the most recent developments in national and international forums attempting to regulate space resource exploitation.

BACKGROUND[1]

The legal regulation of the exploitation of natural resources in outer space has been the subject of intense discussions ever since the first rules regarding space activities were adopted over half a century ago. Though some legal developments of note have taken place in recent years[2], we are still no closer to answering some of the most fundamental questions regarding space resource exploitation in general, and asteroid mining in particular. This is mainly due to the fact that the key legal document that sets out the basic principles of the international regime on outer space, the 1967 Outer Space Treaty, does not contain the notions 'natural resources' and 'exploitation'. (OST, 1967) Specific rules on the exploitation of the natural resources of celestial bodies have been included in another text, the 1979 Moon Agreement, but this convention is not binding on any one of the major spacefaring countries due to lack of ratification.[3]

Since the only widely accepted international texts on space law do not mention the exploitation of natural resources, applying the key principles of these texts to this activity presents something of a challenge. In particular, the application of the fundamental space law principles may differ according to one's views of how natural resources should be defined in the context of space law, and how the activity of exploitation compares to the other types of activities that have been explicitly mentioned in, and

DOI: 10.4018/978-1-5225-7256-5.ch013

regulated by, the existing space treaties. Finally, the regulation of asteroid mining may well depend on the type of use that is made of the resources once extracted from their location on the celestial body. The biggest issue in this respect is the possibility of establishing property rights over asteroid resources, or indeed the necessity thereof, in order to commercially exploit resources.[4]

TYPES OF RESOURCES AND THEIR SCARCITY

Asteroid mining has become particularly salient in light of the recently announced plans of a number of companies, established under US and Luxembourg law, to appropriate valuable resources from a number of celestial bodies, which may be operationalized as early as in 2025 (Wall, 2015). These plans refer to the following natural resources in space:

- The extraction of rare minerals from asteroids at a distance no further than the Moon for use on Earth and the processing of water found on these asteroids to fuel the in-space economy (Planetary Resources);
- The excavation of precious ores and minerals on asteroids, and the use of solar energy (Deep Space Industries); and
- The processing of ice deposits on the Moon for establishing a network of refuelling stations in low-Earth orbit and on the Moon to provide fuel and consumables for commercial and government customers (Shackleton Energy). (Planetary Resources, 2018) (Shackleton Energy, 2018).

The exact legal regulation of all of these different types of use of natural resources from celestial bodies may differ depending on the legal relevance of their classification as scarce, depletable or renewable; in other words: on the impact States' exploitation of these resources may have on their supply. In the end, however, it is the interaction between these characteristics and two fundamental principles of international space law that will determine whether the activities planned by the above companies will present any legal difficulties.

As noted, the basic instrument containing the most fundamental principles of international space law, and which has been ratified by all major spacefaring countries, is the Outer Space Treaty. Every type of use by any State or its nationals has to abide by two key rules. First, outer space, including the Moon and other celestial bodies, are free to be explored and used by all States, without discrimination of any kind, on a basis of equality (Art. I, para. 2 OST). Obviously, this does not mean that every part of every celestial body must be available at all times to every State or every private company, for otherwise no one would be able to exercise its freedom to use outer space. Indeed, every activity in outer space is accompanied by a certain degree of exclusivity, and this exclusivity is protected for the duration of the actual use (Art. IX OST). A satellite launched by one telecommunications company occupies a spot that cannot simultaneously be used by another company; on the contrary, other companies have to respect the fact that this satellite has been placed in orbit and cannot interfere with it. Likewise, a base established by one State in a certain area on the Moon occupies a stretch of land that cannot be used by another State.

What the above principle does mean, however, is that every user in outer space has to take into account the fact that all other States of the international community, and their nationals, have the same freedom to engage in spacefaring activities of exploration and use. In other words, the principle is a guarantee against monopolistic uses of outer space that would be so intrusive as to deprive others from engaging in comparable activities. Applied to asteroid mining, this means that activities of exploitation must be planned so as to minimize interference with the spacefaring activities of others, and may not involve the exploitation of such amounts of natural resources as to deprive, in effect, other States or companies from using these or other, comparable resources. Of course, this is where the difference between depletable and renewable resources will become a relevant factor, because it will determine the scarcity of the resource concerned. It is not the only relevant factor, however, as scarcity is also a product of the strategic relevance of the resource and the practical feasibility of its exploitation, among other things.

As we have seen in the first section of this chapter, resources in space may take many forms, from regolith to water to solar energy and even immaterial riches such as valuable orbits. The use of orbits as natural resources, though not free of controversy, has generally raised few concerns in terms of fundamental lawfulness, since orbits are not depleted upon use; a satellite using a valuable orbit can, in theory, be replaced by another satellite belonging to a different country or company, without the characteristics of the resource having been substantially altered. Like orbits, solar energy, too, has been used in outer space ever since the first satellite was launched. Though it is depletable, the actual use of solar rays by one user does not diminish the overall future supply of solar energy, because it is a renewable energy source. Hence, this form of exploitation of resources from celestial bodies in outer space has not raised too many concerns, either.

However, things appear to be very different for the exploitation of depletable, non-renewable material resources such as water, because their use by one consumes a definite portion of the resource supply that can no longer be used by any other interested party. The issue becomes even more pressing when considering the case of rare minerals on asteroids or Helium-3 on the Moon. Though arguably more abundant than on Earth, the total supply of these resources in space may be still be so small that their use by a couple of States or companies may raise concerns for the broader international community, taking into account the strategic relevance of these resources and the practical difficulties in reaching new mining sites. This is an issue of supply and demand, however, that may rise regardless of the physical characteristics of the resource. Ultimately, the basic rule still holds for these types of consumable resources as well: as long as their supply is sufficiently large as to accommodate all prospective users, the application of the principle of equal freedom to use these resources does not raise immediate problems.

RESOURCE EXPLOITATION V. APPROPRIATION

The second key principle of international space law that needs to be taken into account when considering the regulation of asteroid mining is the fact that outer space, including the Moon and other celestial bodies, are not subject to national appropriation (Art. II OST). This prohibition applies to all types of property rights, whether they have been established on the basis of sovereignty, occupation, use, or any other means. International law recognizes a number of means through which a State can establish sovereignty over new territories, including, for example, occupation. Sovereignty signifies ownership and possession over a territory, granting a State the authority to exercise exclusive jurisdiction. Article II OST bans these traditional means for acquiring sovereignty and expands the list to 'any other means'. Hence, States cannot establish ownership over outer space or the celestial bodies therein.

Though the prohibition only refers to 'outer space, including the Moon and other celestial bodies', it also applies to parts of outer space and celestial bodies. Otherwise, the principle would have little to no effect, because it is practically impossible to appropriate the entirety of outer space, or a large celestial body such as the Moon. Moreover, the non-appropriation principle already defines celestial bodies as a part of outer space. Finally, even though the text of the legal provision refers to 'national' appropriation, it is accepted that the prohibition applies to States as well companies. States are responsible for the activities of their nationals and have to supervise them (Art. VI OST). This means that States cannot allow companies under their supervision to appropriate celestial bodies or parts thereof. Such rights could only be granted if the State adopted national legislation that bestows property rights over outer space to their nationals. The State cannot grant these rights, however, since it does not have them itself.

The fundamental freedom to use natural resources in space and the ban on property rights creates a tension that is at the very heart of the controversy surrounding many of the plans related to asteroid mining (Lee, 2012). As we have seen, the use of depletable, non-renewable resources such as water and mineral ores on an asteroid by one company may not directly affect the freedom of other users to exploit comparable resources on another part of the same asteroid, or on another celestial body altogether. However, any consumption of such a resource does make it permanently unavailable for use by others, because it is destroyed after a single instance of use. Hence, the key issue regarding asteroid mining is that, though the natural resources of asteroids are free to be used, exercising this freedom appears to either require or entail the creation of property rights (Fasan, 1987). Indeed, destruction is commonly considered to be the ultimate form of appropriation (Fasan, 1998; Viikari, 2002). And since appropriation of celestial bodies and parts thereof is not allowed, this appears to also prohibit companies from engaging in the exploitation of asteroids.

The application of the two most fundamental principles of international space law to asteroid mining hence creates a paradox. There are two possible ways to resolve this in order to arrive at a workable regime. Most commonly, it is argued that natural resources simply escape the application of the non-appropriation principle, meaning that they can be excavated, processed and transferred in a commercial environment on the basis of property rights (Jenks, 1965; Gorove, 1984; Christol, 1984). Though it is the majority opinion, the solution appears a bit arbitrary, because it introduces a distinction that is not supported by the text of the legal principles of free use and non-appropriation, and runs counter to logic. To define their scope of application, both provisions in the Outer Space Treaty refer to 'outer space, including the Moon and other celestial bodies'. Neither mentions natural resources. Whether this means that resources are implicitly included or excluded from the scope of these provisions is impossible to tell. However, it is clear that our interpretation of the scope should be the same for both provisions. This means that natural resources should either (1) be free for use and not subject to appropriation; or (2) not free for use but appropriable. If neither option appears to make much sense at first sight, only the latter clearly runs contrary to logic: you cannot appropriate resources without also being able to use them, because property rights also include a right to use.

The only sensible conclusion is hence that resources from asteroids can be used, though not on the basis of property rights. In other words, the solution to the paradox does not lie in distinguishing between natural resources and everything else in outer space, but between the activities of exploitation and appropriation. As we have seen, States are free to use outer space and celestial bodies therein. Though exploitation is not mentioned in the Outer Space Treaty, 'use' includes 'exploitation', and, indeed, appears difficult to distinguish from it (Böckstiegel, 1981; Goedhuis, 1981; Jakhu, 1983). What is not allowed is that States, and their nationals, appropriate celestial bodies. The crucial question then becomes: when does exploitation become appropriation, and is it possible to engage in ISRU or other forms of asteroid mining without appropriation?

Answering this question requires a closer look at the concepts 'natural resources' and 'property rights'. Natural resources can be described as natural phenomena with an intrinsic economic value that is only realised upon transformation through human activity, *i.e.* the act of processing.[5] It is only when such transformation has taken place that a part of a celestial body actually manifests itself as a natural resource; before processing it, ice deposits and other asteroid resources are merely a part of the celestial body, indistinguishable from it and subject to the same legal regulations. The act of exploitation is hence the activity of using natural resources through a series of actions that typically include excavation and some type of processing, as we have seen in the first section of this chapter, for their economic potential.

Property rights can be described in a lot of ways. Typical components of property rights are the rights to use, exclude, and transfer (Hohfeld, 1913, 1917; Honoré, 1961). Hence, ownership over natural resources typically includes the right to use and exploit them. This does not mean, however, that the exploitation of resources always implies or, indeed, requires the existence of property rights. Appropriation is not only, or even primarily, defined by a right to use, but by a right to exclude (Cohen, 1927; Munzer, 1990). To be sure, using something means that others cannot use the same thing at the same time. Therefore, property rights only exist when it is established that your exclusion of others is based on your authority as an owner, and not a logical consequence of your use. For example, as long as I am reading a book that I loaned from the library, others cannot use it. This does not mean that I own the book, however, because the exclusion of others is merely factual, and is not based in any form of authority derived from property. Applied to natural resources, this means that I own resources when I can exclude others *because I have sole authority over them*; exploitation occurs when others are excluded *only as a natural consequence of me exercising my freedom to use these resources.*

International space law allows States and companies to use natural resources, since every user in space has an equal freedom to do so. Once this freedom is exercised, it is protected from interference. It has to be, because otherwise no one could exercise their freedom to engage in spacefaring. The freedom to operate satellites can only exist when others cannot interfere with it by placing a satellite in the same orbit; a Moon base can only be established if you can use the soil on which it is placed to the exclusion of others (even if the law imposes certain obligations to open up Moon bases to others, on a basis of reciprocity). Likewise, the freedom to engage in ISRU requires a certain level of exclusivity to be actually exercised. The fact that the exploited resource ceases to exist after use does not mean it has been appropriated, however, because the exclusivity is merely factual, it is the natural and inevitable consequence of a lawful form of use of outer space, not the legal proof of ownership. This means not only that States and private users are free to exploit resources, but also that such exploitation does not have to be based on ownership rights.

This subtle distinction between exploitation and appropriation is not always followed by lawmakers, who rather tend to agree with the space mining industry's assertions that property rights over asteroid resources are necessary for reassuring investors and creating a climate of legal security. This is confirmed by recent developments in the law of the United States, as will be discussed at the end of this chapter. Whether an industry has to be based on property rights to prosper is unclear, however, as is demonstrated by the telecommunication sector. Though worth over a trillion dollars, this industry has bloomed in the absence of property rights over the resources on which it so crucially relies. Indeed, law-making history at all levels is replete with explicit rejections of any form of property rights over radio frequencies and orbital positions.[6] To be sure, the renewable nature of these resources has been acknowledged to be legally relevant. We have argued that this characteristic only impacts the amount of resources one

State or company can exploit, however; it does not, as such, impact the legality, or necessity of property rights over them.

TYPES OF USE: ISRU AND OTHER FORMS OF EXPLOITATION

What does all of this mean for the regulation of spacefaring activities that rely on the exploitation of resources from asteroids and other celestial bodies? The following segments will cover the types of activity of asteroid mining that are most likely to occur in the near future, as covered in the first section of this chapter, based on their consumption by personnel, the construction of bases and other installations on celestial bodies, and commercial trade.

In-Situ Resources Utilisation

Resources can be used for a variety of purposes. Though the fundamental legality of the exploitation of resources from asteroids and celestial bodies has been established in the previous segment of this chapter, the specifics of their legal regulation will depend on the goal for which the resource is being exploited. First, the most obvious application of resources in space, in particular water, is to be used as sustenance for personnel on lengthy space missions and fixed bases on celestial bodies. This is closely related to a second application, which pertains to the construction of structures and installations on celestial bodies and in outer space as such, by using resources found on asteroids. The existing treaties do not specifically cover any of these activities in particular, but a number of provisions give guidance as to the limits of ISRU in this respect.

Most of the relevant provisions on ISRU are contained in the Moon Agreement. We already noted that this treaty is rather ill-ratified and most of the controversy surrounding it relates to the exploitation of natural resources. However, the provisions of the Agreement concerning the use of areas on celestial bodies and the resources in them for the support of space missions are relatively uncontroversial. Moreover, recent years has seen a rise in support for the Moon Agreement, with new ratifications and increased efforts of the international community to highlight the benefits of adherence to the treaty. Finally, a number of major spacefaring nations such as France and India have signed the Moon Agreement without ratification, meaning that they cannot do anything that would directly contravene or impede the aims and goals of the treaty.

Pursuant to the Moon Agreement, States may establish manned or unmanned stations on the Moon and other celestial bodies (Art. 9 MA). No limitations are imposed on States establishing such stations, apart from the requirement that they "shall use only that area which is required for the needs of the station and shall immediately inform the Secretary-General of the United Nations of the location and purposes of that station" (Art. 9 MA). Moreover, the Moon Agreement emphasizes that "neither the surface nor the subsurface of the Moon, nor any part thereof or natural resources in place, shall become property of any State, international intergovernmental or non-governmental organization, national organization or non-governmental entity or of any natural person. The placement of personnel, space vehicles, equipment, facilities, stations, and installations on or below the surface of the Moon, including structures connected with its surface or subsurface, shall not create a right of ownership over the surface or the subsurface of the Moon or any areas thereof" (Art. 11 (3) MA).

These provisions have several implications for the regulation of ISRU. First of all, Stations erected on celestial bodies are likely to be established on a (quasi-)permanent basis, for reasons of cost-effectiveness and strategic purposes. Exercising the freedom to establish stations on celestial bodies may thus well result in the permanent exclusion of all others from using the area on which the station is built. Nevertheless, it is considered a lawful activity, despite the explicit reiteration of the non-appropriation principle in the Moon Agreement. This supports the interpretation that resources, too, can be used even if they result in the permanent exclusion of others, and that such use does not require or result in property rights. Establishing such a station on a celestial body does not mean that the owner or operator of the station owns the resources underneath it or surrounding it. This does not mean, however, that these resources cannot be used. However, such exploitation should be limited to the needs of the station. This follows from the general requirement that stations should only use the area that is necessary for the needs of the station.

Hence, ISRU that is connected to the sustenance of personnel for the duration of the mission is, most likely, a lawful activity, as long as it does not exceed the functional needs of the mission. The provision also implies that ISRU is lawful when the exploited resources are integrated into the structure of the basis, station, or other form of installation on a celestial body, as long as this integration is required for the continued functioning of the installation. Once more, such integration would likely be permanent, but, due to its functional connection with the lawful exercise of the freedom of States and companies to use outer space, this integration should not pose problems. Legal issues may arise, however, when the integration of resources of a celestial body in an artificial construct is of such an invasive nature that it becomes difficult to distinguish between the natural body and the artificial installation (Fasan, 1984; Gorove, 1993). This is the case when States would engage in activities of terraforming celestial bodies (Roy, Kennedy & Fields, 2009). The issue that would arise in that case, however, is one of distinction between the celestial body as such and a space object. Indeed, both notions are subject to widely differing rules in space law, in particular regarding property rights.[7] These issues do not concern us, however, because they are not specific to the use of natural resources.

Removing Resources

As the name implies, ISRU refers to those forms of asteroid mining that occur on or near the mining site. The natural resources remain in place on the celestial body but are merely repurposed after transformation. Most plans of private space mining companies go further, however. Their plans are based on commercial aspirations to either distribute fuel processed from ice deposits among orbital stations or bring valuable minerals to Earth. There does not appear to be, as of this moment, much difference between the legal regimes for ISRU and other types of asteroid mining, considering the absence of provisions in international space law dealing with asteroid mining as such.

However, one provision in the Moon Agreement does raise some questions in this respect. Article 11 of the Moon Agreement, as we have seen, provides that "neither the surface nor the subsurface of the Moon, nor any part thereof or natural resources in place, shall become property" of anyone, including States, international organizations, and private individuals or companies. The phrase is a peculiar addition to the text of the non-appropriation principle in the Outer Space Treaty, which simply refers to 'outer space, including the Moon and other celestial bodies. True, the Moon Agreement does not apply to most of the States currently planning ISRU missions. However, the States that have ratified the Agreement insist that most of its provisions are simply a clarification of the rather vague and general principles of the Outer Space Treaty, with which they are consistent. (UN COPUOS Joint Statement,

2008) If the addition to the non-appropriation principle in Article 11 of the Moon Agreement should therefore be read as a clarification of the Outer Space Treaty, it might reveal something about the way the latter should be interpreted.

In particular, the question rises whether the formulation of Article 11 MA introduces a distinction between the legal regime of natural resources that are used in place, and other resources. Opinions are, once more, heavily divided on the importance and meaning of the reference to 'natural resources in place' in the non-appropriation principle of the Moon Agreement. Does the fact that natural resources 'in place' cannot be appropriated mean that natural resources that are no longer in place can be appropriated? And if so, does the 'in place' reference mean that resources can be appropriated only when they are removed from the celestial body on which they were initially found? Or does it simply mean that natural resources have to be excavated from their location on the asteroid before they can be used by any one in particular? This may be important when considering natural resources that are used for the construction of installations that are physically integrated in the celestial body on which they were found. Are these resources still in place? What if future processing technologies would allow resources to be used without their prior physical excavation from an asteroid? Does the non-appropriation principle then still apply?

Many authors argue that the simple act of removing a natural resource from the surface or subsurface of a celestial body is sufficient to remove it from the scope of application of the non-appropriation principle because the Moon Agreement refers to 'natural resources in place' (Hobe, 1992; Gál, 1996; Wassenbergh, 1996). However, some arguments make this assumption unlikely, to say the least. First, it is unclear why the Moon Agreement simply does not say that natural resources that are not in place can be appropriated, if this really was the intention of its drafters. Second, excavation is always the first step of any type of exploitation of asteroid resources. If this action is sufficient to remove the ban on property rights, then that fundamental principle loses much of its meaning (Viikari, 2002; Nyman-Metcalf, 1999). Third, the Moon Agreement also contains specific rules on the removal of samples from celestial bodies for scientific purposes (Art. 6 MA). It is generally assumed that scientific exploration is subject to fewer legal limitations than commercial exploitation, considering its less exclusive and invasive impact. However, the Moon Agreement stops short of granting full property rights over samples that have been removed from their celestial body. This makes it unlikely that such rights should be granted over commercially mined resources.[8] Finally, we have seen that property rights are not, in our view, absolutely necessary to allow exploitation to take place.

Without further indications, it is therefore unclear whether the Moon Agreement changes anything to the rules that already apply to the States that have ratified the Outer Space Treaty. This view is reinforced if we look at the other parts of Article 11. Even though the Moon Agreement was adopted largely in order to shed light on the contentious issue of asteroid mining, it ultimately refuses to adopt specific rules. Instead, the treaty simply notes that "States Parties to this Agreement hereby undertake to establish an international regime, including appropriate procedures, to govern the exploitation of the natural resources of the Moon as such exploitation is about to become feasible" (Art. 11 (5) MA). The plans announced by US companies in recent years surely seem to indicate that asteroid mining is 'about to become feasible'.[9] Nevertheless, no negotiations have been undertaken so far to adopt an international treaty on asteroid mining. Considering this vacuum in international space law, States have recently started to adopt their own national laws to fill this void.

RECENT LEGAL DEVELOPMENTS

The most concrete steps toward developing a legal regime for asteroid mining have been taken by the United States and Luxembourg. In direct response to demands from the American space mining industry, US Congress passed, on 18 November 2015, the Commercial Space Launch Competitiveness Act. (Congress, 2016) The act mainly deals with streamlining the US legal regime for commercial space launch activities. However, it also includes, at the very end, a brief title on Space Resource Commercial Exploration and Utilization. The historic significance of this Space Resource Title lies in the fact that it is the first legal instrument, at any level of governance, to explicitly grant property rights to private enterprises over resources extracted from asteroids and other celestial bodies. In particular, the act provides that

[a] United States citizen engaged in commercial recovery of an asteroid resource or a space resource under this chapter shall be entitled to any asteroid resource or space resource obtained, including to possess, own, transport, use, and sell the asteroid resource or space resource obtained in accordance with applicable law, including the international obligations of the United States.[10]

As we have seen, the requirement to obtain the mined space resources 'in accordance with the international obligations of the United States' may very well pose difficulties considering the current state of international space law. Indeed, as noted during congressional hearings for one of the previous incarnations of the adopted text, resource extraction represents "a very volatile and contentious issue at the international level". (Gabrynowicz, 2014) It should not surprise us, then, that the Space Resource Title was heavily criticized upon its adoption by space scholars and the specialized media. (Jakhu & Lee, 2015) (Oduntan, 2015) In light of this controversy, the requirement to take into account the international obligations of the US when mining asteroids becomes rather crucial, as the experience of the Luxembourg initiative two years later would illustrate.

On 1 August 2017, Luxembourg became the second country in the world to provide a domestic legal framework for the regulation of prospective space mining activities with the entry into force of its Law on the Exploration and Use of Space Resources. More declaratory than its US counterpart, the Luxembourg law explicitly proclaims that space resources are 'capable of being appropriated' (Art. 1). Interestingly, an earlier version of the law specified that these resources were capable of being appropriated 'under international law'. (Luxembourg Gouvernment, 2018) However, when the Luxembourg Council of State remarked that "the exploitation of space resources does not currently benefit from an international normative framework to regulate them", the legislator was advised to suppress the entire provision declaring space resources capable of appropriation under international law. (Conseil d'Etat, 2017) (De Man, 2017) Rather than deleting the entire provision, however, it was decided to merely drop the reference to international law, thereby leaving only a rather hopeful declaration of appropriability with unclear legal status, since international obligations trump domestic law in the Luxembourg legal order.

Though momentous developments, the US and Luxembourg laws ultimately leave unanswered the question of how to relieve the tensions between existing space treaties and the practice of space mining, including the matter of orderly management of competitive claims to the same resources. In this respect, we may note that previous versions of the US Act did contain some additional provisions on the way disputes relating to asteroid mining would be settled. Most importantly, they stated that "any assertion of superior right to execute specific commercial asteroid resource utilization activities in outer space shall prevail if it is found to be first in time, derived upon a reasonable basis, and in accordance with all existing international obligations of the United States". (Congress, 2018) Even though these rules were ultimately not adopted, they appear to be in line with the rules that currently exist for the use of orbits in space by satellites, and which have been developed by the International Telecommunication Union. (ITU Radio Regulations, 2016) It follows that these rules should not be ignored, because they will likely inform the further development of the legal regime on asteroid mining in the United States and beyond.

Ultimately, any State adopting domestic legislation on the exploitation of resources from celestial bodies will only be able to fully realize its objective of creating legal certainty for its citizens if it complements these national efforts with an attempt to clarify the application of the existing legal principles of space law to this contentious activity. An international market trading space resources will only emerge if States share a common understanding of the legal regime. To this end, the Legal Subcommittee of the United Nations Committee on the Peaceful Uses of Outer Space, the prime international forum for the development of international space law, has recently started discussing potential legal models for space resource exploitation. In its 2017 session, the Subcommittee has embarked on a general exchange of views on this topic that will hopefully increase our understanding of the current state of play in space mining, and the regulatory efforts that can be undertaken to govern this emerging activity in a controlled, efficient and equitable manner.[11]

CONCLUSION

Whether companies can appropriate resources from asteroid will depend on the interpretation of the international obligations of the United States and other countries that have ratified the Outer Space Treaty and, possibly, the Moon Agreement. What is clear, however, is that the activity of asteroid mining as such should be considered lawful under the current constellation of space law treaties. A distinction might be made between ISRU and other forms of asteroid mining, depending on the location of the actual exploitation and the purpose for which it occurs. This distinction does not affect the lawfulness of the act of mining itself, however, as long as three key rules are taken into account.

First, mining activities may not interfere with the authorised activities of other space actors. Second, the plans of one actor may not result in a substantial reduction of the supply of a particular resource so as to deprive other states or companies from engaging in comparable mining activities. Third, the plans of mining companies should clearly indicate the mining site and identify the legal purpose for which they wish to exploit the resources that are found there. The amount of resources should not exceed the functional needs of the mission. For commercial exploitation, this means that no more resources should be exploited than can reasonably be expected to be used by the international community in the near future.

Though we have argued that these rules are already present, at least in inchoate form, in the existing space law treaties, further efforts need to be taken to regulate the activity of space mining in a manner that satisfies the interests of all States. The first steps toward the regulation of space resources are currently only being taken at the national level, by a small number of intrepid spacefaring nations. However, the environment of outer space has been designated an international area, the use and exploration of which is the province of all mankind. (OST, 1967) It follows that any domestic space resource legislation that may adversely affect the interests of other States must be accompanied by a multilateral clarification of the limits of the freedom to use outer space by all States, as it applies to space mining.

During these discussions, the question of whether or not asteroid mining activities may be accompanied by the creation of property rights will likely remain an issue of grave contention. Its practical relevance may be limited, however, as we have tried to argue that the exploitation of asteroids does not require or necessarily result in appropriation. Every form of use of outer space, including the exploitation of asteroids, is, once actually planned and implemented, protected by a right not to be interfered with by others. The exclusive use of asteroid resources is therefore lawful, even if it permanently destroys these resources in the process. In the context of such a forgivable legal regime, the only thing that the ban on property rights would add, is a right to exclude others that is not based in use. In other words, it would simply introduce a license to waste, which hardly seems like something worth adding to the existing regime.

REFERENCES

Böckstiegel, K. H. (1981). Legal implications of commercial space activities. *Proc. Coll. L. Outer Space*, *24*, 1–17.

CBC News. (2015). US space-mining law seen leading to possible treaty violations. *CBC News*. Retrieved on 30 April 2018 from http://www.cbc.ca/news/technology/space-mining-us-treaty-1.3339104

Christol, C. Q. (1984). Article 2 of the 1967 Principles Treaty revisited. *Annals of Air and Space Law*, *9*, 217–264.

Cohen, M. R. (1927). Property and sovereignty. *Cornell L.Q.*, *13*, 8–30.

De Man, P. (2015). *Exclusive use in an inclusive environment: the meaning of the non-appropriation principle for space resource exploitation*. Berlin: Springer.

De Man, P. (2017). Luxembourg law on space resources rests on contentious relationship with international law. *The Space Review*. Retrieved on 30 April 2018 from http://thespacereview.com/article/3355/1

Fasan, E. (1984). Large space structures and celestial bodies. *Proc. Coll. L. Outer Space*, *27*, 243–246.

Fasan, E. (1987). Celestial bodies and the exploitative use of outer space. *Annals of Air and Space Law*, *12*, 227–234.

Fasan, E. (1998). Asteroids and other celestial bodies - some legal differences. *J. Space L.*, *26*, 33–40.

Foust, J. (2014). Hearing raises questions about asteroid mining bill. *Space News*. Retrieved on 30 April 2018 from http://spacenews.com/41825hearing-raises-questions-about-asteroid-mining-bill/

Gál, G. (1996). Acquisition of property rights in the legal regime of celestial bodies. *Proc. Coll. L. Outer Space*, *39*, 45–49.

Goedhuis, D. (1981). Some recent trends in the interpretation and the implementation of the rules of international space law. *Colum. J. Transnat'l L.*, *19*, 213–234.

Gorove, S. (1993). Toward a clarification of the term «space object» - an international legal and policy imperative? *J. Space L.*, *21*, 11–26.

Gorove, S. (1984). International space law in perspective - some major issues, trends and alternatives. *Rec. Cours*, *3*(181), 3-12.

Hobe, S. (1992). *Die rechtlichen Rahmenbedingungen der wirtschaftlichen Nutzung des Weltraums*. Berlin: Duncker und Humblot.

Hobe, S., & Tronchetti, F. (2013). [Moon Agreement:] Article 6 (scientific investigations/samples/minerals). In *Cologne commentary on space law, Volume II: Rescue Agreement, Liability Convention, Registration Convention, Moon Agreement* (pp. 369-371). Cologne: Carl Heymanns.

Hohfeld, W. N. (1913). Some fundamental legal conceptions as applied in judicial reasoning. *The Yale Law Journal*, *23*(1), 16–59. doi:10.2307/785533

Hohfeld, W. N. (1917). Fundamental legal conceptions as applied in judicial reasoning. *The Yale Law Journal*, *26*(8), 710–770. doi:10.2307/786270

Honoré, A. M. (1961). Ownership. In Oxford essays in jurisprudence: a collaborative work (pp. 107-147). Oxford, UK: Clarendon.

Jakhu, R. S. (1983). *The legal regime of the geostationary orbit*. Montréal: McGill University.

Jakhu, R. S., Pelton, J., & Nyampong, Y. (2017). *Space Mining and Its Regulation*. Berlin: Springer. doi:10.1007/978-3-319-39246-2

Lee, R. J. (2012). *Law and regulation of commercial mining of minerals in outer space*. Berlin: Springer. doi:10.1007/978-94-007-2039-8

Munzer, S. R. (1990). *A theory of property*. Cambridge, UK: Cambridge University Press. doi:10.1017/CBO9780511609138

Nyman-Metcalf, K. (1999). *Activities in space: appropriation or use?* Uppsala: Iustus Förlag.

Oduntan, G. (2015). Who owns space? US asteroid-mining act is dangerous and potentially illegal. *The Conversation*. Retrieved on 30 April 2018 from http://phys.org/news/2015-11-space-asteroid-mining-dangerous-potentially-illegal.html

Roy, K. I., Kennedy, R. G., & Fields, D. E. (2009). Shell worlds: An approach to terraforming moons, small planets and plutoids. *J. Br. Interplanet. Soc'y*, *62*, 32–38.

Tronchetti, F. (2008). *The exploitation of natural resources of the moon and other celestial bodies: a proposal for a legal regime*. Leiden: Nijhoff.

Viikari, L. (2002). *From manganese nodules to lunar regolith: a comparative legal study of the utilization of natural resources in the deep seabed and outer space*. Rovaniemi: University of Lapland.

Wall, M. (2015). Asteroid mining may be a reality by 2025. *Space News*. Retrieved on 30 April 2018 from http://www.space.com/30213-asteroid-mining-planetary-resources-2025.html

Wassenbergh, H. A. (1996). The international regulation of an equitable utilization of natural outer space resources. *Proc. Coll. L. Outer Space*, *39*, 138–143.

Williams, S. M. (1987). The law of outer space and natural resources. *Int'l & Comp. L.Q., 36*(01), 142–151. doi:10.1093/iclqaj/36.1.142

ENDNOTES

1 This chapter is an updated and revised version of De Man, P. (2017). The exploitation of natural resources in outer space. In Failat, Y.A. and Ferreira-Snyman, A. (eds.). *Outer space law: legal policy and practice (243-256)*. Surrey: Globe Law and Business.

2 See *infra*, section 2.4.

3 Agreement governing the activities of states on the Moon and other celestial bodies of 18 December 1979, entered into force 11 July 1984, 1363 *U.N.T.S.* 21. See UNOOSA, Status of international agreements relating to activities in outer space as at 1 January 2015, UN Doc. A/AC.105/2015/CRP.8 of 8 April 2015. The agreement has been ratified by some Member States of the European Space Agency (ESA), such as Austria, Belgium and the Netherlands, meaning that these States have to abide by their obligations under the Moon Agreement when participating in relevant ESA activities. Moreover, the Moon Agreement has been signed by important space players such as France and India, meaning these States have to refrain from acts which would defeat the object and purpose of the treaty (Art. 18 of the Vienna Convention on the Law of Treaties of 23 May 1969, entered into force 27 January 1980, 1155 *U.N.T.S* 331).

4 As we are dealing here with legal issues that are in part subject to different interpretations of existing principles that are rather general in their formulation, the reader may wish to consult additonal reading material for further information on the contentious topic of the regulation of space mining in general, and *in situ* resource utilization in particular. See, among others, F. Tronchetti, *The exploitation of natural resources of the moon and other celestial bodies: a proposal for a legal regime*; R.J. Lee, *Law and Regulation of Commercial Mining of Minerals in Outer Space*; R.S. Jakhu, J. Pelton and Y. Nyampong, *Space Mining and Its Regulation;* De Man, P. (2015). *Exclusive use in an inclusive environment: the meaning of the non-appropriation principle for space resource exploitation.*

5 See the few definitions of natural resources in international law in Art. 2 Convention on biological diversity of 5 June 1992, entered into force 29 December 1993, 1760 *U.N.T.S.* 79; Art. 2 International treaty on plant genetic resources for food and agriculture of 3 November 2001, entered into force 29 June 2004, 2400 *U.N.T.S.* 303.

6 See, for example, United Nations Committee on the Peaceful Uses of Outer Space, Letter dated 16 October 1985 from the Secretary-General of the ITU addressed to the Secretary-General of the United Nations, UN Doc. A/AC.105/360 of 5 November 1985.

7 Space objects remain the property of their State of registry, even if they are launched into outer space or land on a celestial body: see Art. VIII OST.

8 Some authors argue that the rights granted by the Moon Agreement should be interpreted as property rights, though no convincing arguments are given that would justify this departure from the language of the Agreement. See Hobe, S. & Tronchetti, F. (2013). [Moon Agreement:] Article 6 (scientific investigations/samples/minerals). In Hobe, S., Schmidt-Tedd, B. & Schrogl, K.-U. (Eds.). *Cologne commentary on space law, Volume II: Rescue Agreement, Liability Convention,*

Registration Convention, Moon Agreement (369-371). Cologne: Carl Heymanns; Williams, S.M. (1987). The law of outer space and natural resources, *Int'l & Comp. L.Q., 36*, 142-151.

9 See the examples mentioned in the text *supra*.

10 §51303. Asteroid resource and space resource rights.

11 See the single issue/item 'General exchange of views on potential legal models for activities in exploration, exploitation and utilization of space resources', to be discussed by the Subcommittee at its 56th Session from 27 March to 7 April 2017 in Vienna. The item was renewed for discussion in 2018 and 2019.

Chapter 14
Human Planetary Exploration: Technical Aspects

Sophie Gruber
Space Generation Advisory Council, Austria

ABSTRACT

The human exploration of planetary bodies started with the Apollo missions to the Moon, which provided valuable lessons learned and experience for the future human exploration. Based on that, the design of hardware and operations need to further be developed to also overcome the new challenges, which arise when planning crewed missions to Mars and beyond. This chapter provides an overview about the environment and structure of the Red Planet and discusses the challenges on operations and hardware correlated to it. It further provides insights into the considerations regarding the hardware development which need to be investigated and defined before launching a crewed mission to Mars.

INTRODUCTION

By the year 2018, 12 astronauts have set foot on the Moon as a result of NASA´s Apollo space program. Preceded by various robotic missions and hardware demonstrations, the first landing on the Moon was realized within only few years between 1961 and the Apollo 11 mission where Neil Armstrong became the first human to set foot on the Moon in 1969, following president John F. Kennedy´s dedicated goal to land a man on the Moon by the end of that decade. Two years and six missions later, the era of human exploration of the Moon came to a halt after the Apollo 17 mission. Since then, over four decades have passed, without any human setting foot on an extra-terrestrial body. Astronauts later found their workspace aboard space stations like the MIR or the International Space Station (ISS) to investigate the influence of a low gravity environment on the human physiology and the hardware.

Meanwhile, robotic exploration missions advanced to new regions of space. Mars Pathfinder´s rover Sojourner was the first rover to explore the Martian surface in 1997 and was followed by rovers and landers which continued to investigate with the aim to gather insights about Mars. As of June 2018, six orbiters and the two rovers Opportunity and Curiosity are delivering scientific data from the Red Planet back to Earth together with several orbiters, which deliver data from the planet´s orbit. Recent

DOI: 10.4018/978-1-5225-7256-5.ch014

discoveries of water ice deposits, seasonal changes in the level of Methane in the Martian atmosphere and even organic molecules represent driving factors for sending humans to Mars, in order to look for traces of life to understand how life could arise on Earth. As seen with those discoveries, robotic missions significantly contribute to the exploration of Mars. However, humans would be able to investigate the surface of the Red Planet more efficiently. Especially due to the fact that a crewed mission would be more efficient when it comes to the coverage of scientifically interesting areas, together with the ability of humans to take fast decisions on the spot.

The goal to understand our place in the Universe and search for the origins of life on Earth is also stated in the Global Exploration Roadmap (ISECG, 2018), released in February 2018, where the 14 space agencies of the International Space Exploration Coordination Group (ISECG) further agree to "expand human presence into the solar system, with the surface of Mars as a common driving goal."

In the ISECG Mission scenario (ISECG, 2018), the Moon is seen as a first stepping-stone to serve as a test bed and foundation for human exploration of our Solar System. The next steps include robotic missions, the establishment of a deep space gateway, Lunar surface missions and a sample return mission towards the end of the 2020´s to Mars in order to gain in-depth knowledge from in-situ studies of various sites prior to a crewed mission to the Martian surface. These steps shall finally lead to crewed missions to Mars´s orbit and surface.

The significance of space exploration and its benefits for human kind were also expressed in the joint statement, resulting from the 2nd International Space Exploration Forum, which was held in March 2018 in Tokyo. Furthermore, the statement names the following reasons for national investments in space exploration through recognizing "the importance of investments in space exploration, which contribute to economic growth and societal well-being for all citizens on Earth." (ISEF2, 2018, p. 1). The 40 participating nations and intergovernmental organizations further affirmed that "extending exploration deeper into the Solar System, form Low-Earth orbit through the Moon to Mars and beyond, is a goal widely shared by the international community." (ISEF2, 2018, p. 2) and thereby also reflect a common understanding of the importance to extend the exploration of our Solar System towards Mars.

This chapter aims to provide insights on the environmental conditions on Mars, which astronauts would have to face. Furthermore, it discusses the technical and operational challenges of crewed surface exploration on a planetary body and the means to prepare for such a mission through analogue research.

ENVIRONMENTAL CONDITIONS AND CHALLENGES ON MARS

Mars is the fourth planet from the sun, which makes it Earth´s neighbouring planet in our solar system. One of the first observations humans could make about Mars was the planet´s reddish surface colour, which yielded the naming after the Roman god of war.

Nowadays, data obtained from instruments aboard Mars orbiters or rovers on the Martian surface, have revealed some more insights about the Red Planet. By these means have for instance instruments from the Thermal Emission Spectrometer of the Mars Global Surveyor (MGS) shown that the variations in the surface appearance, ranging from red-ochre to dark grey, result from different grain size distributions as well as a variation in their composition. Surface areas on Mars with a high albedo are linked to dust deposits and darker regions (i.e. low-albedo areas) are correlated to basaltic rich materials (Barlow, 2008).

Starting from these observations, further investigations by orbiters about the structure and composition of the Red Planet have contributed to the knowledge researchers have today about Mars. Although Mars is on average only 75 million kilometres away from Earth, the planet has an essentially different environment than Earth, which as it is, does not necessarily favour human settlements. Thus, a crewed mission to Mars will have to face environmental challenges and come up with technical solutions to cope with the harsh Martian conditions.

This aspect is already prominent when comparing the Martian atmosphere to the Earth´s. Where Earth´s atmosphere is breathable for humans and mainly consists of Nitrogen and Oxygen, the Martian atmosphere consists of 95% Carbon dioxide, which is poisonous for humans and further has only 1% of Earth´s atmospheric pressure. Furthermore, the average temperature on Mars is only -63 °C (Williams, 2016b). Hence, the habitat and space suit, which are going to be deployed and used for surface operations on Mars will have to provide a countermeasure against these conditions, in a way that humans are never going to actually feel these differences compared to Earth.

This differs when it comes to gravitational acceleration on the surface. Measurements from the Mars Orbiter Altimeter aboard the MGS provided additional information about the planet´s shape, contributing to a more complete picture about the structure of Mars. For crewed missions to the Red Planet, the gravitational acceleration is particularly interesting, as any difference to the Earth´s surface acceleration of 9.81 m/s^2 can trigger physiological changes in the human body. Mars` equatorial radius is about 0.53% of the Earth´s, which inter alia leads to the fact that the surface acceleration is with 3.71 m/s^2 only about a third of what humans are used to on Earth. This leads to the challenge and need for proper training for future astronauts to act against muscle loss due to the low gravity environment, which especially for long term stays on Mars will take a significant amount of the astronauts´ time. Research aboard the International Space Station is currently looking into this issue, trying to understand the effects of a low gravity environment on the human physiology (i.e. fluid shifts, visual alterations, balance disorders or even cardiovascular deconditioning) and providing sufficient countermeasures to be taken pre-flight, in flight and on the planetary surface itself.

Despite this challenge, reduced gravity means also that astronauts will be able to carry heavier loads. For instance, a space suit, weighting 120 kg on Earth, would only weight about 40 kg on Mars. Hence, the smaller Martian gravitation facilitates the engineering challenge when it comes to light weight design of the equipment used during surface operations. This consideration, however, does not take into account that for launching equipment from Earth, still every additional kilogram also means an increase in the amount of fuel necessary to go beyond Earth´s orbit.

When going beyond Earth´s orbit towards Mars, one also leaves behind the protection of the Earth´s magnetosphere. This leaves a space craft without the natural shield of the magnetosphere against space radiation. Issues correlated to the exposure to space radiation can either occur acute in-flight, after a short amount of time under a high dose of radiation, or after long-term exposure (for instance leading to an increase in the risk of cancer) and represent a considerable risk to the astronauts´ health. Since measurements from the Magnetometer/ Electron Reflectometer aboard the MGS reviled that Mars currently has no global magnetic field, caused by an internal dynamo (i.e. a moving layer in the Martian subsurface, yielding a magnetic field due to electromagnetic induction), the planet itself would also not provide sufficient shielding against space radiation after landing. There have merely been located regions of weak fields up to 0.5 nano Tesla (nT) (Acuña et al., 2001). compared to Earth´s magnetic field strength of 24,000 to 66,000 nT (Williams, 2016a). These results indicate that radiation protection is posing yet another challenge, which has to be solved before humans can start their journey to Mars.

Despite the challenging surface and environmental conditions on Mars, a crewed mission would also need to be able to have a secured water supply. There are basically three different approaches that can be taken into account for solving this issue:

1. In-situ utilization of resources, or
2. Recycling of used water and
3. Bringing water along from Earth.

Especially, when thinking of long-term settlements on Mars, all three approaches will have to be considered.

With this aspect in mind, the perception of Mars has changed significantly over the last years, as several findings lead to a better understanding of aqueous activities on past and present-day Mars. Scientist concur that Mars can no longer be considered the dry desert-like planet as it was used to be seen as. This is inter alia based upon findings from the OMEGA spectrometer aboard the Mars Express spacecraft from Poulet et al. (2005), which provided evidence for the presence of liquid water on early Mars by identifying Phyllosilicates. Such silicates are known to exist at locations with features typical for long-term contact with liquid water. Furthermore, shows the comparison of pictures from the Mars Orbiter Camera, taken only few years apart, the formation of recent gullies on steep terrain, as for instance along craters or canyons. Such gullies are proposed to have formed due to the melting of water-rich snow deposits, thus indicating present water flows on Mars (Christensen, 2003).

By far one of the most explicit evidences of water on Mars however was published in January 2018. Imagery from the HiRISE instrument on the Mars Reconnaissance Orbiter shows buried ice deposits which are now exposed due to erosion at steep slopes (Dundas et al., 2018). This also proofs that water ice is accessible at locations which have more moderate climate conditions than the Martian poles, which makes those deposits potentially a valuable water resource for future crewed missions to Mars. Additional to locations where water ice deposits are exposed and thus more easily accessible, cave systems on Mars are suggested to harbour water ice. In particular lava tubes or basaltic cavities would offer natural protection against radiation and more stable physico-chemical conditions (e.g. temperature, humidity) than the Martian surface. Hence, lava tubes are interesting from an astrobiology point of view as they could potentially harbour bio signatures (Léveillé & Datta, 2010) as well as they could provide a possible shelter for astronauts.

HUMAN EXPLORATION OF MARS

Preparing for Mars Through Analogue Research

Before sending humans to Mars, technologies, experiments, hardware, Standard Operating Procedures and workflows have to be tested in order to find weak spots and identify issues beforehand, to prevent them from occurring on Mars. Analogue research aims to investigate the test case of a human mission to Mars on Earth under conditions which are as similar as possible to those which can be found on Mars.

By these means can regions as for instance in the Dhofar desert in Oman or the Mauna Loa and Kilauea volcanoes in Hawaii, serve as an environment with features analogous to Mars, especially in terms of for instance geomorphology (e.g. volcanoes, dried river beds) or geochemistry (e.g. mineralization). In addition to laboratory test, where for instance the atmospheric pressure on Mars can be simulated, such regions can be used to test the hardware and experiments, which are designed for missions to Mars with the aim to further develop them or test their functionality under conditions similar to their future use case. Despite the test case of the mere hardware deployment, analogue research also offers the means to develop the procedures for conducting the experiments. This goes together with the opportunity to train and prepare astronauts for the EVAs, in the course of which they have to conduct the experiments, as it was already done during the Apollo program or during other analogue missions (Mahoney, 2009; Dorizon, Ciarletti, Plettemeier, & Benedix, 2016). Despite the hardware tests which are focusing on the engineering and geoscience aspects, remotely located analogue sites and long duration deployment of teams further provide the opportunity to study human factors, in particular with regards to isolation and team cohesion in extreme environments.

Due to the fact that the distance between Mars and Earth causes latencies in the communication from 4 to 20 minutes, it is important to establish a proper communication protocol, procedures and means through which the crew on Mars can communicate with the Earth support team. It is evident that voice communication is not practicable, as it can take up to 20 minutes until a question from Mars reaches Earth and it then again takes up to 20 minutes until Mars can earliest receive an answer. This has an impact on the situational awareness of the Earth support team, as it is difficult to follow the progress of an EVA. The latency can further lead to confusions in the sequence of questions, in case another question is sent, before the answer to the previous one can reach the recipient. This issue can partly be resolved by introducing text-based communication, which makes it easier to follow the communication sequence. However, the challenge hereby is that emotions and feelings are not easily transmitted via text-based communication. Hence, misunderstandings can occur more easily which can lead to an (emotional) separation of the teams. This would endanger a mission on Mars and can be avoided by proper training between those two teams already on Earth, as for instance through analogue research by artificially installing communication delays between the team operating in the Mars analogue region and the team supporting the mission remotely.

A deployment of a team in a Mars-like environment with the aim to conduct a series of experiments, analogous to the way astronauts would on the Martian surface, also requires a schedule for their activities to use the time efficiently and make sure the experiments can be conducted properly according to the requirements from the scientists. In particular, the establishment of a remote mission support team and infrastructure (comparable to a mission control centre), offers the opportunity to simulate the workflow and mission elements behind an experiment. This includes also the necessary system to develop an activity plan for the crew at the Mars-analogue site, while considering the time-delay in communication as well as in the data transmission, as conducted by Hettrich et al. (2014), during the MARS2013 mission simulation of the Austrian Space Forum (Groemer, 2014). Few research teams worldwide also implement analogue space suits in their operations, which enable the operating person to experience similar workload and constraints while conducting an experiment as they would on a planetary surface other than Earth.

Despite the operational and hardware related test cases of analogue missions, similar studies are performed to assess the adaptability of the human physiology and psychology to long duration isolation and confinement. This is of great importance to limit the risk of mental issues as well as physiological problems for interplanetary space flight.

A prominent example for such research is the Mars500 ground simulation experiment (Wang, 2014), which emulated the interplanetary journey between Earth and Mars. The experiment involved six multinational crewmembers, who experienced 520 days of confined isolation in a spaceship mock-up with a simulated EVA on Mars in the middle of the mission. The experiment yielded amongst other the tendency that the crew members assigned positive ratings to negative stimuli, which were presented in the course of the experiment. Hence indicating that the crewmembers tend to avoid severe mental feelings and issues, the longer the mission lasts. This positive bias on negative stimuli could be seen as an indication of the psychological status of the crewmembers (Wang, 2014). Events like the simulated landing on Mars offer new stimuli which can influence the feelings of the crew in a positive way. The lack of novel stimuli, however, can enhance negative feelings and sleep issues. What also became evident is a significant decrease of psychological adaptation of the crew, comparable to the third-quarter phenomenon, where the realization that almost half of the mission is still remaining has an impact on the crew communication, which during Mars500 decreased, as well as the motivation. Thus, this respective time-period of a mission has to be paid attention to when schedules are prepared for the activities for crewed missions. This way the Mars500 experiment and other analogue simulations provide valuable insights into psycho-social effects of long duration confinement to humans and help to prepare for future crewed missions to Mars and beyond.

Surface Operations on Mars

The activities to be conducted during surface operations and Extra Vehicular Activities (EVAs) of a crewed mission on Mars depend on various factors like the designated scientific goal(s) of the mission, as well as the landing site and the duration of the stay on the surface. Furthermore, several technological aspects have to be considered when planning a crewed mission to Mars.

Extra Vehicular Activity Suits

In order for humans to be able to investigate the Martian surface, the use of EVA suits is unavoidable, due to the need for protection against the harsh environmental conditions as mentioned in section 1 of this chapter. In addition to that, EVA suits for planetary surface operations must be designed in a way that they are easy to maintain by the astronauts with regards to long term usability and are as lightweight as possible, but at the same time they have to be sturdy and resilient against abrasion due to the fine-grained Martian soil. Despite being light weight, EVA suits also should not restrict the movement of the astronauts. Especially for tasks involved with the conduction of experiments or taking geological samples, the astronauts need to be able to bend over easily or extensively move their arms, without the losing balance or the risk of falling over.

The suits´ Primary Life Support System (PLSS) further needs to provide a safe and reliable environment for the astronauts. This also includes the provision of breathable Oxygen, water, nutrition, and proper thermal control. The thermal control system has the function to ensure adequate temperatures for humans. On the one hand, the temperatures on Mars are with -63 °C on average by far too low for humans to survive. On the other hand, the low atmospheric pressure prevents the heat generated by the astronauts during extensive physical work to be radiated out as there is almost no air, which could transfer the heat like it does on Earth. This can in the worst case even cause a heat stroke. Using material with a high thermal conductivity can contribute to mitigate this risk.

As soon as there is any kind of interaction with a planetary surface, planetary protection regulations have to be taken into account, as outlined in chapter 15. This dictates that an EVA suit also has to act as a barrier between the astronaut and the Martian environment. Along these lines, the suit design needs to ensure that no dust or other material can reach the astronauts, whereas at the same time it must prevent contamination of the Martian environment caused by leaking organic material either from the astronauts or which was brought from Earth. In order to compensate for the low atmospheric pressure, EVA suits also need to be pressurized, which entails a high risk of a leakage. Even a minor leak could release microbes or airborne particles which were in contact with the astronauts, yielding the contamination of the Martian environment. A possible solution to mitigate this risk would for instance be the implementation of enhanced seals and mechanical connections or sterile covers at joints, where the risk of a leakage is especially high (Drake & Watts, 2009).

Surface Mobility (Rovers)

In order for astronauts to be able to extend the exploration radius on a planetary surface around their landing site, rovers represent essential assets. As found during the Apollo 15, 16 and 17 missions on the Moon, the Lunar Roving Vehicle (LRV) provided substantial support to the astronauts by carrying equipment and reducing crew fatigue during the seated traverses from one site to another. Such an unpressurized system would have to be operated in combination with an EVA suit. Hence, a rover, similar to the LRV, could also share responsibilities with the EVA suit as for instance power supply, navigation or communication. Further considering the fact that significant EVA time during the Apollo missions were spent on the rover during traverses, a rover could also take along supplies as for instance nutrition or oxygen and enable the EVA suit systems to recharge and generally extend the suits´ capabilities.

A pressurized rover could further allow for even more extended exploration EVAs, both in terms of distance as well as time, by functioning in a similar way than a habitat, despite being smaller and mobile. The pressurization of the rover would also entail the engineering challenge that a transition between vehicles (or habitats) is not trivial and should be feasible as fast as possible in case of (medical) emergencies. One idea to face this challenge is to implement a suit port, where the astronaut can dock the suit backwards to either a rover or the habitat and can exit or enter via an opening in the back. This approach is supposed to limit a) the donning and doffing time of the EVA suits and b) minimize the time needed to enter or exit a pressurized vehicle or habitat, which operates under the same pressure as the EVA suit. Hence, suit ports would also eliminate the need for depressurization prior to exiting the habitat or rover and by that also reduce the risk of abrasive material from the Martian surface entering the habitat.

From an operational and safety point of view, the "buddy system" would have to be applied during EVAs. The buddy system requires that at any time during an EVA, two astronauts have to be within reach of each other in case of an emergency. Furthermore, another team located in an outpost or habitat should be within rescue range of the astronauts on EVA. Along these lines, a rover should be able to carry at least two astronauts (up to four astronauts in case of a rescue mission) and a second rover would need to be available at the outpost or habitat in case the rover used for the EVA breaks down or the astronauts need to be rescued.

Surface Habitats

The design and requirements of a Mars habitat depend for the most part on the duration, crew size of the mission and the general mission architecture. Various scenarios are currently being investigated by space agencies, including long duration stays of up to 500 days on Mars. Adding the crew transfer time from and back to Earth to that, the total duration of a crewed mission to Mars can be up to 28 months (Drake & Watts, 2009). Most scenarios further consider a crew size of four to six astronauts to cover all the different skills and expertise necessary. Bases upon these decisions, estimations about the equipment and supplies can be made which lead to the mass and design requirements for the surface habitat. In addition to these decisions about the general mission architecture, it is evident that the resulting requirements also entail the engineering challenges as for instance of minimizing the mass and volume. Further key technologies and engineering concepts that need to be developed for habitat concepts include inter alia the deployment mechanism, which also could work autonomously prior to the crew´s arrival on the surface, food and waste processing, power supply and a life support system providing a safe environment for the crew.

Some strategies and approaches can be derived from designs made for lunar habitats and adapted to the needs for a Mars mission in terms of crew size, human factors and the demand for a higher rate of spares. The latter also implies that a Mars habitat needs to be easy to maintain by the astronauts, as well as it should endure long-term usage over several missions, while ideally being on a stand-by mode in between missions.

In the course of the Lunar Architecture Team Phase 2 (LAT-2) Study (Dorsey, Chauncey, & Smith, 2008), several major design drivers have been identified. Based on those and the preceding arguments, the following design drivers can be identified, which are applicable for a Mars surface habitat:

- **Payload Mass:** The mass of the habitat represents one of the major constraints of the habitat design and should be minimized as much as possible. As an example, estimations from NASA`s Human Exploration of Mars Design Reference Architecture 5.0 state a payload mass of 16,500 kg for the habitat (Drake & Watts, 2009).
- **Habitable Volume:** Sufficient living space must be provided to the crew to also satisfy the psychological needs correlated to long duration isolation in a hostile environment.
- **Interfaces:** Due to the low pressure of the atmosphere on Mars, a habitat must be pressurized and interfaces between the Martian surface or other pressurized systems as for instance rovers or EVA suits must be designed in a way, that they are long-lasting, enabling a fast ingress and egress as well as they must prevent dust from reaching the interior of the habitat. One approach of achieving those criteria are suit ports as described in the previous section as an alternative to airlocks.
- **Consumables:** Supply missions from Earth are estimated to take around 6 months until they reach Mars, depending on the constellation of the two planets. Hence, ad hoc re-supply missions are not feasible. Technologies involved with 3D-printing or In-situ resource utilization (ISRU) can be used to produce spare parts or life support consumables on site and with that reduce the overall payload mass and/or volume.
- **Deployment:** In case a cargo mission is realized prior to the arrival of the crew on Mars, the autonomous deployment of surface infrastructure would enable a verification of the system functionality before the crewed mission launches from Earth. Despite that consideration, it further has to be decided whether the habitat shall be removed from the lander or stay connected to the lander.

Adding a system, which enables mobility of the habitat would present a design option for extending the exploration radius on the Martian surface.

- **Protection from Martian Environment:** The habitat must offer protection from the Martian environment, especially with regards to radiation protection, extreme temperatures, low atmospheric pressure and dust to provide a safe environment for the crew.

Furthermore, design decisions about the structure (e.g. inflatable or hard-shell), the geometry (e.g. cylindrical barrels, domes or combinations) and the number of habitat elements have to be made. Those decisions also impact the mass and volume estimations. By these means provides for instance an inflatable structure a higher packing efficiency when loaded on the launcher than a hard-shell component. Depending on the approaches chosen for the habitat design the following input parameters are necessary to determine the structural sizing, which later yield the mass estimation: material properties (e.g. density, tension modulus), minimum gage and environmental factors (e.g. pressure).

Power Supply

The power supply for surface operations on Mars can be divided into two categories: stationary and mobile power systems. Where the stationary power systems are required to provide sufficient power for the habitat and ISRU plants, which could be used to generate amongst other the O_2, N_2 and Ar, necessary for astronauts during EVAs, the mobile power supply is needed for rovers, which are also used for long-range exploration EVAs up to several days.

One possible power source is nuclear fission power, as for instance provided through the Plutonium-238 isotope, which was already used for the Multi Mission Radioisotope Thermoelectric Generators (MMRTG) on the Mars Science Laboratory Curiosity. When using nuclear fission power, shielding against the radiation is needed, which additionally adds to the mass. Another option to ensure radiation protection for the crew would be to bury the reactor beneath the surface, which would involve a high-risk remote operation. An additional concern regarding nuclear fission power is again planetary protection, where contamination of the Martian environment through radioisotopes must be avoided.

Complementary to the stationary power supply near the habitat, a mobile power supply system is required to power the rovers. For this application, a combination of photovoltaic arrays and batteries can be used (Drake & Watts, 2009). This solution however needs for an optimization between the array size, which is still transportable on a rover and the mass of the batteries. The more mass is loaded onto the rover, the more energy is again needed for the rover to drive, thus the desired speed of the rover is also an important component which has to be factored into the calculation, when trying to find the optimum relation between array size, battery mass and time used for a single recharge cycle. Combinations of photovoltaic systems and for instance radioisotope power systems are also an option to optimize the before mentioned calculation. Such combinations would further limit the risk of outages of the photovoltaic systems during sandstorms or reduced efficiency caused by sand in the atmosphere. Due to the low atmospheric pressure, it is furthermore evident that wind driven power sources are not efficient when considering the power output compared to the mass and volume of the technology needed to provide power this way.

Scientific Surface Operations

Driving factors for the first crewed missions to Mars are inter alia scientific rationales and the aim to investigate the Red Planet´s potential to host long-term human settlements. Part of a scientific rationale for crewed missions to Mars is due to the diverse and complex geological features we have already discovered on Mars through robotic missions. These features are, amongst others, indicating that Mars could once have been habitable. In particular, the former presence of liquid water on the surface and a denser atmosphere could potentially have allowed for life to exist on Mars. The question is thus not whether Mars was habitable, but rather whether it was inhabited and along these lines whether it still hosts forms of life. Furthermore, it is not known why the climate on Mars has changes so drastically. The answers to those questions could help understand how life arose on Earth and determine the potential for a sustained human presence on Mars, while providing insights about the evolution of the Martian environment, from which parallels to Earth´s environment could be drawn.

Robotic missions have been providing scientific data from the Red Planet since the Mariner 4 mission successfully conducted the first Flyby in 1965 and delivered the first close-up photographs of Mars. Although several missions and rovers followed up on that, robotic missions to the Martian surface do have their limitations, especially when it comes to dexterity or speed. Both characteristics are essential for conducting (time) efficient surface operations and exploration on Mars. Those skills, however, can be provided by human explorers. Furthermore, a crewed mission could bring expertise in the fields of Geology, Astrobiology and Geophysics along, which would enable a real time assessment of the conducted experiments and samples taken from the Martian surface. This also leads to the advantage of being able to adapt to the findings from the experiments and further improve the scientific operations to maximize the efficiency and scientific output. Due to the fact that a signal from Mars to Earth can take between 4 and 20 minutes, real time support from Earth is not feasible. This, together with the increased need for adaptability of a crewed mission, imposes a challenge on defining and scheduling the activity plan for the crew on Mars. It has to be determined in advance of the mission launch, how much autonomy the crew should have in terms of deviating from the scheduled activities when they deem necessary. The issue of timeline deviations during EVAs was already experienced in the course of the Apollo 14-17 lunar surface EVAs, where only three out of 11 EVAs were finished ahead of schedule. Out of those three EVAs, which were finished ahead of time, two EVAs, the last ones of Apollo 16 and 17 respectively, were rescheduled (Miller, Claybrook, Greenlund, Marquez, & Feigh, 2017). The most significant deviations were observed during traverses and the initiation and execution of science objectives, whereas the purely engineering tasks during the initial overhead, where typically conducted well within the scheduled timeline. This leads to the question, which time increments are feasible and reasonable for planning the surface activities on Mars. Especially, given the fact that real time consultation between the support teams on Earth and the crew on Mars is not possible. When questions regarding the execution of the activities arise, which have an impact on the timeline and fast decisions have to be made, crew autonomy is an important consideration for the scientific operations on a planetary surface with significant interplanetary communication transit time.

Another aspect related to optimizing the scientific output of a crewed mission on Mars is the human robotic interaction. As of now, robotic instruments on Mars are instructed via time delayed commands from Earth and also have to function autonomously for a significant amount of time to use the time on the planet efficiently. The autonomous activities provide the chance of additional scientific findings on one hand, whereas they are also depending on the limited ability of a camera-based system to autonomously identify targets of scientific interest on the other hand. By these means has the AEGIS (Autonomous Exploration for Gathering Increased Science) autonomous targeting system aboard the Curiosity rover significantly increased the rate of ChemCam observations. This intelligent targeting system further ensures a better target selection than through blind targeting (Francis et al., 2017). Nevertheless, a human operator, who is able to take decisions about which target to investigate on the spot while also contributing scientific expertise would further increase the scientific output.

However, rovers are able to explore areas, which are not accessible or too risky for astronauts, undergo long distance exploration while utilizing less resources than astronauts and can take on sampling tasks. Hence the scientific exploration of Mars will benefit from a cooperation between humans and robotic instruments.

Amongst the general aim to search for life on Mars, several goals for the scientific exploration of Mars have been identified by NASA´s Mars Exploration Program Analysis Group (MEPAG) (2015):

- Determine if Mars ever supported life;
- Understand the process and history of climate on Mars;
- Understand the origin and evolution of Mars as a geological system;
- Prepare for human exploration" (p.5).

Together with several sub-goals, these definitions provide an exemplary base for the scientific rational towards the human exploration of Mars. When taking into account that the first crewed mission to Mars is still several years away, those goals might have to be adapted according to findings provided by robotic missions in the meanwhile. By these means could results from the ExoMars or InSight mission influence the search for life by delivering data about the Martian subsurface.

Landing on Mars

The scientific goals of a mission also influence the landing site selection. Additional to that, engineering requirements are limiting the options for potential landing sites, together with the need for regions which can be exploited for ISRU. Based upon these considerations several criteria for potential landing and exploration sites can be derived. It is evident that there might not be a single area which fulfils all criteria and requests at once. Thus, it will be necessary to identify several regions of interest, where each meets one (or more) of the requirements from a scientific, engineering and operational point of view and which lie within a reasonable radius while being a) traversable and b) reachable with a surface mobility unit (rover).

From an engineering perspective, it is not possible to determine the exact longitude and latitude, where the space craft is going to land on the surface beforehand. Environmental conditions like the low density of the atmosphere and its variable dust content, or complex terrain structures (e.g. craters, rocks, slopes, …) pose challenges when narrowing down the exact landing coordinates. Furthermore, the atmosphere is on the one hand not dense enough to significantly lower the descent velocity (especially for large masses), but on the other hand still creates heat through friction. Computer simulations and calculations, which include such variables, yield normally distributed possible landing points following a bell curve. The standard deviations ("sigma") for each of the two directions in the 2-dimensional case, can be interpreted as the axis of an ellipse, with the mean value of this distribution determining the centre of the so-called landing ellipse. Depending on the accuracy of the calculations, the size of the axis of the ellipse can be one, two or three times sigma. The bigger the landing ellipse, the less precise it is, but also the more likely it is that the final landing point lies within it.

A further requirement due to the atmospheric composition is that the landing site shall not lie above +2 km altitude, as the atmospheric density is too low above that altitude to deploy landing mechanisms. The data required for this restriction can be obtained from the Mars Orbiter Laser Altimeter. Furthermore, it is only possible to land systems with a ballistic coefficient below 50 kg/m^2 under subsonic conditions (Braun & Manning, 2007), the necessity for landing systems that can cope with that, represents another issue to overcome when considering landing massive structures on Mars. Compared to the 1 metric ton mass capability of Curiosity, the estimated payloads up to 100 metric tons for human missions to Mars (Drake & Watts, 2009) require the development of new technologies for entry, descent and landing.

In order to limit the potential risk imposed by rocks on the surface, thermal emission measurements in combination with high-resolution imagery offer data on the surface structure and the potential landing sites. Such data is for instance provided by images from the High Resolution Imaging Science Experiment of the Mars Reconnaissance Orbiter, which can already identify features of about 1 m in size (National Aeronautics and Space Administration, 2005). In addition to that, the Thermal Emission Imaging System (THEMIS) aboard the Mars Odyssey spacecraft takes images in the infrared spectrum to determine physical properties of surface features. Besides rocks, surface features like craters and slopes also have to be avoided as they can lead to the spacecraft tipping over or impede post landing deployments.

Based upon the scientific goals for the exploration of Mars, the scientific requests to the exploration zone on the planet can be derived with the same goals in mind. As a starting point for the *First Landing Site/ Exploration Zone Workshop for Human Missions to the Surface of Mars,* the Lunar and Planetary Institute has amongst others formulated the following criteria which are exemplary for the scientific rationales to select a landing site on Mars, which need according to Bussey and Davis (2015):

- *Access to 1) deposits with a high preservation potential for evidence of past habitability and fossil bio signatures and/or 2) sites that are promising for present habitability*
- *Noachian and/or Hesperian rocks in a stratigraphic context [for investigating rock layers] that have a high likelihood of containing trapped atmospheric gasses*
- *Exposures of at least two crustal units that have regional or global extents, that are suitable for radiometric dating, and that have relative ages that sample a significant range of Martian geological time*
- *Access to outcrops with morphological and/or geochemical signatures (with preference for sites that link the two) indicative of aqueous or groundwater/mineral interactions. (p.4)*

Especially regions with remnants from the Noachian period of Mars (beginning from the formation of Mars until 3.5 billion years ago) could provide useful information about the past hydrological processes. Furthermore, the Noachian period was characterized by extensive geological processes (e.g. volcanism) and impacts from the late heavy bombardment, which abated during the Hesperian period and thus can help understanding the geological origin and evolution of Mars.

The final landing site for the first crewed mission to Mars will have to have as many sites as possible within reach from the landing site, which are interesting from a scientific point of view, as well as there must be sites, which can be exploited for ISRU. Furthermore, a long-term exploration strategy needs to be implemented to accommodate the search for areas which could potentially host long-term human settlements. Although such an endeavour still needs substantial research in terms of resource utilization and policy, the search for potential sites for settlements also goes along with the long-term goal to explore the Red Planet. Such areas would need an increased amount of available resources and ideally provide protection from radiation. Lava-tubes or caves would offer natural protection from radiation, but, as of 2018, too little information about the subsurface structure of Mars is known, to determine whether this could be a solution for radiation protection and long-term human presence on Mars.

CONCLUSION

The human exploration of a planetary body requires different reliability and safety precautions for the equipment, technologies and planning operations than a purely robotic mission. This is especially with regards to the requirements for providing a safe environment for the astronauts, which withstands a hostile environment like the one on Mars and planetary protection issues, which should prevent contamination. When planning for a long-duration mission, additional constraints about ISRU, power supply and human physiology have to be accounted for and solutions yet need to be identified.

The Apollo missions in the late 60´s and early 70´s of the last century have provided substantial knowledge and experience from which human spaceflight and exploration can still benefit today. Developments and design studies as for instance for rovers and surface habitats can be used as a starting point for designing the hardware to be implemented for crewed missions to Mars. In addition to that, provide the lessons learned from the activity planning for the EVAs during the Apollo missions to the Moon key aspects to be considered for Mars. However, the time-delay in interplanetary communication poses a new challenge from a planning as well as an operational point of view. Studies conducted through analogue research efforts on Earth can help to overcome this challenge by testing alternatives to voice communication as for instance text-based conversation. Furthermore, through analogue research, the robotic instruments and experiments can be tested prior to their deployment on Mars. This is important as human surface operations on Mars will also be supported by roving vehicles and robotic instruments to enhance the exploration area and efficiency.

From what we know about Mars, its environmental conditions and surface structure today, it is evident that further research is necessary before a crewed mission to Mars can be launched. In particular assessments about the subsurface structure and composition of the Red Planet are still missing in order to prepare the scientific operations and experiments to be conducted by astronauts on Mars. Such assessments are further contributing to the development of the technology, which will be needed for crewed missions to Mars, as for instance the Mars surface habitat, entry, decent and landing systems and launching systems, as well as the EVA suits. What scientist however concur upon today is that a crewed mission to Mars will be feasible in the near future, once the technological and policy issues are resolved.

REFERENCES

Acuña, M. H., Connerney, J. E. P., Wasilewski, P., Lin, R. P., Mitchell, D., Anderson, K. A., ... Ness, N. F. (2001). Magnetic field of Mars: Summary of results from the aerobraking and mapping orbits. *Journal of Geophysical Research*, *106*(E10), 23403–23417. doi:10.1029/2000JE001404

Barlow, N. (2008). *Mars: An introduction to its interior, surface and atmosphere.* Cambridge, UK: University Press Cambridge. doi:10.1017/CBO9780511536069

Braun, R. D., & Manning, R. M. (2007). Mars exploration entry, descent, and landing challenges. *Journal of Spacecraft and Rockets*, *44*(2), 310–323. doi:10.2514/1.25116

Bussey, B., & Davis, R. (2015). *First landing site/ exploration zone workshop for human missions to the surface of Mars: Second announcement.* Retrieved from https://www.nasa.gov/sites/default/files/atoms/files/hls2-2nd-announcement-7-8.pdf

Christensen, P. R. (2003). Formation of recent martian gullies through melting of extensive water-rich snow deposits. *Nature*, *422*(6927), 45–47. doi:10.1038/nature01436 PMID:12594459

Dorizon, S., Ciarletti, V., Plettemeier, D., & Benedix, W. S. (2016). Performance validation of the ExoMars 2018 WISDOM GPR in ice caves, Austria. *Planetary and Space Science*, *120*, 1–14. doi:10.1016/j.pss.2015.10.008

Dorsey, J. T., Chauncey Wu, K., & Smith, R. (2008). *Structural definition and mass estimation of lunar surface habitats for the lunar architecture team phase 2 (LAT-2) study.* Retrieved from https://ntrs.nasa.gov/search.jsp?R=20080013395

Drake, B. G., & Watts, K. D. (2009). *Human exploration of Mars design reference architecture 5.0.* Retrieved form https://www.nasa.gov/sites/default/files/files/NASA-SP-2009-566-ADD2.pdf

Dundas, C. M., Bramson, A. M., Ojha, L., Wray, J. J., Mellon, M. T., Byrne, S., ... Holt, J. W. (2018). Exposed subsurface ice sheets in the martian mid-latitudes. *Science*, *359*(6372), 199–201. doi:10.1126cience.aao1619 PMID:29326269

Francis, R., Estlin, T., Doran, G., Johnstone, S., Gaines, D., Verma, V., ... Bornstein, B. (2017). AEGIS autonomous targeting for ChemCam on Mars Science Laboratory: Deployment and results of initial science team use. *Science Robotics*, *2*(7), 1–12. doi:10.1126cirobotics.aan4582

Groemer, G., Soucek, A., Frischauf, N., Stumptner, W., Ragonig, C., Sams, S., ... Sivenesan, C. (2014). The MARS2013 Mars Analog Mission. *Astrobiology*, *14*(5), 360–376. doi:10.1089/ast.2013.1062 PMID:24823799

Hettrich, S., Dinkelaker, A. N., Sejkora, N., Pfeil, I., Scornet, Q., Moser, L., ... Luger, U. (2014, May). Efficiency analysis of the MARS2013 planning strategy. *Astrobiology*, *14*(5), 377–390. doi:10.1089/ast.2013.1076 PMID:24766437

International Space Exploration Coordination Group (ISECG). (2018, February 2). *The Global Exploration Roadmap.* Retrieved from https://www.globalspaceexploration.org/wordpress/?cat=3

Léveillé, R. J., & Datta, S. (2010). Lava tubes and basaltic caves as astrobiological targets on Earth and Mars: A review. *Planetary and Space Science*, *58*(4), 592–598. doi:10.1016/j.pss.2009.06.004

Mahoney, E. (2009). *Past and present: Field testing for the Moon.* Retrieved from https://www.nasa.gov/exploration/analogs/then-and-now.html

Mars Exploration Program Analysis Group. (2015). *Mars scientific goals, objectives, investigations, and priorities: 2015.* Retrieved from https://mepag.jpl.nasa.gov/reports/MEPAG%20Goals_Document_2015_v18_FINAL.pdf

Miller, J. M., Claybrook, A., Greenlund, S., Marquez, J. J., & Feigh, K. M. (2017). *Operational assessment of Apollo lunar surface extravehicular activity.* Retrieved from https://ntrs.nasa.gov/archive/nasa/casi.ntrs.nasa.gov/20170007261.pdf

National Aeronautics and Space Administration (NASA). (2005). *NASA Facts: Mars Reconnaissance Orbiter.* Retrieved from https://mars.nasa.gov/mro/files/mro/MRO-060303.pdf

Poulet, F., Bibring, J. P., Mustard, J. F., Gendrin, A., Mangold, N., Langevin, Y., ... Gomez, C. (2005). Phyllosilicates on Mars and implications for early martian climate. *Nature*, *438*(7068), 623–627. doi:10.1038/nature04274 PMID:16319882

The 2nd International Space Exploration Forum (ISEF2). (2018, March 7). *Joint Statement.* Retrieved from http://www.mext.go.jp/isef2/pdf/1-e.pdf

Wang, Y., Jing, X., Lv, K., Wu, B., Bai, Y., Luo, Y., ... Li, Y. (2014). During the long way to Mars: Effects of 520 days of confinement (Mars500) on the assessment of affective stimuli and stage alteration in mood and plasma hormone levels. *PLoS One*, *9*(4). PMID:24695321

Williams, D. R. (2016a). *Earth fact sheet.* Retrieved from https://nssdc.gsfc.nasa.gov/planetary/factsheet/earthfact.html

Williams, D. R. (2016b). *Mars fact sheet.* Retrieved from https://nssdc.gsfc.nasa.gov/planetary/factsheet/marsfact.html

ADDITIONAL READING

European Space Agency. (2017). Bulletin. *Astrobiology*, *172*. Retrieved from https://esamultimedia.esa.int/multimedia/publications/ESA-Bulletin-172/Bul172.pdf

Explore Mars, Inc. (2018). The humans to Mars report 2018. Retrieved from https://www.exploremars.org/the-humans-to-mars-report

Hartmann, W. K. (2003). *A traveler's guide to Mars.* New York, NY: Workman Publishing.

Kranz, G. (2000). *Failure is not an option.* New York, NY: Simon & Schuster Paperbacks.

Olson, J., Craig, D., Gresham, E., Hay, J., Graham, R., Mullins, C., . . . Callahan, J. (2011). *NASA's analog missions: Paving the way for space exploration.* Retrieved from https://www.lpi.usra.edu/lunar/strategies/NASA-Analog-Missions-NP-2011-06-395.pdf

Preston, L., Grady, M., & Barber, S. (2012). *Concepts for activities in the field of exploration: The catalogue of planetary analogues.* Retrieved from https://gsp.esa.int/documents/10192/43064675/C4000104716ExS.pdf/733a9456-a578-4ffb-8158-0d6af7f32767

Price, H., Baker, J., & Naderi, F. (2015). A minimal architecture for human journeys to Mars. *New Space*, *3*(2), 73–81. doi:10.1089pace.2015.0018

Race, M. S., Kminek, G., & Rummel, J. D. (2008). Planetary protection and humans on Mars: NASA/ ESA workshop results. *Advances in Space Research*, *42*(6), 1128–1138. doi:10.1016/j.asr.2008.03.003

Soucek, A., Ostkamp, L., & Paternesi, R. (2015). Suited versus unsuited analog astronaut performance using the Aouda.X space suit simulator: The DELTA experiment of MARS2013. *Astrobiology*, *15*(4), 283–290. doi:10.1089/ast.2013.1067 PMID:25811713

KEY TERMS AND DEFINITIONS

Bio Signature: An indication for a biological process which could be linked to the presence of life forms.

Exploration Radius: Maximum range which can be reached for investigations.

Gully: Trenches in the earth occurring from water flows, typically on steep terrain.

Orbiter: A spacecraft designed to operate in orbit around a planetary body.

Rover: A robotic vehicle, which can traverse rough terrain, in particular planetary surfaces.

Space Radiation: High-energy, charged particles present in the outer space environment.

Chapter 15
Human Planetary Exploration:
Legal Aspects

Anja Nakarada Pecujlic
University of Cologne, Germany

ABSTRACT

This chapter examines the legal implications of the ambitious plans to send manned missions to Mars and establish permanent human presence in outer space announced. It addresses the concept of "colonizing planets" and its potential consequences. It offers an overview of the applicable existing legal norms and presents a number of potential legal scenarios concerning the governance of permanent human habitant in outer space. Even though technology is not completely there yet, we should not wait for the space shuttle to be ready to board in order in the nearest future to start discussing potential legal frameworks required to govern multi-planetary existence.

INTRODUCTION

It is the year 2017 and Elon Musk, the CEO of SpaceX presented his ambitious plan to take humans to Mars, that is, for "Making Humans a Multi-Planetary Species". (Musk, 2017) According to Musk, the time has come to "colonize" Mars. However, as exciting and appealing as this plan may sound, there are a number of issues that need to be resolved beforehand in order to get humans to Mars. The issues are above all of a technical nature, but they are not the only challenges that need to be addressed. Far-reaching ethical considerations and vital legal questions require concentrated attention as well.

In relation to a permanent planetary habitat, ethical dilemmas cover a very broad spectrum. They range from environmental reflections (e.g. would it be permissible to perform terraforming of Mars, especially using nuclear bombs to warm it up? What are the environmental repercussions of geo-bioengineering?) to humanity considerations (e.g. how are the first inhabitants to be chosen and who should decide this? Should there be a universal referendum or should private companies have the freedom of choice?). These ethical questions lead to the next logical step, that is, to pragmatic deliberations including, among others, the following questions: what is more desirable, developing a self-sufficient, independent Mars society or one that is Earthly bound? And how would these different scenarios e.g. affect human employment

DOI: 10.4018/978-1-5225-7256-5.ch015

and the distribution of natural resources on Earth? How would it be able to trade and under which conditions? These examples serve only to indicate the range of dilemmas that surround ambitions to make human presence interplanetary and that need to be discussed before such a mission takes place rather than afterwards, particularly since the space mission to Mars raises more fundamental questions than any previous space mission.

While humanity and society considerations are highly relevant and require reflection, they exceed the scope of this book. Therefore, the following chapters will present some of the substantial technical and legal issues related to the envisaged space mission to Mars.

LEGAL BACKGROUND AND NEW LEGAL UNCERTAINTIES

In the first phase of the Space Age, space technology was developing at a slow pace, there were only a few actors and possible space activities were very limited. However, nowadays, things are different. There are numerous new space applications that influence new business models and challenge the current legal framework. Furthermore, in recent years, the rise of private space actors is such that they have achieved a predominant role in the space domain. (Bank of America Merill Lynch, 2017) This dominance initiated a second space race, the race between various players in the private sector competing for national contracts and tenders. In addition, private actors are also competing for media attention. Therefore, they often make "sensationalistic" statements that run contrary to the well-established legal principles governing space activities.[1]

In addition, rivalry tends to spill over from the commercial sector to the state, forcing governmental space agencies to enter the race with the private industry regarding break-through missions, i.e. human space exploration missions. The question as to who is going to be the first to deploy a manned mission to Mars has become another axis of the competition. This fear that the governmental sector has been left behind, has led to several public declarations announcing the refocusing of national space policies, increasingly towards human space exploration. This is exemplified by the United Arab Emirates' (UAE) commitment to accelerate the global development of space exploration and space habitation through projects like UAE Mars City and the Mars 2117; as well as by the US Space Policy Directive 1, which provides for a government-led, integrated program, with private sector partners, for a human return to the Moon, followed by deep-space and missions to Mars. (Science Alert, 2018) (NASA Press Release, 2017)

Even though it is uncertain if private ventures will actually be able to make it to Mars before governmental space agencies, as it is they portrayed in the media, one should keep in mind that the required technology is currently primarily being developed by the private sector. In order for this technology to be utilized, it has to comply with all the legal requirements presently governing the conduct of space activities: meaning that operational features and materials used both need to comply with existing principles. In addition to these legal requirements, break-through missions, whether conducted by national agencies or private companies, often entail numerous legal questions that are not covered by the existing space law treaties (e.g. liability for space debris, ownership of extracted resources, refusal to cooperate etc.). Hence, these type of legal uncertainties may open a dangerous door to a legal grey area that can be misused. For example, some entrepreneurs are claiming that the 'non-appropriation' principle is only applicable to governmental activities, but not to private ones.

For this reason, it is important to identify legal voids from the outset and initiate discussions on how to solve them before any misconduct occurs and injured parties appear. In order to demonstrate what responding to legal uncertainties entails, the following paragraphs will discuss questions related to the usage of the concept of colonialism, *erga omnes* obligations, the applicability of existing international law principles to planetary habitats and the possibility to use other legal regimes in analogy.

IS THE EXPRESSION "COLONIZING PLANETS" LEGALLY APPROPRIATE AND APPLICABLE?

The term "colonizing planets" is exploited considerably in media and it has been mostly used to describe the plan of sending humans to the Red planet. The concept has been used in a rather neutral, value-free, historical manner as if it is a modern invention. The tendency toward historical de-contextualization is the reason why the first question to be taken up in this chapter will be to define the concept "colonizing", its legal implications and applicability in regard to planets.

How Is the Term "Colonization" Defined?

The abundance of existing definitions in theory, encyclopaedias/dictionaries[2] demonstrate that colonialism is a "phenomenon of colossal vagueness". (Osterhammerl, 1997) Vagueness is caused by the historical extent of the colonization practices, the variety of "colonial situations", and the contrasting perceptions of the colonizer and the colonized.

A broad definition considers colonialism to be a "constant in world history". From the ancient Greeks to the peak of European colonial expansion in the nineteenth century, world history is full of examples where one nation gradually enlarged and extended its power by incorporating other territories and moving its people to newly conquered areas.

A more narrow definition pertains to practices of European colonial powers in non-European, overseas territories.[3] This European phase was influenced by the industrial revolution and inspired by the discovery of new markets, new territories where poor migrants could settle, and was driven by the mission to "civilize barbarian nations". It is characterized by the extension of sovereign European power, dominance, rule, to non-European territories lacking sovereignty and is of particular interest to this topic because it problematizes the legitimizing principle of colonialism - the civilizing mission of the European powers and its function in developing the norms of international law. "Colonialism is central to the formation of international law, and, in particular, to its founding concept, sovereignty. International law has always been animated by the civilizing mission, the project of governing and transforming non-European peoples." (Anghie, 2006) Transformation begins with turning sovereignty into a mechanism of exclusion, i.e. proclaiming non-European countries as non-sovereign, which makes them a legitimate object of colonialism. (Anghie, 2004) Whether a broad or narrow view is taken, colonialism as a practice of domination, territorial acquisition, political subjugation, cultural disregard of the native habitants throughout history had many forms that vary from another, depending on the intent, the impact on subjugated indigenous people and the force that was used to accomplish it. (Loomba, 1998) Variations of dominance are expressed in numerous different typologies, e.g. one typology includes colonies of exploitation (resources, land, labour), settlement colonies, and naval colonies (strategically important territories), another distinguishes settler colonies, dependencies, plantation colonies, trading posts, and another also notes settler colonies, but adds administered territories and maritime enclaves. (Osterhammerl, 1997) (Robert, 2001) Despite different types and typologies of colonization, they show that the following features are common to all forms: supremacy, control, and occupation over a territory outside of national borders. In addition, they demonstrate that colonialism is not restricted to a specific time or geographical place, it is a "fundamental phenomenon of world history". (Osterhammerl, 1997)

Definitions also reflect the definer. If the definer is a European, he may still attempt to salvage the civilizing mission, by pointing out that colonialism was not all bad, there were economic and political dimensions that presented an advancement in relation to, for example the infrastructure, and representative institutions. These views are not shared by the Third World and they see colonialism as an exploitative relation, a practice of political oppression and dehumanization on the basis of race. (Aimé, 2000) These differences also have to be taken in consideration when attempting to describe a tone of ordinariness in the use of the term "colonizing planets".

Concerning concretely the call for colonizing Mars, since it is not inhabited by "indigenous people", applying the concept of colonization would suggest that primary interests would be territorial and economic expansionism, i.e. the exploitation of resources and land. This would open legal issues referring to the right of extending sovereignty, the right to appropriation, just distribution of benefits, fairness and due regard to the less technologically advanced countries. However, if increasingly the context of colonizing Mars becomes the pessimistic assessment that the threat of extinction of life on Earth is real, as e.g. Elon Musk claims and Steven Hawkins warned, then the problem of regulating conditions, procedures, and forms of interplanetary presence will increase in urgency. The challenge is whether the old principles of power and self-assigned civilizational superiority will define the new legal norms? Or, whether International law in addressing the new interplanetary challenges will transcend the very principles that have endangered life on Earth and demonstrate species responsibility.

Is There a Legal Right to Colonize?

Following the implications that the ordinary meaning might have for treaty interpretation, in the next step, it will be examined whether the term "colonialism" has a legal notion and what consequences this might entail.

In international law, the legal term "colony" refers to two types of territorial entities. First, it can be a dependent territory subjected to the sovereignty of an independent country. This dependent territorial entity is considered to be part of the "colonizing" in the eyes of other countries. (Definitions, 2018) Second, it can also be a distant territory where group of emigrants or their descendants have settled, but they still keep a close relationship with their parent country or territory. Seeing that "colonizing" imposes certain rights and obligations on a "colonized" territory, and thus has legal implications, the next question is, is it legal to colonize territories at all?

Established in the Age of Discovery (from the 15th to the 18th century), legal principles have been declared that gave expressive rights to European explorers to conquer and colonize new lands. Based on these middle age norms, the majority of the non-European world was colonized under an international legal principle that is known today as the Doctrine of Discovery. (Miller, 2011) (Miller, Ruru, Behrendt & Lindberg, 2010) This doctrine evolved from Papal Bulls and gave to Christian explorers the right to claim, populate and exploit land that was occupied by "pagan" inhabitants. (Miller, LeSage &lopez Escarcena, 2011) Concerning the legal nature of these bulls, it should be mentioned, first, that they were mostly only binding between the parties involved. Second, they served as a legal confirmation of the valid claims made by discovery and occupation, or as licenses to occupy territory. (Brownlie, 1990)

The legal right to occupy foreign territories under the Discovery Doctrine turned into one of the very first international law principles. It was set forth by the United States Supreme Court in a series of decisions; most notably in Johnson v. M'Intosh in 1823 (this case was further cited and relied upon by Australian, New Zealand, and Canadian courts). (D'erico, 2001) (Johnson vs. M'Intosh case, 1823) The Doctrine provided that "Europeans automatically acquired property rights in native lands and gained governmental, political, and commercial rights over the Indigenous inhabitants without their knowledge or consent". (Johnson vs. M'Intosh case, 1823) Additionally, colonists also imposed sovereign governmental powers over the indigenous peoples and their governments, thereby taking control of their international political, economic, commercial, and diplomatic powers.

Although the basis and justification of this legal principle lay in the idea of the religious, cultural, and racial superiority of Europeans and Christians over other peoples and religions, the Doctrine is still utilized today. In international and national laws across the world, it is being occasionally applied against Indigenous nations. The legal reasoning behind this is that certain rights of parties to a dispute derive from legal acts, or treaties concluded, very long ago. "It can be regarded as an established principle of international law that in such cases the situation in question must be appraised, and the treaty interpreted, in the light of the rules of international law as they existed at the time, and not as they exist today". (Sir Fitzmaurice, 1928) Hence, for example, this Doctrine affects United States Indian Law today and has been cited as recently as 2005[4], and in 2007 and 2010, when Russia and China evoked the Doctrine by planting their flags in the Arctic Ocean and South China Sea seabed, thereby claiming commercial and sovereign rights. (Broad, 2010) (Miller, 2007) Setting aside its application, the Doctrine has been severely criticized for being racist, socially unjust, and in breach of basic and fundamental human rights. Therefore, in May 2012, the United Nations Economic and Social Council Permanent Forum on Indigenous Issues declared the Doctrine to be a fundamental violation of indigenous rights and called for a mechanism to investigate historical land claims.[5] However, the Doctrine is still effective to this day. Thus, it is possible that outdated values and reasoning outlive their era and are still embodied in the wording of the principles presently in force. For this reason, it is extremely important to be careful when choosing concepts that could become the basis for establishing future legal principles: especially if these concepts already have a legal meaning that runs contrary to fundamental human rights and also relevant in this case, to the well-established principles of space law.

To put this into the perspective of the legal framework governing outer space activities – outer space is considered as an international commons- only a limited number of international common spaces exist and are recognized by the international community.[6] (Hobe, 2015) These areas are all governed by a specific international framework.[7] In the case of outer space and celestial bodies, that means that they are also subject to a specific international legal regime, which was negotiated and agreed by the international community and which explicitly prohibits appropriation of any kind.[8] Therefore the main characteristics of the legal regime governing conduct in outer space, i.e. the non-appropriation principle, the principle of the freedom of exploration and use, as well as of the scientific investigation of outer space and the principle of peaceful purposes (see below) – all run contrary to the Doctrine of Discovery and in no way should this Doctrine be applicable in outer space. Therefore, the expression "colonizing" Mars should not be used by perspective CEOs nor by the media as it may lead to the wrong conclusion, that the nationality of the first manned mission to Mars will also signify the extension of this national legislation on Mars and that property rights are automatically acquired. Additionally, it could be understood that inequalities between nations and people are reinforced, which is completely contrary to the spirit and objective of the existing *corpus juris spatialis*.

Is There a Legal Right to Colonize Under the United Nations Charter?

Notwithstanding that the Doctrine of Discovery is still applicable as a legal principle, a short overview of the UN's view on colonies and colonialism will be briefly discussed. The UN is the world's biggest political body, where almost all States are represented, and it is in charge of maintaining peace and security. Furthermore the UN Charter is applicable to space activities[9] and the UN sub-organ, UN COPUOS, is responsible for ensuring the peaceful uses of outer space.

Although the UN Charter does not mention "colonialism" *expressis verbis*, this main UN document has been the focus of controversy regarding the legality of colonization. (Quaye, 1991) The main argument supporting the view that the Charter does not prohibit colonialism but in fact expressly recognizes its legitimacy is the following: the UN failed to allow the use of force to eradicate colonialism, and in addition, through the wording of Chapter XI, it stipulates the duty of accountability to the UN for the administration of colonies. However, this interpretation ignores the fact that the Charter, in Chapter I, grants equal rights and the right to self-determination of people[10], and furthermore, colonialism has not been supported by international practice since the foundation of the UN.

Furthermore, the UN adopted the Declaration on the Granting of Independence to Colonial Countries and Peoples in 1960 and founded the United Nations Special Committee on the Situation with Regard to the Implementation of this Declaration. At present, there is a generally accepted view that decolonization is in line with the principles of the Charter, the Universal Declaration of Human Rights, the International Covenant on Civil and Political Rights and relevant Resolutions of the United Nations.[11]

The UN has declared 2010 – 2020 the Third International Decade for the Eradication of Colonialism and has called upon its Member States to speed up the process of decolonization and to finalize the complete elimination of colonialism. (UNGA Resolution 65/119, 2010)

Summarizing the points elaborated above: colonization stands for control, an extension of one nation's sovereignty over territory beyond its borders, and often the acquisition of all the governing power; the derogation of indigenous people's rights and the displacement of natives from their natural habitat. While, it was developed on the basis of the idea of Western Christian supremacy, now it is considered to be a breach of human rights and the UN, with its Member States, has been actively working on eradicating colonialism. All of this leads to the conclusion that the term "colonizing" should not be used lightly, as if it is not burdened by history, nor should it be used when establishing permanent habitats on other planets is advocated. Colonizing has a legal meaning, which indicates that colonies are dependent upon, and (economically, politically, culturally) subordinate to a distant political entity. Colonization and its eradication pervaded 20th century history with violence, political instability and suffering, forcing the international community to acknowledge that colonization would no longer be accepted on Earth. (Gabrynowicz, 2016) Hence, we should from the start be careful not to encourage use of this term in relation to planetary habitats, as the use of the "colony" concept might bring with it a colonial mindset that can contribute to a repetition of the terrestrially abandoned practice.

WHAT ARE THE EXISTING REGULATIONS GOVERNING A PLANETARY HABITAT?

The first part of this chapter focuses on the importance of utilizing the correct terminology in representing unambiguous values when it comes to planetary habitats and on the possible dangers and implications inappropriate terminology might cause. In the second part, this chapter will not deal with concepts, but will examine the existing regulations that are applicable to planetary habitats.

At the very beginning of negotiations relating to space activities, States were focused on the launching of space objects, their operational-life in orbit and return to Earth. No nation then considered the possibility of humans on celestial bodies. Nevertheless, States, very soon after the first discussions, began to raise considerable concern regarding the normative void governing the exploration of celestial bodies and decided to include these regulations in the debate. (Hobe, Stubbe & Troncheti, 2013) (UN Doc.A/AC.105/58, Annex I, 1969)

Currently we can observe a developing trend to surpass the types of missions performed up to now and a move towards more complex solar system exploration, such as sample return and human missions, to Mars for example. As mentioned in Chapter VI, there are considerable legal issues and challenges when it comes to natural resource extraction and the utilization of celestial bodies. As these have been elaborated upon in the previous chapter, the following paragraphs will focus only on legal questions arising from creating a permanent planetary habitat and permanent human habitants.

Unlike natural resource utilization, where ownership questions are highly debatable when it comes to building planetary habitats which are most valuable assets, ownership is expressly addressed by the OST. The State that builds a construction retains jurisdiction and control over such an object.[12] Jurisdiction means the legislative power to enact and enforce laws and rules to persons and objects, and control means the actual possibility to supervise the objects and personnel thereof. (Schmidt-Tedd & Mick, 2009) Hence, questions relating to the rules applicable to the construction itself are already established under the OST regime and do not pose any new legal issues. According to the established regime, if a scientific base was built on the Moon or Mars, each state would retain jurisdiction and control over the part which it contributed, as is the case with the ISS. However, a set of unanswered legal questions arise in regard to habitats and the surface they are built on: what is allowed to be built and what measures and regulations must be taken into account under the current legal regime? For example, can it have purposes other than a scientific basis, are we allowed to build a permanent habitat?

What Is an Erga Omnes Obligation and How Does It Apply to a Planetary Habitat?

There are certain obligations in international law that are considered to be owed to the whole of humanity, called *erga omnes* (from Latin "towards everyone") obligations. (Duhaime's Law Dictionary) (Shelton, 2014) *Erga omnes* obligations means that there are certain obligations of a broad nature that are of interest to everyone (the whole international community) and not only to one injured State. They unveil that simultaneously to national interests, certain interests and values of the international community of States also exist. Interests protected by these obligations can be broadly divided into two groups. The first group includes obligations aiming at regulating the internal behaviour of the State (e.g. human rights violations, which do not affect other states). The second group encompasses obligations safeguarding such important rights, where all states have a potential legal interest in the case of a breach (e.g. a violation of principles protecting the environment).[13] Considering the latter group of obligations, the International Court of Justice (the ICJ) has confirmed that "there are certain Treaties of such a fundamental and broad nature [...] that [they] are incumbent upon States towards the international community as a whole.[14] This is the case with the OST. The object and purpose of the OST recognize the global public interest in outer space by underlying the notion of free access to outer space for all States without discrimination of any kind. (Jakhu, 2006) The scope of the global public interest in outer space encompasses following principles:

- The "common interest" principle;[15]
- The freedom of exploration and use of outer space;[16]
- The prohibition of national appropriation;[17]
- Respect for the rights of other states.[18]

These four principles, stipulated in the very first articles of the OST, form the interest of the international community in outer space and they are reinforced by other articles in the treaty (e.g. the principle of "due regard" in Art IX OST). "It is generally accepted that these principles are not only legal norms of conventional international law but have also become a part of customary international law binding upon all States. Moreover, the global public interest in outer space imposes international obligations *erga omnes* applicable to, and enforceable by, all States." (Jakhu, 2006)

Therefore, it can be concluded that space law regulates the interests of the international community and entails obligations *erga omnes*. This means that all States have an interest in planetary habitat aspirations and plans. This also encompasses states not party to the OST. Thus, one State cannot build whatever/wherever on its own thus disregarding the global public interest in such an endeavour. More precisely, planetary habitat's plans are limited by the inclusive interest of the international community, and any State could claim a violation of *erga omnes* principles if in some way the habitat would affect its e.g. freedom of the use of outer space.

What Are Space Law and International Law Principles Applicable to the Planetary Habitat?

As elaborated above, the OST grants certain rights and they will now be examined in more detail. Rights in Art I OST entail the freedom of exploration and use and they also encompass scientific activities. Explicit beneficiaries of these freedoms are States. However, Art VI OST obliges national governments to authorize and continuously supervise the actions of private actors, thus indirectly transferring freedoms to private companies and corporations, public private partnerships, IGOs and NGOs and individuals.[19] Therefore, an endeavour announced by e.g. a nation like United Arab Emirates or an individual like Elon Musk to inhabit Mars are both covered by Art I of OST.

Notwithstanding this, certain limitations to these freedoms also exist. In this regard, Art IX OST states that "State Parties to the Treaty shall pursue studies of outer space, including the Moon and other celestial bodies, and conduct exploration of them so as to avoid their harmful contamination...". (OST, 1967) Space activities are per se ultra-hazardous activities and the risk they carry is of substantial proportions. For this reason, they require a higher standard of care in order not to cause any harm on Earth and to preserve celestial environments.[20] From identifying activities that could endanger the outer space environment to taking all appropriate measures to prevent harm and/or minimize risk, States are obliged to perform due diligence. (Marchisio, 2009) On the question of what this due diligence entails and what exactly falls under the definition "harmful contamination", Art IX OST is silent. It has been interpreted as a broad concept, covering both intentionally and accidentally caused harm, however without defining specific criteria. More details are provided under Art 7 of the Moon Agreement, which stipulates that "In exploring and using the Moon, States Parties shall take measures to prevent the disruption of the existing balance of its environment, whether by introducing adverse changes in that environment, by its harmful contamination through the introduction of extra-environmental matter or otherwise."[21] However, due to the fact that a small number of States have ratified it, as already mentioned above, its scope of application is very limited.

In addition to space law, international environmental law is also relevant in regard to planetary habitat. As mentioned above, according to the Art III OST, activities related to the exploration and use of outer space are to be carried out in accordance with international law (which encompasses environmental law as well). Article 21 of the Stockholm Declaration and Article 2 of the Rio Declaration, which are seen by most scholars as international customary law (Shelton, 2008), stipulate that States have to ensure that activities within their jurisdiction and control do not damage areas beyond national jurisdiction, such as outer space.[22]

Furthermore, an international non-governmental organization was already established in 1958 by the International Council for Science, called the Committee on Space Research (COSPAR). Its role is to promote scientific research in space on an international level, to encourage the free exchange of results and information, and to provide an open forum for all scientists, for the discussion of problems that may affect space research. In doing so, COSPAR has developed the Planetary Protection Policy with "associated implementation requirements as an international standard to protect against inter-planetary biological and organic contamination, and later a guide to compliance with Article IX of the OST". (COSPAR, 2018) Despite its position as a consultative body of the United Nations, and the fact that these standards are not legally binding, so far space-faring countries have broadly implemented COSPAR policy, as it is perceived as the interpretation of provisions of the OST. (Science, 2018) In practice, these standards have been a base for State practices over the last fifty years.

Lastly, on a national level, some States have set forth in their national legislation a requirement for private actors to obtain a license for space activities to ensure that their planned activity does not cause the contamination of outer space.[23]

Hence, although not conclusive, under the existing legal regime, a positive obligation for States to protect the environment of celestial bodies exists. So for example bringing microbes from Earth to Mars could jeopardize the future science value of celestial bodies for the future of mankind. Furthermore, as one of the biggest dangers in reaching to Mars is successfully landing on the surface, in the unfortunate event of an accident, a crash of a human spaceship would spread contaminating fragments over the planet. This could cause adverse changes to the Mars environment. Therefore, planetary protection needs to be incorporated into the planning of future planetary activities. As Art I OST mentions the exploration and use of outer space as the province of all mankind, there is a strong notion of the preservation of the celestial environment in the treaty. Inter-generation equity should be secured in order not to rob future generations of the knowledge they could gain from a preserved celestial environment. (Hobe, 2009)

Does National Space Legislation Foresee Licenses for Planetary Habitats?

As elaborated above, plans to inhabit planets have to take into account numerous legal principles and abide them. This is applicable to both governmental and private activities. However, for private activities, there might be additional national regulations and requirements that have to be fulfilled. National space legislation is enacted in order for States to regulate private and commercial space activities so as to ensure compliance with their international and obligations under the space treaties, and to prevent and avoid harmful consequences. For this reason, a growing number of nations are adopting national space legislation.

In the example used in this chapter, Elon Musk's ambitious plan to inhabit Mars would, in addition to the international legal framework, also fall under US national space legislation. According to US national space legislation, the Federal Aviation Administration (FAA), part of the Transportation department, is in charge of reviewing launch applications and for issuing licenses. Essential to the grant of a launch license for a private space activity is "payload review" and there are two types of possible licenses:

- A launch-specific license: "a launch-specific license authorizes a licensee to conduct one or more launches, having the same launch parameters, of one type of launch vehicle from one launch site."
- A launch operator license: "a launch operator license authorizes a licensee to conduct launches from one launch site, within a range of launch parameters, of launch vehicles from the same family of vehicles transporting specified classes of payloads." (51 USC § 50904 and 14 CFR § 415.3)

However, the current licensing authority granted to the FAA does not provide for a positive payload review and a subsequent license for a private mission to Mars entailing the "colonization" plans proposed by Musk. (Listner, 2018) Unless this type of mission is performed as a government mission, and thereby does not fall under the FAA licensing process, the appropriate license cannot be issued. Draft legislation with the modification of the current licensing regime was discussed and the "enhanced payload review" option was introduced. This review option would allow the granting of "mission authorization" to approve or reject non-traditional private missions such as the mission proposed by Musk. (Bridenstine, 2018) However, this bill has not even made it to the Space Subcommittee.

Therefore, according to the current US regulations, one of the requirements SpaceX would have to meet is to demonstrate that the Mars mission would not violate US international legal obligations: more specifically, that NASA's planetary protection requirements[24], which are an application of the US obligations under Article IX OST, are complied with. Musk's proposed Mars mission would not fulfil these requirements and would thereby pose a risk of the US being in violation of Article IX OST. Thus, Musk could not obtain the required license at present in the US.

The United Arab Emirates (UAE) are also in the process of enacting their national space legislation, which is supposed to entail human space exploration regulations. Considering UAE repeatedly announced its plan to reach Mars and build a habitat, it will be interesting to see if the new legislation will have specific license provisions for launching and conducting such a mission.

WHICH EARTH-BASED LEGAL REGIME COULD BE USED FOR PLANETARY HABITATS IN ANALOGY?

In the 1960s, when the need to create an independent legal regime to govern space activities, emerged, lawyers turned to the already existing legal frameworks for guidance. As a result, international space law was modelled upon a number of different legal precedents, varying from the UN Charter, the Antarctica Treaty System, maritime law including the Law of the Sea Treaty, aviation law and various Conventions (Warsaw, Chicago, Montreal, etc.) to those agreements regulating the use of nuclear arms and defense. (Hertzfeld, 2009) Therefore, it is plausible that the future legal regime governing habitats on Mars will be based on some existing legal model. But when and how can a legal analogy be used?

Legal reasoning in search of a new legal system that forms the progressive development of international law, is built on two central pillars. Legal precedents, and analogous legal arguments, constitute these two pillars. (Stanford Encyclopedia of Philosophy, 2018)It is justifiable to search for regulatory answers in other legal areas, which share common characteristics that are relevant to the issue at hand. However, at the same time, one must bear in mind that even when legal issues are similar, they are not exactly the same. This is one of the reasons why different legal frameworks were developed for different areas of activities. Thus, one cannot simply transfer one legal regime to another area. Nonetheless, it is prudent to look at the negotiating and drafting history, lessons learned and the possibility of a similar legal system adapting to new challenges when searching for legal inspiration on how to deal with an unregulated sphere of activities. Based on this logic, the following paragraphs will look at a legal regime that could in analogy serve as a model for a future planetary habitat legal framework.

Could the Antarctic Treaty Be Used in an Analogy?[25]

The Antarctic Treaty, the original international treaty for new territories post-WWII was adopted on 1 December 1959, coming into force 18 months later on 23 June 1961.[26] The Treaty was concluded between twelve States: seven States that have made sovereignty claims over the Antarctic territory - Argentina, Chile, the U.K., Norway, France, Australia and New Zealand - and five other States that had an interest in the Antarctic continent – the US, the USSR, South Africa, Belgium and Japan. Despite these explicitly diverging starting positions, the States managed to reach an agreement. They agreed that sovereign claims over Antarctica would not be impeded but "frozen" for the time being. In other words, this meant that the claims were not resolved – claimant States did not have to revoke their claims, and the other States by agreeing to the Treaty, did not recognize such claims as valid. By putting aside (not-resolving) sovereignty claims, this standstill mechanism enabled the necessary collaboration and linkages between claimant states and others, which was essential to embed the primary value of the Antarctic Treaty – the focus on scientific research.

Following, the Antarctic Treaty, States in the following years negotiated and adopted two more Conventions (the Convention for the Conservation of Antarctic Seals and the Convention on the Conservation of Antarctic Marine Living Resources) and one Protocol (the Protocol on Environmental Protection to the Antarctic Treaty). Together, they constitute the Antarctic Treaty System (ATS). This conglomerate of interlinked conventions and protocol, that balances diverging interests, provides a forum for multinational collaboration, and reinforces the importance of science. It has proven to be a productive international law legal mechanism. Science, as a cornerstone, was acknowledged as a method of diplomacy and indeed for the last 50 years a tenth of the Earth has been preserved and regulated peacefully solely in the interest of scientific research. It has become a model for regional environmental management founded upon agreed common values of cooperative scientific research and peaceful purposes, which facilitated unprecedented collaboration e.g. in understanding the causes of the hole in the o-zone layer and the melting of glaciers. (Triggs, 2013) The ATS even demonstrated itself to be bulletproof against some of the modern challenges, as the member states confirmed their commitment to the prohibition on mineral resource exploitation in 2009, instead of giving in to increasing demand for natural resources. This underlines the ATS's ability to adapt constantly and effectively to new challenges.

A potential celestial body that could be inhabited shares a number of common features with the Antarctic: it provides new territory that does not fall under any nation's sovereignty; multiple States with diverging politics have an interest in it; there will potentially be a possibility to acquire natural resources;

and its environment should be protected in order to ensure the possibility of scientific research and the usability for future generations. Therefore, it is not improbable that the potential new legal regime for the planetary habitat could look to the Antarctic Treaty System.

WHICH NEW REGULATIONS COULD POTENTIALLY BE APPLICABLE TO PLANETARY HABITAT?

Once we have the technology in place and are able to reach another planet, safely land, build a habitat (without destroying the existing environment) and settle, the question arises – what laws will govern the life and conduct in this habitat? The matter of the administrative legal regime[27] of space habitats is another noteworthy issue to be considered.

From a broad perspective, three possibilities are thinkable. Either a habitat is built and managed by a single State, or by multiple States or it is completely autonomous from any existing Earth-based State. In another words, extra-terrestrial human habitats might fall under the direct administrative control of Earth-based governments or become politically independent. Following this rational, three possible scenarios will be examined more closely:

Scenario one, planetary habitat is built by a single Earth-based State. In this case, it remains under the jurisdiction and control of this State and the applicable administrative law is this State's national law.

Scenario two, planetary habitat is built by multiple Earth-based States. In this case, we would most likely have a similar governing regime to the ISS model. There would be an international treaty binding for the involved States, regulating conduct in the habitat and every participating nation would retain jurisdiction and control over the part that it built (as elaborated in the ISS chapter).

Scenario three, the planetary habitat becomes a politically independent entity. According to the Montevideo Convention, a state is considered as a person of international law if it fulfills the following four qualifications – it has a:

1. A permanent population;
2. A defined territory;
3. Government; and
4. Capacity to enter into relations with the other states.

If these four necessary requirements are satisfied, a state does not depend on other nations recognizing it. Therefore, even before recognition the permanent planetary habitat would have the right to *"defend its integrity and independence, to provide for its conservation and prosperity, and consequently to organize itself as it sees fit, to legislate upon its interests, administer its services, and to define the jurisdiction and competence of its courts"*. (Montevideo Convention, 1934) In this case, a planetary habitat would have a status as a legal person, with sovereignty over its territory and general authority over its nationals, it wouldSovereignty over its territory and general authority over its nationals be free to define its legislation and acquire property, and it would be able to enter into relationships with other Earth-based nations.[28]

CONCLUSION

This chapter looked into the far-reaching consequences of inappropriate conceptualization, offered an overview of the applicable existing legal norms and presented a number of potential legal scenarios concerning the governance of permanent human habitant in outer space. Regardless of the fact that these considerations are at the moment by and large in the domain of intellectual debate, it is essential to understand which rules of the existing legal body would be applicable, and what models could serve as a good example in the future. Simply because human aspirations to inhabit planets permanently are still in their infancy, one cannot ignore the fact that legal questions regarding future habitants pose a very real problem that must be dealt prior to engaging in extensive and comprehensive planetary activities. Even though we are not there yet, we should not wait for the space shuttle to be ready to board, in order to start asking ourselves – what kind of regulations would be desirable for a future planetary habitant? Furthermore, bearing in mind the ambitious plans of private enterprises to reach Mars by 2030, (The Colonization of Mars, 2018) the legal community needs to follow this pace and begin discussing potential legal frameworks required to govern multi-planetary existence in the near future.

Indeed, this type of progressive development of the space law regime is in line with the ideas embodied at the very creation of the *corpus juris spatialis*. Namely, from the outset of space activities, regulations were created in order to define standards of desired behaviour in outer space, before space actors had the chance to act otherwise. For example, even though there were already a number of military satellites in orbit, the drafters of the Outer Space Treaty agreed to completely demilitarize the Moon and other celestial bodies while no nation had the capability to reach them. This tendency, that was exhibited at the very beginning of the space law era, should also be followed today as well. Thinking through the implications of establishing a permanent presence in space before irreversible changes are made or irreparable damages are caused is simply sensible.

REFERENCES

Agreement Governing the Activities of States on the Moon and Other Celestial Bodies, entered into force July 11, 1984, 1363 U.N.T.S. 3.

Aimé, C. (2000). *Discourse on Colonialism*. Monthly Review Press.

Anglo-Norwegian Fisheries (U.K v. Nor.), 1951 LC.J. 116 (Dec. 18).

Antarctic Treaty of December 1, 1959, entered into force June 23, 1961, 402 U.N.T.S. 71.

Bank of America Merill Lynch Thematic Research: "To Infinity and Beyond – Global Space Primer". October 30, 2017.

Broad, W. J. (2010). China Explores a Rich Frontier, Two Miles Deep. *New York Times*, p. A1.

Brownlie, I. (1990). *Principles of Public International Law*. Oxford University Press.

Canada Island Visit Angers Danes. (2005). BBC News. http://news.bbc.co.uk/2/hi/europe/4715245.stm

Charter of the United Nations, entered into force Oct. 24, 1945, U.N.T.S. XVI.

Check out this incredible Mars City the UAE is building for training purposes. Last accessed May 8, 2018,

Colony Law and Legal Definition. (n.d.). Last accessed January 16, 2018. http://definitions.uslegal.com/c/colony/

COSPAR. (n.d.). Retrieved from https://cosparhq.cnes.fr/about

Crawford, J. (n.d.). *Responsibility to the International Community as a Whole*. Retrieved from http://lcil.law.cam.ac.ukISnyderlectOO(f).doc

D'Errico, P. (2001). Native Americans in America: A Theoretical and Historical Overview. In F. E. Hoxie, P. C. Mancall, & J. H. Merrell (Eds.), *American Nations: Encounters In Indian Country, 1850 To The Present*. New York: Routledge.

Decision City of Sherrill vs. Oneida Indian Nation of N.Y.

Declaration of the United Nations Conference on the Human Environment, U.N. Doc. A/Conf.48/14/Rev. 1, Princ. 21 (1973)

UN Doc.A/AC.105/58

UN Doc. A/AC.105/C.2/L.69,

Doctrine of Discovery". HR/5088. United Nations Economic and Social Council. May 8, 2012.

Does Elon Musk's Plan Violate The Outer Space Treaty - Planetary Protection For Mars After Human Crashes. (n.d.). Retrieved from http://www.science20.com/robert_inventor/does_elon_musks_plan_violate_the_outer_space_treaty_planetary_protection_for_mars_after_human_crashes-180496

Duhaime's Law Dictionary. (n.d.). Retrieved from http://www.duhaime.org/LegalDictionary/E/ErgaOmnes.aspx

Fanon, F. (1967). *The Wretched of the Earth*. London: Penguin.

Freeland, S., & Nakarada Pecujlic, A. (Forthcoming). How do you like your Regulation – Hard or Soft? *The Antarctic Treaty and the Outer Space Treaty Compared.*

Gabčíkovo-Nagymaros Project (Separate Opinion Weeramantry), p. 118

Gabrynowicz, J. I. (2016). Review of the book: The Meaning of Liberty beyond Earth. *Quest., 23*(1).

General Assembly Resolution 1514 (XV) of December 14, 1960.

General Assembly Resolution 65/119, A/RES/65/119, December 10, 2010.

Hertzfeld, H. (2009). Current and Future Issues In International Space Law. *ILSA Journal of International & Comparative Law, 15*(2), 327.

Hobe, S. (2015). Rights and Obligations in the International Commons: The Case of Outer Space. In *IISL Proceedings*. Eleven Publisher.

Jakhu, R. (2006). Legal Issues Relating to the Global Public Interest in Outer Space. *Journal of Space Law, 32*, 31.

Johnson vs. M'Intosh (1823). 9 21 U.S. 543 (8 Wheat.). The case involved land purchases made by British citizens in 1773 and 1775.

Judge Max Huber in the Island of Palmas Case. Netherlands vs. US (1928). Permanent Court of Arbitration. 2 RIAA 829 at 845-6.

Listner, M. J. (2018). *A Legal Look at Elon Musk's Plan to Colonize Mars.* Retrieved from http://www.thespacereview.com/article/3286/1

Miller, R. J. (2007). Finders Keepers in the Arctic? *L.A. Times*, p. A19.

Miller, R. J. (2010). Christianity, American Indians, and the Doctrine of Discovery. In A. Yong & B. Brown Zikmund (Eds.), *Remembering Jamestown: Hard Questions About Christian Mission* (p. 53). Pickwick.

Miller, R. J. (2011). *American Indians, Manifest Destiny, and the Doctrine of Discovery. Wyo. L. Rev.*

Miller, R. J., LeSage, L., & Lopez Escarcena, S. (2011). The International Law of Discovery, Indigenous Peoples, and Chile. *Nebraska Law Review*, *89*(4), 6.

Miller, R. J., Ruru, J., Behrendt, L., & Lindberg, T. (2010). *Discovering Indigenous Lands: The Doctrine of Discovery in the English Colonies.* Oxford University Press. doi:10.1093/acprof:oso/9780199579815.001.0001

Montevideo Convention on the Rights and Duties of States, adopted in Montevideo in 1933, 165 LNTS 19; 49 Stat 3097.

Musk, E. (2017). Making Humans a Multi-Planetary Species. *New Space.*, *5*(2), 46–61. doi:10.1089pace.2017.29009.emu

New Space Policy Directive Calls for Human Expansion across Solar System. (2017). Retrieved from https://www.nasa.gov/press-release/new-space-policy-directive-calls-for-human-expansion-across-solar-system

New World Encyclopedia. (n.d.). Retrieved from http://www.newworldencyclopedia.org/entry/Colonialism

Newcomb, S. T. (2008). *Pagans in the Promised Land: Decoding the Doctrine of Christian Discovery.* Fulcrum Golden.

Osterhammerl, J. (1997). *Colonialism: A Theoretical Overview.* Markus Winer Publishers.

NASA Policy Directive (NPD) 8020.7G.

NASA Procedural Requirements (NPR) 8020.12.

Protocol on Environmental Protection to the Antarctic Treaty, adopted on October.4, 1991, entered into force on January 14, 1998.

Quaye, C. O. (1991). *Liberation Struggles in International Law.* Temple University Press.

Representative Jim Bridenstine (now NASA administrator) had proposed this: H.R. (discussion draft), F:\M14\BRIDEN\BRIDEN_160.XML, 114th Congress, 2D Session. (n.d.). Retrieved from http://www.spacepolicyonline.com/pages/images/stories/Bridenstine%20discussion%20draft%20sept%2014%20 2016.pdf

Restatement (Third) Of Foreign Relations Law of the United States, § 201 (1987).

1992. Rio Declaration on Environment and Development, UN Doc. A/CONF.151/26 (vol. I)

Shelton, D. (2014). International Law and 'Relative Normativity'. In M.D. Evans (Ed.), International Law (4th ed.). Oxford, UK: Oxford University Press.

Stanford Encyclopedia of Philosophy. (n.d.). Retrieved from https://plato.stanford.edu/entries/colonialism/

The Colonization of Mars: What Legal Issues Will Arise out of a Multi-Planetary Existence? (n.d.). Retrieved from http://www.jhtl.org/blog/the-colonization-of-mars-what-legal-issues-will-arise-out-of-a-multi-planetary-existence/

The Space Settlement Institute. (n.d.). *Lunar Land Claims Recognition*. Retrieved from http://www.space-settlement-institute.org/strategy.html

Third International Decade for the Eradication of Colonialism, Report of the Secretary-General

Traction, B. Light and Power Company, Limited, second Phase, Judgment, ICJ reports 1970, p. 3, para 33

Treaty on principles Governing the Activities of States in the Exploration and Use of Outer Space, Including the Moon and Other Celestial Bodies, entered into force Oct. 10, 197, 18 U.N.T.S. 205.

Triggs, G. (2013). The Antarctic Treaty System: A Model of Legal Creativity and Cooperation. Washington, DC: Smithsonian Institution Scholarly Press.

UK's Outer Space Act of 1986.

United Nations Convention on the Law of the Sea, Montego Bay, done December 10, 1982, entered into force November 16, 1994; 1833 U.N.T.S. 3.

51. U.S. Code § 50904 and 14 CFR § 415.3 (a).

Wolfrum, R. (1984). Die Internationalisierung Staatsfreier Räume. In Veröffentlichungen des Max-Planck-Instituts für ausländisches öffentliches Recht und Völkerrecht (Band 85). Academic Press.

ENDNOTES

1 E.g. the private sector is claiming that "International law bans governments from owning land on the Moon [or Mars], but private entities could legally own such land", The Space Settlement Institute, Lunar Land Claims Recognition. Last accessed January 15, 2018, http://www.space-settlement-institute.org/strategy.html.

2 For example, the Stanford Encyclopedia of Philosophy points out the blurred lines between "colonialism" and "imperialism", and for a better understanding of the two, it refers to "colonialism" as

"a broad concept that refers to the project of European political domination from the sixteenth to the twentieth centuries that ended with the national liberation movements of the 1960s", "Colonialism". Stanford Encyclopedia of Philosophy. Last accessed January 16, 2018. https://plato.stanford.edu/entries/colonialism/; The Cambridge dictionary defines it as "a belief in and [the] support system of one country controlling another", "Colonialism". Cambridge Dictionary. Last accessed January 16, 2018. https://dictionary.cambridge.org/dictionary/english/colonialism; whereas in the New World Encyclopedia and Oxford Dictionary the term "colonialism" stands for "[the] policy or practice of [the] extension of a nation's sovereignty over territory beyond its borders, occupying it with settlers or forming administrative dependencies, ruling or displacing indigenous populations, and exploiting it economically", "Colonialism". New World Encyclopedia. Last accessed January 16, 2018. http://www.newworldencyclopedia.org/entry/Colonialism.; Oxford Dictionary. https://en.oxforddictionaries.com/definition/colonialism. Last accessed January 16, 2018.

3 See further Kämmerer, J.A. "Whereas colonization, a form of conquest usually involving alien dominance and subjugation, has been known to all historic eras, colonialism at least in its narrower and frequently implied meaning, reaches beyond the mere taking of foreign land and peoples. Generally speaking, it refers to a concerted racist of, in most cases, European States dividing up other inhabited parts of the world amongst themselves". Oxford Public International law: Colonialism.

4 Decision City of Sherrill vs. Oneida Indian Nation Of N.Y: "Under the 'doctrine of discovery...' fee title (ownership) to the lands occupied by Indians when the colonists arrived became vested in the sovereign-first discovering European nation and later the original states and the United States."

5 "Preliminary study of the impact on indigenous peoples of the international legal construct known as the Doctrine of Discovery", United Nations Economic and Social Council. February 4, 2010; "Doctrine of Discovery". HR/5088. United Nations Economic and Social Council. May 8, 2012.

6 These are areas of the world that like the High Seas, the Deep Seabed, Antarctica and Outer Space, are not subject to the sovereignty of states. For a general account on common spaces outside national jurisdiction see: Wolfrum, R. (1984). "Die Internationalisierung Staatsfreier Räume". In: Veröffentlichungen des Max- Planck-Instituts für ausländisches öffentliches Recht und Völkerrecht (Band 85), Berlin/Heidelberg.

7 E.g. United Nations Convention on the Law of the Sea, Montego Bay, done December 10, 1982, entered into force November 16, 1994; 1833 UNTS 3 (referred to hereinafter as 'UNCLOS'); Antarctic Treaty of December 1, 1959, entered into force June 23, 1961, 402 U.N.T.S. 71.; Protocol on Environmental Protection to the Antarctic Treaty, adopted on October.4, 1991, entered into force on January 14, 1998.

8 Art II OST provides as follows: "Outer space, including the Moon and other celestial bodies, is not subject to national appropriation by claim of sovereignty, by means of use or occupation, or by other means".

9 Art III OST, provides as follows: "States Parties to the Treaty shall carry on activities in the exploration and use of outer space, including the moon and other celestial bodies, in accordance with international law, including the Charter of the United Nations…"

10 Art I (2) UN Charta provides as follows: "To develop friendly relations among nations based on respect for the principle of equal rights and self-determination of peoples, and to take other appropriate measures to strengthen universal peace.

11 Seventieth session Item 63 of the preliminary list, Implementation of the Declaration on the Granting of Independence to Colonial Countries and Peoples, Third International Decade for the Eradication of Colonialism, Report of the Secretary-General.

12 Art VIII OST provides as follows: "A State Party to the Treaty on whose registry an object launched into outer space is carried shall retain jurisdiction and control over such object, and over any personnel thereof, while in outer space or on a celestial body. Ownership of objects launched into outer space, including objects landed or constructed on a celestial body, and of their component parts, is not affected by their presence in outer space or on a celestial body or by their return to the Earth."

13 Barcelona Traction, Light and Power Company, Limited, second Phase, Judgment, ICJ reports 1970, p. 3, para 33; Gabčíkovo-Nagymaros Project (Separate Opinion Weeramantry), p. 118; Proclamation 2 UN Declaration on the Human Environment, 1972; Principle 7 Rio Declaration on Environment and Development, 1992.

14 Barcelona Traction, Light and Power Company, Limited (Belg. v. Spain), 1970 I.C.J. 3 (Feb. 5). For a detailed analysis, see Ragazzi, M. (2000). The Appearance of the Concept of Obligations Erga Omnes on the Agenda: The Dictum of the International Court in the Barcelona Traction Case. In The Concept of International Obligations Erga Omnes. Oxford University Press; Crawford, J. Responsibility to the International Community as a Whole.

15 Art I para 1 OST provides as follows: "The exploration and use of outer space, including the moon and other celestial bodies, shall be carried out for the benefit and in the interests of all countries, irrespective of their degree of economic or scientific development, and shall be the province of all mankind."

16 Art I para 2 OST provides as follows: "Outer space, including the moon and other celestial bodies, shall be free for exploration and use by all States without discrimination of any kind, on a basis of equality and in accordance with international law, and there shall be free access to all areas of celestial bodies."

17 Art II OST provides as follows: "Outer space, including the moon and other celestial bodies, is not subject to national appropriation by claim of sovereignty, by means of use or occupation, or by any other means."

18 Rules of general international law are applicable to the activities in outer space according to Art III OST, which provides as follows: "States Parties to the Treaty shall carry on activities in the exploration and use of outer space, including the moon and other celestial bodies, in accordance with international law…", and hence States may exercise their rights in such a way so as not to infringe upon similar rights of other States (see further Anglo-Norwegian Fisheries (U.K v. Nor.), 1951 LC.J. 116 (Dec. 18), and Brownlie, I. (2003). Principles of Public International Law 301. Oxford University Press.

19 Art VI OST provides as follows: "States Parties to the Treaty shall bear international responsibility for national activities in outer space, including the moon and other celestial bodies, whether such activities are carried on by governmental agencies or by non-governmental entities, and for assuring that national activities are carried out in conformity with the provisions set forth in the present Treaty."

20 This type of contamination is called "forward contamination" and it signifies the transfer of viable organisms from Earth to another celestial body. There is also "back contamination", which means the transfer of "Extraterrestrial life" organisms, if such entities exist, back to the Earth's biosphere. However, this is out of the scope of this chapter.

21 The Moon Agreement is also pertinent to other celestial bodies according to Art 1, Moon Agreement.

22 Declaration of the United Nations Conference on the Human Environment, U.N. Doc. A/Conf.48/14/Rev. 1, Princ. 21 (1973) [hereinafter Stockholm Declaration]; 1992 Rio Declaration on Environment and Development, UN Doc. A/CONF.151/26 (vol. I), Art. 3 (June 14, 1992) [hereinafter Rio Declaration].

23 E.g. Section 5 of the UK's Outer Space Act of 1986.

24 Compliance with planetary protection is mandatory for solar system missions, as stipulated in the NASA Policy Directive (NPD) 8020.7G. A brief summary of the requirements for orbiters, flyby missions, landers, and rovers, along with many of the implementation practices required to achieve compliance, are further detailed in the NASA Procedural Requirements (NPR) 8020.12.

25 This paragraph is a part of a paper by Freeland S. & Nakarada Pecujlic, A. How do you like your Regulation – Hard or Soft? The Antarctic Treaty and the Outer Space Treaty Compared. Forthcoming. National Law School of India.

26 The Antarctic Treaty was signed in Washington on 1 December 1959 by the 12 countries whose scientists had been active in and around Antarctica during the International Geophysical Year of 1957-58. As at October 2015, the total number of State Parties to the Treaty was 53; Secretariat of the Antarctic Treaty.

27 Administrative law is the body of rules and procedures that organizes government and provides mechanisms for redress of grievances as a result of decisions or the actions of government.

28 Restatement (Third) Of Foreign Relations Law of the United States, § 201 (1987).

Chapter 16
Space Policy:
Understanding Space Activities, Motivations and Programs

Arne Lahcen
European Space Agency (ESA), Italy

Djordje Andrijasevic
Federal Ministry of Defence (BMLV), Austria

ABSTRACT

This chapter explores the concept of space policy and sheds light on its importance in understanding the nature and evolution of space activities and programs. More specifically, this chapter seeks to offer an introduction on how space policy can be defined and understood, followed by an overview of the main current trajectories of space policies of both national actors as well as international and intergovernmental institutions. Finally, this chapter formulates some conclusions regarding the future potential avenues of space policy and describes how these scenarios could impact the future of human activities and behaviour in space.

INTRODUCTION

Evolved in Africa over the course of millions of years, humans have ever since ventured into new territories, pushing boundaries wherever they would go. At some point in the distant past, humanity invented vessels to traverse rivers, seas and eventually oceans, allowing us to venture beyond the first environment – the land – into the vast open second environment which makes up over 70% of our planet's surface. Breakthroughs like this have paid off, because they expanded our options, thus increasing our chances of survival dramatically. In the beginning of the 20th century, we added a third environment to that list. With the invention of the airplane, we finally succeeded in realising the dream of flying, enabling us to venture into skies which for so long has been so close yet so far out of reach. Just a few decades after that, in the middle of the 20th century, an even bolder dream was realised, when we achieved the 'unachievable': venturing beyond the confined surface of our planet. At that memorable moment of the first satellite launch, we entered a new era: the Space Age.

DOI: 10.4018/978-1-5225-7256-5.ch016

Today, more than 60 years since the start of the Space Age, the world is witnessing an ever-growing interests of states and private actors to become engaged in space activities. The club of spacefaring nations and its set of capabilities is constantly expanding. At the same time, strong innovation dynamics are drastically reshaping the way in which many space activities are conducted. All these developments are changing the playing field for space activities, programmes and capabilities, bringing space ever closer to the heart of a vastly inter-connected society. As a result, we are relying ever more on space assets, increasing their strategic importance for humanity. As space activities become increasingly strategically important, however, the risk and potential impact of disagreement is on the rise too. All these recent trends are expected to continue – if not intensify – in the foreseeable future.

This chapter explores the concept of space policy and sheds light on its importance in understanding the nature and evolution of space activities and programs. More specifically, this chapter seeks to offer an introduction on how space policy can be defined and understood, followed by an overview of the main current trajectories of space policies of both national actors as well as international and intergovernmental institutions. Finally, this chapter formulates some conclusions regarding the future potential avenues of space policy and describes how these scenarios could impact the future of human activities and behaviour in space.

Throughout this chapter, particular attention is given to the reasons why space policies have been and are evolving. In doing so, the diverse driving forces that give rise to these changes – which range from geopolitics, over economics to technological innovation – are illustrated and explained.

Considering that space activities are typically conducted in a highly specialised and segmented fashion, the authors hope that the interdisciplinary nature of space policy can help readers in getting a better overview and insight into the historic development, current trends and future directions in space.

DIMENSIONS OF SPACE POLICY

Before this chapter focusses on the specific contents and historic developments of space policies around the world, it is crucial to characterise and define the generic nature of space policy. Explaining the lenses through which space policies should be understood is crucial in order to grasp the true meaning and rationale behind them.

Since the term 'policy' is widely used in contemporary society, it is important to look at its generic substance before elaborating on the specificities of space policies. Generally speaking, 'policy' can be defined as "*a set of ideas or plans of what to do in particular situations which has been agreed upon officially by a group of people, a business organization, a government, or a political party*"[1]. Policies formulated by the executive or legislative branch of government – as is typically the case for space and other policies at political level – are therefore a very important basis for making decisions. In fact, policies serve multiple purposes at once, thanks to their different 'dimensions'.

Firstly, policies have a very important 'communicative' dimension: they are typically used to communicate and motivate a cohesive set of principles or vision adhered to by a government in devising and implementing its decisions and resources in a particular field. In this regard, space policies are only a very small field of national policies throughout countries, especially compared to other fields such policies on energy, poverty, mobility, etc. In fact, many countries in the world do not have dedicated space policies and among those who do have one, most are fairly recent. Since space activities are gaining importance throughout the world in the last couple of years – as will be described in more detail further on – so has increased the need for governments to devise policies to express their understanding, vision and roadmap regarding their space activities.

Since policy documents are most of the time readily accessible for anyone to read, their communicative nature serves a binding function. This plays out both on national and international level. On national level, policies are intended to converge the understanding of what actions to take and which roads to travel for all relevant parties involved, such as: public space administration, space industry, other governmental branches, universities and research institutes, NGOs, citizens, etc. On the international level, public space policies allow international partners and actors to take note of a state or region's plans regarding future space activities and programmes. This creates clarity on states' future intentions in space, but it also opens up opportunities for states to react, for instance by looking into mechanisms of cooperation and joint actions, activities or programmes. Especially now that more actors are establishing space capabilities and, given that we are witnessing the dawn of many operational services which will be around for decades to come, the incentives for international cooperation are fundamentally altered for the better. This will be addressed further in this chapter.

Secondly, space policies typically have a strong 'strategic' dimension. This strategic nature of space and other policies finds its roots in the various elements they seek to incorporate and reflect. Evidently, governments want their policies to set out a coherent, sound and sensible vision for the future, preferably one that is based upon the specific needs and requirements within the field in question. In other words, governments want their space policies to be instrumental in achieving rational, effective and efficient outcomes concerning their actions and investments. This is one of the reasons why many (space) policy documents will make ample reference to recent trends, relevant achievements and upcoming challenges. It is also one of the reasons why policy documents are updated on a fairly regular basis, typically every few years. New governments often seek to reorient existing objectives or set new ones. Alternatively, dynamic changes within the playing field for space activities can lead a government to re-formulate its policies, allowing them to keep their grip onto a rapidly changing domain and its activities and investments.

Policy documents often include objectives, such as goals and targets, for the various elements of the policy domain in question. Usually these objectives are not binding, but they are rather set out as desired milestones or future achievements, and in the case of space policy, they can be expressed fairly vaguely. One of the reasons for this is that in the case of space, objectives can be very divergent because of the diversity both in space activities and the national interests that need to be served.

Firstly, what is referred to as 'space activities' is actually a set of diverse activities with differing objectives, including: space science and exploration, human spaceflight, satellite navigation, Earth observation (EO), space transportation (including but not limited to launchers), Space Situational Awareness (SSA), etc.

These activities, though all space-related, have different requirements and needs. Certain activities are very much focused on innovation and novelty, such as science and explorations, which continuously seek to venture into new terrains to make new discoveries. Certain activities in other fields, such as Earth Observation and satellite communication, tend to rely on continuity and therefore have stronger operational needs. Some activities like space transportation were part of the space endeavours at the outset of the Space Age, other activities, such as Space Situation Awareness, are fairly novel. Domains such as space resource exploitation are likely to arise as new activities in the future, provided the required technologies mature to a point where the activity becomes economically viable. Moreover, the technology involved in all these domains is very divergent too. The different stakeholders within society

Table 1. The Main space activities domains

Human Spaceflight	Space Science	Space Exploration
Earth Observation	Satellite Communication	Satellite Navigation
Space Situational Awareness	Space Transportation	Space Resource Exploitation (in the future?)

Source: Authors' visualisation

for each of these fields thus often end up competing for resources, especially given that resources from space activities often come from the same source.

In space, size does matter. Depending on the nature of a space programme and the resources available, decision makers will thus prioritise certain fields of activities in their space policy and describe which direction they want to go in the various space application domains. Countries that do not have the budget to finance a comprehensive space programme will typically make a trade-off and clearly prioritise which achievements are to be reaped in the near future. This is typically the case for smaller countries, or countries with a small space programme. Factors that might steer their decision in one way or another might include the presence of expertise or specialisation they have built up nationally or new application areas they want to develop. Large countries with substantial space programmes, such as the U.S., Russia and China, do not necessarily have to make very distinct choices for one domain over the other, but they often use their policies to express a change of direction or approach, or to signal a changed focus of one particular space activity.

In addition to their diversity, space activities also seek to serve various national strategic interest. One of the primary objectives of any functioning state is to ensure its existence and continuity in the future. States will go to any length to achieve this and will therefore pursue activities which are in their self-interest, especially when these can ensure and further boost their future success and continuity. These activities include the development of capabilities and power tools that give them a competitive edge or advantage over other (rivalry) states. Space is a field, or perhaps rather a domain, which serves and supports many crucial strategic national interests, as summarised in Table 2.

In the 21st century, space activities are crucial to the functioning of everyday life and many societal activities would not be able to continue to function as smoothly as they do without them. Space activities are very reliant on strong R&D and thus offer a vast technological potential for state who develop them, both in terms of direct use for future space applications, as well as for the spin-in and spin-out of

Table 2. National strategic interests served by space

	Science	**Technology**	**Autonomy**	**Commerce & Economy**	**Safety & Security**
Main Actors	Scientists	Engineers	Politicians	Business	Defence & Intelligence
Main Fields	Space Science, EO and Space Exploration	All fields	All but in particular Launchers and Space Applications	All, but in particular Space Applications	Launchers, Space Applications and SSA
Objectives	Increased knowledge	Innovation	Power and Prestige	Profit and Growth	Safety & Security

Source: Author's visualization

technologies. The latter allows states to innovate their industrial and knowledge sectors, thereby increasing competitiveness and boosting domestic capabilities of various nature. The vast amounts of scientific data generated by space research and exploration is instrumental in putting countries at the forefront of international scientific research, especially in the highly knowledge-driven economy of today. Together, the space application domains (satellite telecommunication or 'satcom', Earth observation and satellite navigation) generate and support the functioning of modern commerce, agriculture, environmental management, urban infrastructure and, logistics and mobility. In addition, space assets are crucial in operating today's safety and security infrastructures, both for civilian and military purposes.

All in all, this makes that space assets have an incredible value in terms of national autonomy. Any large industrial state genuinely interested in ensuring its ability for self-preservation will therefore pursue a comprehensive space programme, making it independent from others for accessing all these capabilities. This is one of the reasons why there are so many parallel capabilities in space today: many players have their own launching capabilities, their own space research missions, parallel but independent observation and navigation systems, etc.

Together, all these benefits give space programmes a certain allure, one that is associated with power, modernism, progress, and wealth. This reinforces the image of space activities as the ultimate display of national development, and it is an important factor in the motivation and will of many smaller countries that seek to get involved in space activities today. Furthermore, small countries are attracted by space because it offers them a potential to find a niche in a global economy which is highly reliant upon specialisation, added-value, knowledge and technological innovation.

Space policies have to make sense of all this and can be seen as a useful and easily accessible tool to look into the motivation of states for developing space capabilities and investing in space programmes.

As with nearly any field, there has been a huge historical development in space policies over the years. In the European Space Agency context, humanity's relationship with space is often delineated into four distinct eras each characterised by a dominant paradigm. The changing of dominant paradigms over time in this categorisation has led humanity to conduct and think about space activities, programmes and policies in a different way: from Space 1.0 all the way to the current Space 4.0 era. More specifically, Space 1.0 is defined as the era of astronomy, when human gauged into the universe through their eyes and later on through telescopes. The dawn of the Space Age coincides with Space 2.0 at a time when the Soviet Union launched the first spacecraft ever into orbit: Sputnik in 1957. During that time, space activities worldwide were conducted in a great sense of rivalry between the major power blocks of the time, the US and the USSR, with the understanding that a weaponization of the space environment had to be avoided in the interest of humanity as a whole. Instead, the two power blocks competed each other by mean of the Space Race, which culminated into humanity's first ever visits to the Moon. As the Berlin Wall fell and the Cold War ended in 1991, the dynamic within space policies shifted from rivalry to increased cooperation and multilateralism, and eventually the Russians also started participating in what so far has been the most expensive object ever build in the history of humanity: the International Space Station (ISS). In more recent times, the major paradigm has shifted once again. In recent years, space utilisation, applications and technologies have diversified immensely, and space activities are becoming a crucial and indispensable activity within the interconnected global society of the 21st century.

As space technology and capabilities have expanded vastly since the launch of the Sputnik in 1957, so have the motivations for nations to be active in space altered and broadened. Initially, nations and their policies were very concerned with human spaceflight, space exploration and security aspects. In the early days of spaceflight, there was a great wave of optimism and believe in the future ensuing from

Figure 1. The four eras of humanity's space activities

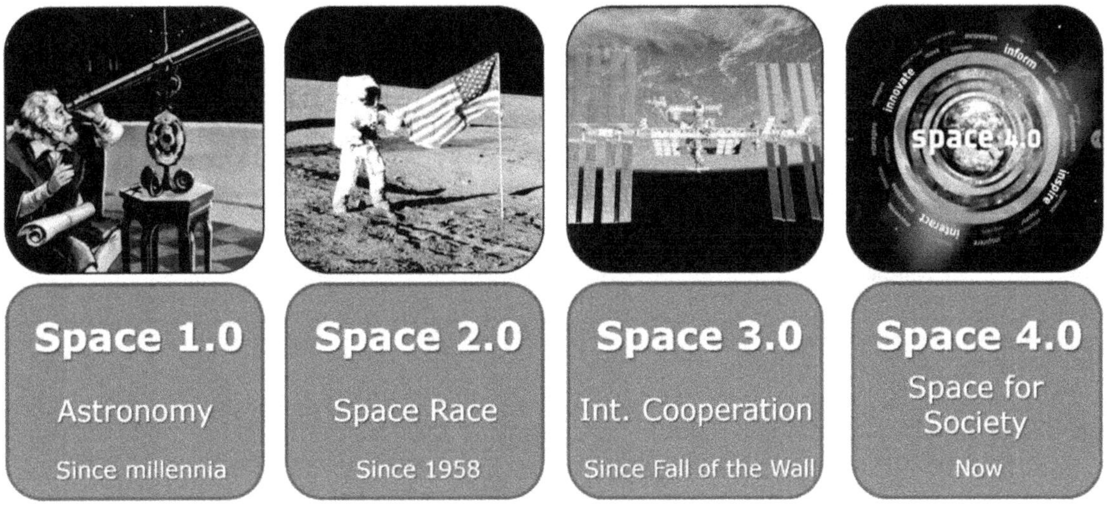

what was believed to be unprecedented technological progress at the time.[2] However, this optimism was accompanied by a great deal of concern over security matters, as launchers were in fact spin-off technology that resulted from the advancements in InterContinental Ballistic Missile (ICBM) capabilities able to deliver nuclear warheads or conventional weapons anywhere on the planet within a matter of hours.[3]

As the space pioneers at the time[4] clearly understood the vast potential offered by this new environment, technologies quickly matured, and space programmes became more diverse and comprehensive in terms of applications and interests served throughout the 1960s, 1970s and 1980s. This trend was reflected in space policies, which expanded in scope and approach and gradually started to emphasise more the role of government in spurring R&D and in providing new experimental missions and later on operational systems. In this regard, space policies, and in particular the space programmes they overarched, have made a strong distinction between different space applications. These distinctions, which go back on the economic features that characterise the different activities, are gaining ever more traction and are leading many space nations to heavily diversify and tailor their approaches, investments and policies for a given field.

In economic theory, goods and services can be classified into four categories based upon two characteristics: rivalry and, excludability. Rivalry refers to the declining benefit or availability of a good or service as a function of the number of consumers. Schoolbook examples of rival goods include food and clothes, i.e. items that can only be consumed by one person at the time. Non-rival goods, for instance, include radio, movies and fireworks – good that can be enjoyed or consumed by many people at once. Accessibility, on the other hand, refers to the possibility of restricting access to a certain good or service. Together, these two parameters give rise to four types of goods, as displayed in the table down: common-pool goods, public goods, private goods and, club goods. Space activities, as well as the space environment, can now be mapped onto the existing categories based upon their characteristics, as illustrated in Table 3.

Certain applications, such as satellite communication, have a strong disposition towards commercial viability. Once these capabilities were proven feasible and demonstrated – through public funding which had financed the early and riskiest phases of technology development – the market largely took over

Table 3. The economic classification of space activities and the space environment

		Rivalry	
		Rival	**Non-rival**
Accessibility	**Non-excludable**	Common-Pool Goods	Public Goods
		Near-Earth space (e.g. commonly used Earth orbits and orbital slots, frequency spectrum)	Space at large (e.g. for planetary and solar system exploration)
	Excludable	Private Goods	Club Goods
		· Satellite communication · Launchers · Human Spaceflight	· Satellite Navigation · Earth Observation data · Space Science

this activity from the public sector. Indeed, public agencies still fund for R&D regarding telecom and certain related activities, but the public sector (i.e. space agencies) is no longer solely responsible for the provision of satcom and broadcasting services to society, as market mechanisms have largely taken over this function.

For satcom, and to a certain extent also for launchers[5], this has been possible because they are private goods: their use or consumption is both rival and excludable, making it possible for market mechanisms to function based upon supply and demand. Human spaceflight, although a private good, has been historically funded by governments, because the market dynamics could not work due to the extreme high risks and related cost involved. Recently, however, this has slowly started changing, as several entrepreneurs and companies are developing private Human Spaceflight capabilities which they seek to offer on the market. These entrepreneurs hope to significantly reduce the cost of these activities through technological innovation, thereby allowing the mechanisms of supply and demand to start functioning – at least for the market segment of dollar billionaires and multi-millionaires. In many space policies, these market dynamics are implicitly or explicitly recognised, and governments actively pursue private sector involvement and development in these areas.

For the so-called 'Club Goods' of space activities, governments have typically followed a very different policy path. Satellite navigation signals, Earth Observation data and space science data are technically all excludable through restricted distribution and encryption. At the same time, they are non-rival, since digital data can be copied infinitely at very low cost, while GNSS signals can be used by billions of users simultaneously. However, this has not led governments to commercialise or restrict access to these services. In fact, these services have historically grown to be treated like public goods considering that governments have not enacted their excludability. This 'decision' was not so much a one-time event, but rather has grown historically over time. In the 1980s, the realisation grew that governments have a strong role to play in spurring general-purpose R&D, since it was then understood that these innovations would lead to long term productivity growth through society and the economy. More recently, this idea has further growth into the realisation that governments are actually 'market creators'. Public investments made in fields characterised by market failures[6] will eventually create new products and services that in turn will constitute entirely new markets. In this spirit, it was realised that the societal benefits of investments made in space science, Earth Observation and satellite navigation systems would be far greater if their services would be made available quite freely, rather than restricting access through price-setting or administrative burdens. That is the reason why today space science data is available to universities and researchers and why we can use GNSS signals free of cost. It is also the reason why many EO missions and constellations have a fairly free and open data policy and why so many EO data exchange agreements and practices are in place worldwide.

In brief, it is fair to say that space policies today strongly reflect the vast complexity of space in terms of applications, domains and national interests. This includes motivations in terms of geopolitics, economics and science and technology. These elements should be kept in mind when reading or analysing space policy documents. Governments have increasingly given more weight and importance to more comprehensive policies, as it allows them to ensure the right degree of sector support and interference. Furthermore, space policy facilitates an evolving approach of intervention and guidance over time, as market dynamics change thanks to technological and societal changes.

Awareness of these dynamics is important because they are crucial to assess future potential courses of evolution by taking into account current trends and developments. If we understand how and why space policies are formed in the way they are, then we can anticipate how they might look like and change in the future and get a glimpse of what this might mean for future space programmes.

SPACE POLICIES OF MAJOR SPACEFARING NATIONS

National space policies are undoubtedly the major catalyst of all space activities whilst states continue to be the key players in shaping global space governance mechanisms. The dominance of the United States (US) and the Soviet Union has been severely shaken up with the emergence of new dominant players such as China and India as well as Europe which is continuously striving to act as a unified actor on the global space scene.

Most spacefaring nations have either conducted or expressed their interest in lunar missions as well as in Mars missions. To be able to handle larger undertakings in the field of space exploration, nations as well as private actors such as the American SpaceX corporation have been heavily investing in the development of heavy launchers which would permit the transport of advanced spacecraft necessary for both lunar as well as deep space exploration missions.

This section offers an overview of the major space policies currently in place. Although the list and descriptions are not completely exhaustive, they should offer insight into the major current agenda and objectives of space policies around the world.

USA

The United States are widely regarded as the most advanced space nation and currently the only nation to have sent humans outside of the Low Earth Orbit (LEO) at the time of the successful Apollo Programme.

The objectives of the US Space Policy are being set by the US government under the direction of the President of the United States which detains a crucial role in the policy setting and are subject for approval and funding by the United States Congress. (History NASA, 2018). It is standard practice for newly elected Presidents to modify objectives of the National Space Policy. This has been also the case following the inauguration of the new Administration under President Donald Trump that was swift to emphasize the strategic importance of space for the United States. On December 11th 2017, President Trump signed Space Directive #1 instructing the National Aeronautics and Space Administration (NASA) to return American astronauts to the Moon, followed by human missions to Mars. (NASA Press Release, 2017). This Space Directive is not the formulation of an entirely new space policy but a modification of the space policy adopted under the Obama Administration which, instead of the Moon, had emphasised asteroids as intermediate step towards further solar system exploration.

Prior to the signing of Space Directive #1 Trump has also reinstalled the *National Space Council,* an institution within the Office of the President headed by the Vice-President and responsible for drafting the direction of America's activities in space. The Council was initially founded in 1989 but disbanded shortly after in 1993. Vice President Mike Pence is the current chair of the Council which includes furthermore the most relevant Secretaries such as the Secretary of State and Secretary of Defence as well as other relevant high-ranking officials such as the NASA Administrator and the Chairman of the Joint Chiefs of Staff.

Space policy in the US is generally a unifying and bipartisan theme in US politics and is closely linked to national security issues. Shortly before the signing of the new directive, Vice-President Mike Pence announced that "under President Trump's leadership, America will lead in space once again on all fronts". The new space policy directions approved by Trump and which foresee a return to the Moon as intermediate stop towards Mars are also in line with Trumps "making America great again" logic and reflect the US's more assertive stance on global issues.

Back to the Moon

Lunar missions have played an important role in US Space policy ever since former President Kennedy's famous "We choose to go to the Moon" speech of May 1961. America's lunar endeavours have led to moments of greatness that have defined the US space program despite the fact that no US astronaut has ever returned to the moon since the last Apollo mission in 1972. The first Moon landing occurred under the Presidency of Richard Nixon despite serious cuts to NASAs budget which were already undertaken under the previous Administration under President Lyndon Johnson given the exorbitant costs of the Vietnam War and massive domestic spending programmes of Johnson known as *Great Society.* It was only Nixon in 1972 to have announced what would become the new milestone of US Space exploration efforts, NASA's Space Shuttle program. The first launch of a crewed Space Shuttle Orbiter, STS-1, was achieved only in 1981 under the Administration of Ronald Reagan. Until the programme's retirement in 2011, Space Shuttle missions have achieved important accomplishments for the US Space Program such as in the construction of the *ISS.*

The loss of Space Shuttle Columbia in 2003 was a serious downdraft for US space policy which had found itself without the means to autonomously transport its astronauts to the ISS and to conduct manned spaceflight. Shortly following the report published by the *Columbia Accident Investigation Board,* President George W. Bush announced a new space policy with the publishing of the *Vision for Space Exploration* in 2004, which set out the new policy directives for future space exploration programmes. One of the fundamental goals of the vision was extend human presence across the solar system starting with a relaunching of the Space Shuttle missions for manned missions towards the ISS, as well as returning to the Moon by no later than 2020 in preparation for human exploration of Mars and other destinations. The manned spaceflight mission aimed at a crewed lunar landing became known as CxP, the Constellation programme and was planned to be executed in 2005-2009 but got cancelled under the Obama Administration.

The 2010 speech of President Obama outlining the new US Space Policy has, however, dubbed the Bush administrations endeavour to return to the Moon. The policy drafted under the Obama administration envisaged to land humans on Mars by 2030 by bypassing the Moon and landing on asteroids as an intermediate step and was known as NASA's Journey to Mars. (NASA Space Policy, 2010)

Towards Deep Space Exploration

A further important milestone of the new Space Policy drafter under Obama was the development of a new launch vehicle which became known as the Space Launch System (SLS) intended as a follow-up of the cancelled Constellation program. SLS is designed as a single launch vehicle usable for both crew and cargo and shall be the backbone of NASA's deep space exploration endeavours given the immense load capacity which is designed to outperform the launch capacity of Saturn V and is set to be launched in December 2019. Other progress towards accomplishing the sophisticated vision for Mars exploration are the successful tests of NASA's Low-Density Supersonic Decelerator LDSD, a re-entry vehicle capable of landing large payloads of up to 3 tons as opposed to the current operational limit of one ton as well as the MAVEN spacecraft, a space probe aimed at studying the Martian atmosphere which was launched in 2013 and currently orbiting the red planet.

Commercialization of Space Activities

On 24 May 2018, President Trump signed off an additional space policy regulative, Space Policy Directive #2 which adopts a new set of measures aimed at providing more support to commercial space activities. The set of measures was elaborated by the National Space Council and seeks to provide a new, less bureaucratic and more stimulating regulatory framework for private investors. "This directive will encourage American leadership in space commerce by creating more certainty for investors and private industry, while focusing on protecting our national security and public safety", Vice-President Mike Pence, Head of the National Space Council, said in a statement after the signing. (Pence, 2018)

Russia

The current Russian space policy is built upon the pillars of the successes in space exploration of the Soviet Union. Cosmonautics developed solid basis already during the times of the Russian Empire with the pioneer works of Konstantin Tsiolkowsky whose works - especially in the fields of rocketry - have inspired the Soviet Union to develop its space programme. Space was and continues to play a fundamental role in the Russian society which is, to a certain extent, also reflected by the fact that the Russian space industry employs over 250.000 people.

Unlike the US which had established its space agency NASA, already in 1958, Russia only officially founded its space agency Roscosmos in 1992, shortly after the dissolution of the Soviet Union. The Soviet space programme did not have a dedicated agency coordinating the country's space activities but was rather more characterised by the presence of a series of *design bureaus* that were responsible for developing launch vehicles, satellites as well as spacecraft.[7]

Despite the absence of a central space agency the Soviet Union was one of the two superpowers that dictated the rhythm of global space exploration endeavours during the Cold War. Soviet satellites in orbit were joined by the first man in orbit, Yuri Gagarin on 12 April 1961 after which the country had started shifting its focus on lunar exploration. Soviet spacecraft "Luna 2" and "Luna 3" were the first man-made objects to reach the lunar surface, delivering also the first pictures of the "hidden" side of the Moon. Much of these successes are to be attributed to the works of the leading Soviet rocket engineer Sergei Pavlovich Korolev, a key figure of Soviet cosmonautics and the development of the R-7 rocket whose death in 1966 constituted a serious setback to the Soviet space programme and lunar exploration in particular. Despite having lost the race towards the Moon to the Americans, the Soviet Union managed to send its lunar space probe "Luna-16" to the Moon and back to Earth returning lunar soil samples. Space exploration efforts were subsequently directed towards Venus and under the "Venera" programme the country conducted "the most detailed investigation of the Venusian atmosphere and surface" following a series of failed attempts in the 1960s. (Galimov, 2015)

The first years of Roscosmos in the harsh post-Soviet context were characterised by a deep state financial crisis which led to severe budget cuts limiting the Russian space activities to the most basic activities. Space exploration missions were almost non-existent. Following the new millennium and the presidency of Vladimir Putin, space activities gradually augmented yet inefficiency, corruption and, as a result, a large failure rate continued to trouble the Russian space sector. In 2013 the space governance system of the country experienced a major reorganisation as ordered by President Putin which led the government to having a substantially better oversight over the space activities.

The current Federal Space Program (FSP) for the period from 2016 to 2025 was approved on 26 March 2016 and disposes over a total financial commitment of 20.5 billion USD subject to a triennial revision. Lunar exploitation is one of the priorities defined in the programme. Russia aims to boost the presence of currently 49 operational spacecraft to 73 by the end of the projected period in 2025. Further, Russia planes to launch a series of lunar missions after more than 40 years have passed since the last Soviet robotic explorations on the Earth's natural satellite.

Parallel to the announcement of the new FSP, Russia also officially inaugurated its new cosmodrome Vostochny in the country's Far East. The first lift-off of a Soyuz-2 rocket with an astronomical satellite on board occurred on 28 April 2016 and was followed by a large governmental delegation headed by President Vladimir Putin. It was Putin in 2010, back then as Prime Minister of the Russian Federation, to have stressed the importance of the new cosmodrome characterising it as "one of modern Russia's biggest and most ambitious projects". (ROSCOSMOS, 2017) Talks regarding a new space port were already launched in 2005 despite the lease agreement for the Baikonur space port, located in the Central Asian Republic of Kazakhstan, running up to the year 2050. The main reason for this has to with the desire and need for Russia to be completely autonomous and non-reliant upon others to manage its space activities.

The development of the new Angara rocket family whose design and construction were entrusted to the Moscow-based Khrunichev State Research and Production Space Center will possibly be a game-changer for the Russian space industry. In December 2014 Russia successfully tested Angara A5, a heavy-load rocket able to put considerable payloads into different orbits. Dmitry Rogozin, former Deputy Prime Minister in charge of space activities and current head of Roscosmos announced that Russia would only continue producing the Proton rockets under the currently signed contracts before using exclusively Angara rockets, including re-usable versions which are still under development.

Europe

Space policy in Europe is quite complex as it is developed by different actors, which different interests, competencies and needs. Three types of players can be identified: the European Space Agency, the European Union and, the individual European states on national level.

The European Union with its 28 members states[8] compromises all major European spacefaring nations such as Germany, France, Italy and the UK and has centred its space policy around three important pillars:

1. **Meeting Key Societal Challenges:** EU space programmes provide public services to EU public authorities, companies and citizens. Space data is essential to answering societal challenges such as the sustainable consumption of natural resources, safety and security, and climate change.
2. **Jobs and Industrial Growth:** The space sector provides over 230,000 jobs in the EU from manufacturing to space operations and downstream services. It is worth between €46-54 billion to the EU economy.
3. **Ensuring EU Autonomy:** Europe's access to space underpins the implementation of many EU policies, the competitiveness of European industry and businesses, as well as its security, defence and strategic autonomy. Space reinforces the role of Europe as a stronger global actor. (EU Space Policy Summary, 2018)

Despite the European Union being a rather new player in space, it is currently managing three major space projects which it regards as its flagship missions in space: Galileo, EGNOS and Copernicus. Galileo is Europe's Global Navigation Satellite System (GNSS) and was created to provide an independent high-precision positioning system in Europe. The programme was conceived during the late 1990s. The major motif behind Galileo was to ensure Europe an autonomous access to satellite navigation at a time where the American GPS system still had a very military character and its availability for civil purposes and other states was not 100% guaranteed under all scenario's. US officials were initially hostile towards the launching of Galileo as they feared that the signals could interfere with GPS. Chinese involvement in Galileo was a further problem but EU officials assured their American counterparts that Galileo would only be intended for civilian uses. As of December 2017, the system is offering global coverage and is the only civilian controlled Global Satellite Navigation System. EGNOS, the European Geostationary Navigation Overlay Service augments the functioning of Galileo and other GNSS systems thus providing precise navigation services for aviation, maritime and land-based users over most of Europe.

Copernicus, formerly known as the Global Monitoring of Environment and Security or 'GMES', is a further flagship mission of the EU and is also the world's largest integrated EO programme. It is conceived as an integrated system drawing from its own dedicated Sentinel satellites and related ground segments, as well as satellite data from third parties and in-situ data sources on a great variety of environmental parameters such as land, atmosphere, ocean and ice monitoring.

Given its technological skills and experience in operating complex space programmes most of the designing as well as many of the satellite operating responsibilities are being managed by the European Space Agency. The European Space Agency, ESA, is an autonomous intergovernmental organisation which is not part of the EU and currently has 22 Member States all of which all, except Switzerland and Norway, are member states of the European Union.

ESA was officially established in 1975 headquartered in Paris as the result of the fusion of the two precursor organisations, ELDO (European Launch Development Organization) and ESRO (European

Space Research Organisation) which had, at a joint European level, conducted space activities already since the early 1960s. The founding treaty of the Agency states that *ESA's purpose shall be to provide for, and to promote, for exclusively peaceful purposes, cooperation among the European States in space research and technology and their space applications, with a view to their being used for scientific purposes and for operational space applications systems*[9]. As a first tool to achieve the ESA Convention states as tool *"by elaborating and implementing a long-term European space policy, by recommending space objectives to the Member States, and by concerting the policies of the Member States with respect to other national and international organisations and institutions*". In practice, however, ESA's programmes, activities, direction and policies are decided upon by its Member States. In this sense, the policy stance of ESA clearly reflects the Agency's intragovernmental structure, whereas the space policies defined by the EU have a stronger supranational character.

ESA is a very comprehensive space agency, and it covers nearly all fields of space activities, including a mandatory science programme and optional programmes and activities on the other space activities including launchers, Earth Observation, satellite navigation, satcom, technology development, SSA etc. The member states contribute to both the mandatory programmes as well the optional ones. Mandatory programmes constitute approximately 25% of the overall ESA budgets and are the technology research and science core technology programmes, general study programmes as well as the European Component Initiative. The remaining 75% of the budget is managed through the optional programmes.

In 2007 twenty-nine European states adopted a new European Space Policy in which they agreed to unify the ESA approach with the ones of the national states creating thus a more harmonised approach and, for the first time, a common political framework for space activities in Europe. This move can be seen as a desire of Europe to become a strong actor in space using all possible space resources of its Member States.

Maintaining Europe's autonomous access to space is one of the main legacies present in the European Space Policy. Currently, ESA relies on three launchers which are being used to cover a large part of the domestic institutional European launch market, as well as for international partners. Currently, Europe relies three types of launchers guaranteeing autonomous access to space. The Ariane rockets cover the heavy-lift launch vehicle segment and development has already begun in the 1980s. The medium-launch segment instead is currently covered by the Russian Soyuz-2 rocket which is the result of a cooperation between ESA and Roscosmos in order for Europe to have access to reliable and cost-efficient medium-sized launchers. The agreement between ESA foresaw the construction of a launch site for Soyuz at the Guyana Space Centre in Kourou as well as the adaptation of the launch vehicle to be launched from Kourou. Vega, the smallest launcher, is a joint project of seven ESA member states under the leadership of Italy.

China

China made remarkable progress in spacefaring since the launch of its first satellite, Dongfanghong I, in 1970 in midst of the Cultural Revolution, a period in which China has laid the groundwork for the future development of its launch as well as satellite capabilities and which has seen the country becoming the fifth nation to achieve independent launch capability. (Chen, 1991) Today, China is regarded to be one of the key players in the space sector disposing over an extensive space industry which, despite recent private initiatives, remains to be state-owned as well as characterised by a strong dual-use nature. The dual-use approach and the strong leadership of the Armed Forces especially in the initial development of ballistic missiles became the standard framework for research and development of the Chinese space industry.

Research in the field of technology took its beginning in the late 1950s with the foundation of the China Academy of Launch Vehicle Technology (CALT) which had the lead in the development of the Long March rocket family which has successfully launched Dongfanghong I. The development of launch capabilities in the late 1950s included considerable technical assistance from the Soviet Union until, following political tensions between the two countries, Gorbachev suspended all scientific interactions with China. Gorbachev's move did, however, not hinder China to further develop its rocket industry. The China Academy of Launch Vehicle Technology (CALT), the country's major developer and constructor of rockets made serious progress developing the Long March series culminating in the transportation of the manned Shenzhou spacecraft as well as the both Tiangong space stations aboard a Long March 2F. China entered the commercial launch market in the late 1980s providing launch capabilities even to the Americans following the 1986 Challenger disaster.

Currently, Chinas major policy goals are focused around lunar exploration as well as manned spacecraft. The Chinese leadership has also put much attention in the last years on its manned space missions and the government has announced to launch a large modular space station into LEO by 2020. (BI, 2014) An important milestone in this context was achieved in 2003, China became the third country in the world to have launched a human into space in when Yang Liwei, aboard a Shenzhou 5 spacecraft, spent almost 22 hours in orbit. In 2011 the country achieved a further landmark with its first space station, Tiangong 1, literally *Celestial Palace 1,* having been successfully injected into orbit and being a precursor mission for a large modular space station.

Chinese space efforts are not only on a sign of the country willing to advance its space capabilities for strategic and economic purposes fulfilling civilian and military requirements, but also have a considerable symbolic effect. Chinese successes in space such as the first manned mission or the Tiangong space station contribute to the overall prestige of the country. The planned large modular space station China will launch will be open for all countries to participate and conduct in-orbit experiments. In 2018, The United Nations Office for Outer Space Affairs (UNOOSA) and China's Manned Space Agency (CMSA) announced an open tender for state and non-state actors from all UN member states. This can be, similar to the prestige China generated having developed its first rockets even following the pull-out of Soviet scientists, seen as a response to the American policy of excluding China from any sort of participation at the ISS. Chinese model open for all countries – new role of leadership.

India

Despite being classified as a developing country, India can claim a long and solid tradition of space activities with scientific missions and visions dating back to the 1960s. Currently, India is one of the most advanced spacefaring nations in Asia and it is likely to soon enter the human spaceflight arena with the first crewed mission scheduled to be launched in 2020.

Space activities in India were historically spurred by the need to address socio-economic needs. Dr Vikram Sarbhai, widely regarded as the father of India's space programme, is often quoted for having said that India *must be second to none in the application of advanced technologies to the real problems of man's society* indicating the importance for India to work on a predominantly civilian space program focussed on societal issues. (Nagai, 2015)

In fact, India's first satellite, Aryabhata, was an astrophysical research satellite and was launched in 1975. Despite the satellite having been launch aboard a Soviet rocket, India had committed itself to developing autonomous domestic launch capabilities. Satellite Launch Vehicle-3 (SLV-3), India's first experimental launch vehicle was successfully launched from the Indian Satish Dhawan Space Centre in 1980 delivering Rohini RS-1, an experimental satellite into LEO making India the sixth country to have autonomously launched its own satellite. As of today, India is guaranteeing its autonomous access to space as well as offering commercial launch services with its Polar Satellite Launch Vehicle (PSLV) as well as Geosynchronous Satellite Launch Vehicle (GSLV). The heaviest version of the GSLV, the GSLV MkIII, was successfully tested in 2017 injecting the communications satellite GSAT-19 into geostationary orbit.

Remote sensing is further an important issue of the Indian Space Policy which helps addressing issues on agriculture and development, an ever-important driver for space projects in the country. The first remote sensing mission was launched in 1988 with the launch of IRS-1A, aimed at providing imagery for various land-based applications. At present, India is operating one of the biggest EO satellite fleets worldwide with 17 operational spacecraft. In January 2018, ISRO launched Cartosat-2F, the eight satellite of the Cartosat series raising concerns in Pakistan, India's regional rival neighbour given the possible dual-use nature of the mission.

Space exploration missions have rapidly gained importance and become one of the core activities of the Indian Space Research Organisation (ISRO). Chandrayaan-1, the country's first lunar orbiter was launched in 2008 and completed its mission in 2009. Chandrayaan-2, the second moon mission consisting of a lunar orbiter, lander and rover, all of which were produced in India, is set to be launched in 2018. Furthermore, India has also launched its first interplanetary probe in 2013 with the lift-off of a PSLV-XL carrying India's sophisticated Mars Orbiter Mission (MOM) which has been orbiting Mars since September 2014. MOM has made ISRO the fourth space agency to reach Mars and the first to do so in its first attempt.

ISRO has expressed its desire to return to Mars with a lander by 2021. The Indian government has announced that it will allocate budgetary resources for the mission to be completed in joint cooperation with France which is also providing lift services for India's heavy launch necessities such as communication satellites in geostationary orbit.

Future policy issues for India are the finalisation of its heavy-launch vehicles, as well as the definition of a coherent space policy which the country is still lacking. (Rajagopalan, 2018)

Japan

Japan is one of the key space actors in the Asia Pacific region looking back on a long history of space activities that took their beginnings in the early 1950s and that have seen Japan becoming the fourth nation to launch a satellite, Ohsumi into orbit by own means in 1970. Research and Development of launch vehicles and satellites was a major priority for Japanese Space Policy and Japan's space policy in the 1970s and 1980s with a mere peaceful commitment rooted in the country's pacifist constitution.

The University of Tokyo had the initial lead in developing the first carrier rockets which became known as Lambda and which were operated until 1977. As of today, JAXA, the Japanese space agency relies on the Epsilon rockets for small launches such as smaller scientific satellites. Development of the Epsilon has started in 2007 and all launches so far have been success. Larger payloads are being by H-II series launchers developed and constructed by Mitsubishi which is also jointly with JAXA developing a new, low cost launcher denominated H3 planned to be first launched in 2020 with a capacity of 6.5 tons to GTO.

2008 marked a crucial census in the space policy development of Japan, whose parliament's resolution of 1969 had limited any involvement of the Japanese self-defence forces in the country's space programme. In August 2008, the Japanese Parliament had enforced a new legislation on space activities entitled the "Basic Space Law" setting out the policy framework for the future space activities. The new law consists of three contents, (1) Basic Concepts & Measures of Space Development, (2) Establishment of Headquarters for Space Policy as well as a (3) Review of Space related Organisations. (UNOOSA, 2018) Within the first concept, Japan commits itself once again to the peaceful use of outer space as well as to the promotion of international cooperation and the consideration of environmental preservation utilising space infrastructure. Furthermore, the country established a new Strategic Headquarter for Space Policy under which the Prime Minister serves as Director-General and the Chief Cabinet Secretary and Minister of State for Space Policy serve as the Vice Director-Generals thus elevating the countries space policy to the highest levels in government.

Emerging Actors

Defining emerging space actors is a complex undertaking. For the sake of an easy understanding the authors classify a state as emerging space actor which at least in the last two decades has demonstrated a growing interest in participating in space activities, despite having space programmes that might date back long ago. The state-of-art of national space programmes must not necessarily reflect the country's overall technological as well as economic capacity. Many Eastern and South-eastern European countries for instance, despite being classified by the OECD as high-income countries and disposing over a solid industrial basis have only rudimental space programmes whilst some countries classified as low-income countries might already have progressed space programmes.

In the current context, emerging actors mainly strive to acquire technological skills that enable them to conduct certain space activities. The form of assistance received is manifold and ranges from the full capacity development by a developed space actor to the provision of launch services for space assets which have been developed either in an entirely autonomous way or with additional assistance from a more developed space actor.

Middle East

The announcement of the United Arab Emirates (UAE) in 2014 to launch its planned *Hope Mars Mission*, a space exploration probe to be sent to mars in 2020, clearly indicated the desire of the country to join the club of advanced space nations. This announcement can be seen in a context of the entire region's governments bids of diversifying their oil-reliant economies as well as spurring technological progress in line with massive investments in human capital development.

The UAE's government announced its first space policy document entitled *National Space Policy* in 2014 which came together with the establishment of the UAE Space Agency. The country's main space policy objective has been defined as to "position the United Arab Emirates among the world's leading nations in space sciences before 2021" which is aimed to be achieved to generate the development of new specialised competencies, diversify the economy as well as bring along national pride and is fully aligned with the *UAE Vision 2021* which sets the key themes for the Socio-economic development of the UAE. (National Space Policy, 2018) This can be seen as an attempt of the Emirate to encompass space activities in a bid to diversify its heavy oil-export oriented economy.

Turkey, despite being one of the key players in the region and an economic as well as political heavyweight has, so far, demonstrated only little progress in developing a space industry. In 2017 the Turkish Parliament has, however, paved the way for forming a Space Agency and Turkey has published plans in which it outlines its desire to construct a spaceport and develop indigenous spacecraft manufacturing capabilities.

Israel remains the most advanced space nation in the Middle East region and its space policy has ever since been mainly evolved around addressing the security needs of the nation. It had become the eight country in September 1988 to have achieved space launch capabilities following the launch of Ofeq 1, a first in a series of Israeli reconnaissance satellites. Ofeq 1 was launched aboard a Shavit rocket which have been developed by Israel based on knowledge matured during the construction of the Jericho-II long-range ballistic missile most likely constructed in the framework of the country's nuclear weapons programme. Further important space programmes of the country are the operation of the Amos series of communication satellites, the Earth Observation capabilities and the development of nano- and microsatellites.

Iran, which became the ninth country to gain orbital-launch-capability in 2009, founded its Space Agency in 2004. The main aim was to boost the countries space activities, including the development of more capable launch vehicles as well as a manned space programme which would potentially include Iranian astronauts aboard the planned Chinese Space station.

Africa

The lack of a significant industrial capacities and educational networks capable of supporting the formation of a viable space sector have hindered most African countries in developing advanced space programmes. Nevertheless, domestic capabilities and programmes exist on the continent. Following the launch of a Kenyan nano-satellite in April 2018[10], Kenya has become the sixth African country to have procured the launch of a satellite demonstrating the growing appetite of African countries to leap forward in the space sector. Besides Kenya, Egypt, Ethiopia and Algeria have advanced their space programmes in the last decade whilst Nigeria and South-Africa remain the two countries with the by far the most advanced space programmes in Africa.

Nigeria became the first African country to have founded a dedicated governmental agency dealing with space, the National Space Research and Development Agency (NASRDA) – established in 2001. Its plans to engage in the space sector date back to 1976 but its first satellite, NigSat-1, was launched only in 2003. The country has currently five satellites in orbit out of which one is a telecommunication satellite and the remaining four Earth Observation satellites which have been widely recognised as a great success with tangible benefits in the field of agriculture as well as fighting terrorism. The country's government also announced ambitions build and launch a satellite of its own from a Nigerian launch site by 2020.

Algeria soon followed track after the establishment of NASRDA, by founding its own dedicated agency, the Algerian Space Agency, in 2002. The country has been working on the development of space policies since 1987 when it founded its National Center for Space Technology (CNTS) and has since made solid progress. So far, six satellites have been injected into orbit, five of which are part of the Alsat programme, an EO programme with a major focus on risk management, as well as one telecommunication satellite. A Centre for Satellite development has been founded in the city of Oran with Algeria now being capable of autonomously designing and constructing Earth Observation satellites.

South-Africa is a key player on the continent despite having founded its national Space Agency only in 2010. The country has a strong focus on developing an home-grown satellite industry and launched, two nanosatellites in 2017 which will be used to record factors in the Earth's lower thermosphere. Furthermore, the country continues to play an important role in fostering academic cooperation in the space field among universities in Africa thus taking over a role of *leadership* in the region.

Latin America

The Latin American scenario is characterized by the strong dominance of Brazil which has matured advanced space technologies that range from independent launch capabilities with two domestic spaceports, one at Alcântara and one at Barreira do Inferno to satellite manufacturing. Brazil currently operates a series of satellites, mainly EO satellites for scientific and reconnaissance purposes. Its space programme dates back to the 1960s with the development of the first sounding rockets Sonda I, II, III and IV and was under military dominance until the foundation of the Brazilian Space Agency in 2014. Currently, Brazil is putting a strong emphasis on developing global partnerships with leading space nations such as Russia and China to gain greater autonomy from the US it once relied heavy on.

Argentina is a further Latin-American country that is expanding rapidly in the space field. The Argentinian space programme is led by the National Space Activities Commission which dates back to 1960. Main priorities of the Argentinian space programme include Earth Observation, the peaceful exploration of Outer Space as well as technological developments for space use.

A series of other Latin American countries such as Venezuela have demonstrated their interest in developing indigenous satellite manufacturing skills mainly aimed at addressing socio-development goals. (Sarli, Cabero, Lopez, Cardoso & Jimenez, 2015)

Asia-Pacific

Space dynamics in the Asia Pacific region are much influenced by the activities of the three regional leaders, India, China and Japan who are massively using their advanced space technologies to assist their regional allies in developing and operating space assets. Malaysia is an exception in this context as the country has sought for cooperation with private companies in order to develop domestic microsatellite manufacturing capabilities rather than seeking cooperation with state actors. TiungSAT-1, Malaysia's first microsatellite was developed in joint cooperation with private UK space company and was launched in 2000. The country further has a series of communication satellites in geostationary orbit and satellite-linked ground stations and is

Similarly, also Singapore has been working on developing and indigenous satellite manufacturing industry. It launched TeLEOS-1, a commercial EO satellite and is looking to further develop more spacecraft. Indonesia disposes over a sophisticated space programme that includes a rocket launch site as well as plans for developing full-scale spaceports and launch vehicles.

Australia became the most recent country to have founded a space agency which its regional ally, New Zealand, had already done in 2016. Both countries are focusing on private investments in the space field. American company Rocket Lab opened the first commercial spaceport, Rocket Lab Launch Complex 1 in New Zealand, launching its first rocket in May 2017. The first attempt was a failure as the spacecraft did not reach the designated orbit but on a second attempt in January 2018 two satellites were successfully injected in the correct orbit.

South Korea established its national space agency, the Korea Aerospace Research Institute (KARI) in 1989 and places a strong emphasis on the development of indigenous launch capabilities as well as advancing its aerospace technologies. Korea Space Launch Vehicle KSLV1, also known as Naro-1, was the first Korean carrier rocket to have successfully reached orbit in 2013 after two failed launches and was designed in joint-collaboration with Russia. Analysts have described the South Korean rocket programme as a move of the country to develop its rocket industry following decades in which the US has hindered the Koreans to do so. (Brown, 2009) Further goals of the national space policy are a lunar exploration which is set to launch a first lunar orbiter in 2020.

Very little is known, on the other hand, about the North Korean space programme. In 2013 the DPRK established the National Aerospace Development Administration (NADA) as a successor institution of the Korean Committee of Space Technology. The latter was previously in charge of the country's space programme which is mainly focussed on launch vehicle development as well as the manufacturing of Earth Observation satellites of which some were successfully injected in orbit.

INTERNATIONAL SPACE COOPERATION

International cooperation in space is crucial and essential to achieving ambitious space exploration undertakings and it typically brings along a series of immense benefits to each cooperating partner. The benefits and thus reasons for countries and other actors to cooperate reach from political and strategic motifs to scientific, technological as well as industrial and commercial gains. Further particular objectives of engaging in cooperative missions are the desire to contribute to a common project, sharing common initiatives as well as results as well as promoting common interests.

Actors engaging in international cooperation can be international public bodies having international legal personality such as states and international organisations, national entities such as space agencies and other governmental entities dealing with space affairs, private actors such as companies or consortiums as well as other international actors such as, for example, both governmental and non-governmental bodies, foundations, federations and committees.

Cooperation agreements can be signed on a bi- as well as multilateral level but can also be intergovernmental – as for instance the Intergovernmental Agreement (IGA) on Space Station Cooperation governing the ISS – and inter-agency, i.e. between agencies but not on a national, bilateral level. Professor Yun Zhao has summarised four levels of cooperation based on an article on lunar colonization of Ryan Zelnio in an article on bilateral and multilateral agreements in international space cooperation

1. Coordination: while operating separately, the states coordinate on technical and scientific matters;
2. Augmentation: other states contribute to certain part of the project of a leading state, thus augmenting the original project of that leading state;
3. Interdependence: the states contribute to the operation of the project, but still controlling their own part of the contribution;
4. Integration: full cooperation with shared resources and responsibilities. (Zhao, 2016) (Zelnio, 2005)

In the first case, actors coordinate each other on technical and scientific matters representing a rather low-level form of collaboration with no or little reciprocal responsibilities. In the second case, actors share a way more encompassing responsibility as they are in charge of developing, constructing or even operating a certain element of the project run by the leading state. One example would be the China Seismo-Electromagnetic Satellite (CSES) mission in which China was the leading state and Austria, as a contributing actor, was entrusted with developing and constructing the magnetometer, one of the most essential payloads of the mission. In case three, actors are already intertwined in an advanced mechanism of cooperation in which they autonomously develop the various contributions but maintain full operational autonomy in both the development as well as the operation of these. An example would be the ExoMars mission, a joint mission of the European Space Agency (ESA) and the Russian space agency Roscosmos aiming the robotic exploration of Mars. In this case Roscosmos, one of the two contributing agencies, is in charge with providing the appropriate launch vehicle, a Proton rocket, as well as the construction of a landing system and surface science platform for the rover in addition to a series of instruments for both the lander and the rover. ESA on the other hand is responsible for the development of the ExoMars Trace Gas Orbiter (TGO) as well as the Entry, Descent and Landing Demonstrator Module (EDM) named Schiaparelli. The fourth example of cooperation which undoubtedly is the most advanced form is best represented in the case of the International Space Station ISS in which all five participating space agencies share resources as well as responsibilities in operating the station. It should be noted that partners can develop degrees of cooperation intensity on a parallel basis, as for instance the European Organisation for the Exploitation of Meteorological Satellites (EUMETSAT) and US National Oceanic and Atmospheric Administration (NOAA) have done in the field of satellite meteorology. (Lahcen, 2013) This approach allows organisations and governments to tailor their partnership models to the needs and requirements of different situations and space programmes. This also allows them to gradually intensify their partnerships over time, which is sensible from a risk management perspective.

Despite the international community having already achieved a high level of cooperation in space through both bilateral as well as multilateral agreements we still lack cooperation mechanisms capable of sustainably vectoring national space efforts into global and exclusively peaceful missions such as a lunar or even Martian basis. Explicating the case of a lunar outpost, Ryan Zelnio notes that *the United States as well as other nations' space policies make the kind of international cooperation needed to create this infrastructure impossible.* (Lahcen, 2013) So far, only the European Space Agency has achieved a far-reaching framework for its member states to contribute towards achieving common goals.

In this context, a series of important global issues arise. The first is whether a high level of coordination and cooperation amongst spacefaring but also non-spacefaring nations may be an alternative to the current multipolar governance approaches. Secondly, it would be appropriate to assess whether space cooperation can act as precursor for processes and sustainable models on Earth, especially in light of the current tensions between the East and West.

In order to elaborate these questions, the role of global space governance bodies will be explained and, *in primis*, the system of the United Nations dealing with space issues. Subsequently this chapter assesses other bodies and mechanisms in place, and it sheds light on the dynamics of multilateral agreements with a special focus on regional space initiatives as a result of multilateral agreements. Finally, the role of space as a potential *foreign policy tool* in the context of *space diplomacy* will be addressed.

Global Space Governance Bodies

Outer Space Policy Making in the UN System

The launch of the first artificial satellite in 1957 immediately gained a vast political dimension and attracted the attention not only of the major world powers but also of the United Nations as major intergovernmental body conferred with the mandate to ensure global peace. In 1958, the General Assembly of the United Nations established an ad hoc Committee on the Peaceful Uses of Outer Space (COPUOS) with initially 18 members becoming a permanent body in 1959 with 24 members at the time.

As of today, the Committee, one of the biggest of the United Nations, consists of 84 Member States with the participation of both governmental and non-governmental organisations accredited as Permanent Observers. The COPUOS has produced the five sore space law treaties and contributed to advancing a series of Transparency and Confidence Building Measures (TCBMs). The United Nations Office for Outer Space Affairs (UNOOSA) acts as Secretariat to the COPUOS and is tasked with implementing the decisions in space matters of the United Nations General Assembly as well as of the COPUOS. The Office is furthermore responsible for discharging the space-related duties under international law of the United Nations Secretary General as well as for promoting international cooperation in the peaceful uses of outer space and assisting developing countries in gaining access to space technologies.

UNOOSA has signed a partnership with the private American Sierra Nevada Corporation (SNC) to conduct an orbital space mission using SNC's Dream Chaser spacecraft enabling all countries to participate. The mission especially aims at the participation of developing countries since it is supposed to become the first space mission dedicated to the Sustainable Development Goals (SDG). As such, all payloads and experiments must be aimed at supporting the achievement of one or more SDGs. This is in line with the policy of the United Nations to encourage and conduct capacity-building measures for developing states to engage in space activities.

UNISPACE +50

UNISPACE+50 is an initiative dedicated to the 50th anniversary of the first United Nations Conference on the Exploration and Peaceful Uses of Outer Space (UNISPACE) held in 1968. The 2018 conference provided new perspectives and fresh dynamics for the future of space governance which are set around seven thematic priorities. The first thematic priority is entitled *Global partnership in space exploration and innovation* and is aimed at fostering new partnerships to mainly boost new space exploration undertakings. Thematic priority 2 (Legal regime of outer space and global governance: current and future perspectives) intends to promote the universality of the five United Nations treaties on outer space whilst thematic priority 3 (Enhanced information exchange on space objects and events) is set to define and develop requirements for enhanced information exchange and notification procedures under the United Nations Register of Objects Launched into Outer Space with a special focus on Transparency and Confidence-Building Measures. Thematic priority 4 (International framework for space weather services) is aimed at strengthening the reliability of space systems and their ability to respond to the impact of adverse space weather while thematic priorities 5 (Strengthened space cooperation for global health) and 6 (International cooperation towards low-emission and resilient societies) have a special focus on the achievement of a series of SDGs by using space resources. The last thematic priority, Capacity-

Figure 2. UN Framework for Space Activities
(Credit: United Nations)

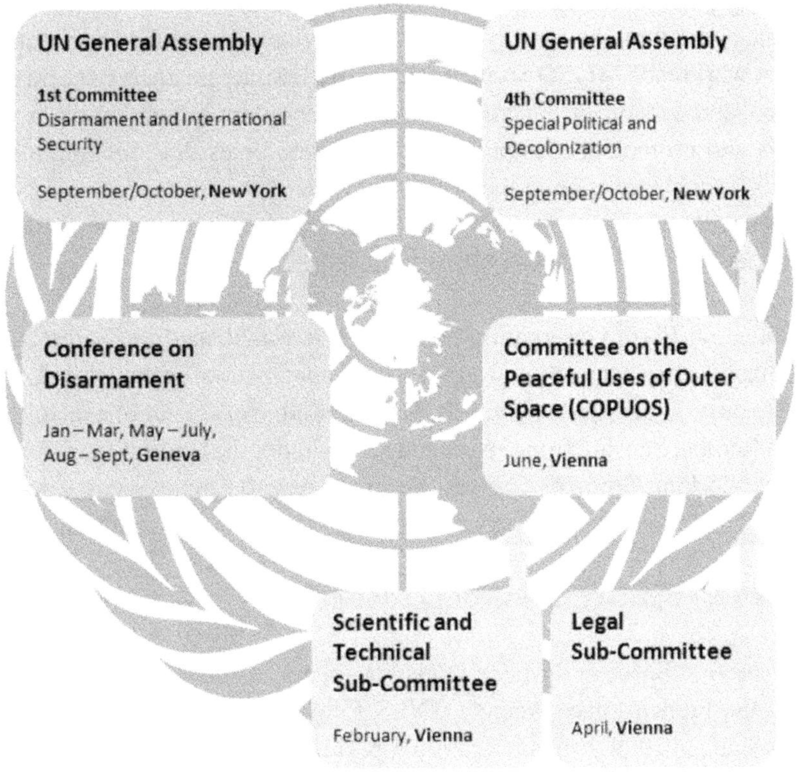

building for the twenty-first century, stresses the need for using space as a driver of and contributor to global frameworks and agendas such as the 2030 Agenda for Sustainable Development, the Sendai Framework for Disaster Risk Reduction and the Paris Agreement on climate change.

Policy Themes Coordinated on a Global Level: A Global Space Governance?

GEO: Group on Earth Observation

The Group on Earth Observation (GEO) is a partnership of currently more than 100 national governments as well as professional organisations that was formed following the Third Earth Observation Summit in Brussels in 2005 with the main scope being the establishment of the Global Earth Observation System of Systems (GEOSS). GEOSS is a unique network of spatial data infrastructure being provided by the various stakeholders with the aim of generating a global, near-real-time Earth observation data set whose resources can be easily disseminated and made available to a wide range of users.

The main policy areas the Group and especially the GEOSS network can assist with and which have been identified in the most recent strategic paper, the GEO Strategic Plan 2016-2025, as objectives are linked to the Sustainable Development Goals (SDGs), a collection of 17 global goals set by the United Nations covering a broad range of social and economic development priorities. (GEO Strategic Plan, 2018)

As set out in the Strategic Plan, the first strategic objective will be to advocate "the importance of Earth observations as irreplaceable resources that must be protected, rendered fully and openly accessible (including through contribution to GEOSS), and integrated to provide maximum value in support of achieving national and international calls for resilient societies, sustainable economic growth, and a healthy environment worldwide." (GEO Strategic Plan, 2018) Strategic goals two and three are linked to the first strategic goal and will aim at engaging various stakeholders in this process as well as delivering data to stakeholders and promoting the uptake of new technologies thus creating also new economic opportunities and fostering innovation in the sector.

CEOS: Committee on Earth Observation Satellites

CEOS, the Committee on Earth Observation Satellites was established in September 1984 and is an international coordinating mechanism involved in the management of international, civil, space-borne missions designed to observe and study the Earth. The committee is a result of a recommendation from a Panel of Experts on Remote Sensing from Space and set up under the aegis of the G7 Economic Summit of Industrial Nations Working Group on Growth, Technology, and Employment. Currently, there are 52 members consisting of both space agencies as well as other national bodies and international organisations.

International Charter "Space and Major Disasters"

The International Charter "Space and Major Disasters" is a joint-initiative launched by the European Space Agency and the French Space Agency CNES following the UNISPACEIII conference at the United Nations in Vienna in 1999. Its main aim is to provide assistance with space data following major catastrophes and currently counts on 17 Member States and 34 contributing satellites. As of today, the charter was activated in a total of 124 countries for 576 times providing valuable disaster relief assistance, in particular for those countries that do not dispose of domestic space capabilities.

ISECG: International Space Exploration Coordination Group

The ISECG is a non-binding coordination mechanism currently composed of 15 space agencies and convened for the first time in 2006. Its main aim is to enable key spacefaring nations to exchange information and define common goals and activities in the field of space exploration. The most important publication of the group, the Global Exploration Roadmap GER was first released in May 2007. The 2007 GER has identified the Mars as ultimate goal for all space exploration scenarios and the Moon as well as near-Earth asteroids as intermediate steps. In 2018, the third and most recent edition of the GER was released by the ISECG. It does not include anymore a near-Earth asteroid as intermediate step but has a rather more extensive focus on building a lunar colony. The 15 participating agencies envisage that, by the mid-2020s, a Deep Space Gateway would be launched in lunar orbit and that between 2030 and 2040 the first humans would reach Martian soil.

Regional Multilateral Cooperation

Multilateral agreements have been crucial in fostering international cooperation on a global level. Binding agreements such as the five major UN space treaties have contributed to creating a global set of mutually accepted norms that have helped prevent expanding the weaponization logic of the Cold War into the outer space environment. Reaching mutual agreements or even non-binding resolutions on a global level is, however, an extremely complicated endeavour as it requires, in the context of the United Nations, the consent of the majority of the member states of the United Nations. For a resolution to be passed by the United Nations General Assembly a simple majority is required. If the topic of the resolution is deemed to be important, a two-thirds majority is required. Any resolution that might be passed by the Security Council of the United Nations can be vetoed by even a single veto of a permanent member state of the Security Council.[11]

In the last two decades the world has witnessed an increase in regional multilateral initiatives that have given birth to a series of agencies, initiatives and organisations as well as other forms of cooperation amongst nations operating on a regional level. The European Union has even lobbied for certain initiatives such as the EU proposal for an *International Code of Conduct for Outer Space Activities* to be adopted on a global level.

Having discussed ESA as a regional, inter-governmental organisation in Europe it is also important to note the existence of similar organisations in other parts of the world. The Asia-Pacific Space Cooperation Organization (APSCO) is an is an inter-governmental organisation that, just as ESA, enjoys a fully international legal status. Its founding convention was signed in 2005 by its founding and current members Bangladesh, China, Iran, Mongolia, Pakistan, Peru and Thailand and is headquartered in Beijing. Unlike ESA, which includes all major European spacefaring nations, APSCO does not count all major spacefaring nations of the Asia Pacific region as its members. This is due to the organisation being considered very much China-dominated thus discouraging other key players such as Japan and India to join.

The Asia-Pacific Regional Space Agency Forum (APRSAF), a similar though not inter-governmental organisation is also active in the region. The forum was founded in 1993 and is, unlike APSCO, also open to non-state actors including entities outside the region which is reflected in the broad participation of a variety of entities from currently 46 countries and regions and 28 international organizations. APRSAF meetings are held annually and are led by the four working groups of the organisation which are dealing with the topics: 'Space Applications', 'Space Technology', 'Space Environment Utilisation' and 'Space Education'. At the APRSAF-24 meeting in November 2017 in Bengaluru, India the four working groups concluded four major points. (ISRO, 2018). First, national space agencies in the Asia Pacific region shall develop small/cube satellites collaboratively in order to provide input to solve various common issue of the Asia Pacific Region. Second, to further promote educational activities, using space technology as a tool that will be beneficial for the human resource development of the next generation. Third, encouraging and accelerating Kibo, the Japanese research module of the ISS, utilisation through feasibility studies in each country. Fourth, to further promote space applications including rice crop monitoring, global rainfall monitoring, fire hotspot, haze monitoring and disaster management, etc.

In Africa, four countries, Algeria, South Africa, Kenya and Nigeria have taken the lead in creating a constellation of satellites denominated African Resources Management Constellation (ARMC) with the aim of providing unrestricted and affordable access to satellite data to support effective environmental and resource management in Africa. The initiative was taken in the margins of a COPUOS meeting in Vienna in 2004 and has led to the signing of a Memorandum of Understanding (MoU) between the four nations. A second intra-African initiative, the African Leadership Conference on Space Science and Technology for Sustainable Development (ALC) was held for the first time in 2005 in Abuja, Nigeria and is a regional conference held every two years with the aim of promoting intra-African cooperation in the uses of space science and technology to support Africa's development as well as creating a forum to shape the evolution of the space arena in Africa. (Martinez, 2012) The regular meeting which have received strong support from the international space community including the UNCOPUOS and UNOOSA have provided important impulses for the shaping of an African space arena. Most notably, in 2011 a special panel of the ACL was tasked with evaluating the feasibility of creating an African Space Agency. Despite the panel not having brought along any concrete steps, the African Union presented a draft statute for an African Space Agency in October 2017 whose policy framework is outlined in the African Space Policy.

Similar initiatives and mechanisms can be observed in Latin-America as countries in the region are recording fast progresses in the space field and as regional integration to address socio-economic but also military issues is continuing to be a trend. The Space Conference of the Americas (CEA) is a regional forum destined at fostering international cooperation in the Latin American & Caribbean region created in the early 1990s. CEAs focus is on the disseminating the use of space technologies in applications related to socio-developmental goals as well as progressing cooperation in the region. A South American Space Agency had initially been proposed by Argentina but has not realised thus far.

FOREIGN POLICY AND SPACE: SPACE DIPLOMACY

One can certainly claim that space programmes generate incredibly valuable economic, scientific as well as technological progress. What is, however, often neglected is the possible use of space resources as a tool of foreign policy. The U.S and other major spacefaring nations have long ago understood that space activities lead to the accumulation of foreign policy capital in a long-term setting. In 1958, one year after the launch of the first Sputnik, Lyndon B. Johnson, then US Senator, warned that whoever would win the space race would gain *"control, total control, over the Earth for purposes of tyranny or for the service of freedom."* (Allen & Husain, 2017) Achievements in space exploration can leverage a country's share in the global space agenda as well as influence the objectives of other players. (Luzin, 2013) International cooperation in the context of space and foreign policy goes well beyond issues such as cost- and risk-sharing and which have been described before in this chapter.

As John Krige points out, NASA's international cooperation activities in the 1950s and 1960s had the effect of "significant political and cultural role to play in projecting a positive image of American power and American democracy abroad." Krige further elaborates in this context that "*those who implemented NASA's mandate had a far broader vision of international cooperation than one that was simply subservient to America's national space needs. From its inception, NASA saw its role as fostering the development of space science and technology in other countries. Its officers, in consultation with other parts of the administration (notably the State department and the department of defense), sought to use American scientific and technological pre-eminence to kick-start and even mould space activities in other countries, notably those of the Western alliance. NASA's international programs were intended to build a world community dedicated to the peaceful exploration of space with American help, under American leadership, and in line with the general objectives of American foreign policy*". (Krige, 2007)

Soon after the establishment of the Committee on Space Research (COSPAR) in 1958 by the International Council for Science (ICSU), NASA offered its Western allies to launch their scientific payloads into orbit. By accepting this offer the United Kingdom was able to launch its first satellite, Ariel 1, in 1962 shortly followed by Canada which launched Alouette 1. In Italy, Luigi Broglio, an aerospace engineer and Italian Air Force General, approached the US delegation during a COSPAR meeting in Italy in 1961 requesting launch services for a satellite to be entirely built in Italy and operated by Italian staff. The project, later officially known as "San Marco" has seen the launch of "San Marco 1" in December 1964 making Italy the third country in the world to operate its own satellite, after the Soviet Union and the United States.

The Soviet Union developed a similar model of cooperation addressing its allies. In 1967 "Interkosmos", a space cooperation programme led by the Soviet Union was launched. The programme aimed at helping the Soviet Union's allies with crewed and unmanned space missions. Soviet allies were quick on capitalising on the offer. Former Czechoslovakia was the first country of the Eastern Bloc to launch its first satellite, Magion 1, into Orbit. The satellite was intended to study the magnetosphere and ionosphere of the Earth and was launched in 1978, the same year the first Czechoslovak cosmonaut, Vladimir Remek travelled to the Salyut 6 space station. Mirosław Hermaszewski, a Polish Air Force officer was the second cosmonaut from the Eastern Bloc to travel into space shortly before Sigmund Jähn from Eastern Germany. Bulgaria was another country that was able to develop a series of space technologies thanks to its involvement in the Interkosmos programme. The country became the second member to launch its first satellite, Bulgaria 1300, also known as "Interkosmos 22" in 1981, three years after its first Cosmonaut, Georgi Ivanov, had departed on Soyuz mission 33 to visit the Salyut 6 space station.

Emerging spacefaring powers such as China and India have been fast to grasp the potential of space systems as substantial tools to achieve diplomatic objectives on Earth. Regional cooperation mechanisms have witnessed a steady propagation in Asia but also in Latin America and are likely to soon encompass the African continent.

This has been especially evident in the case of India which has started directing its space programme – mainly driven by scientific, commercial as well as pragmatic needs addressing developmental issues – towards addressing security issues as a result of geopolitical tensions in the region which are gradually reaching outer space. The uncertain geopolitical developments in the ASEAN region have led to the creation of what could be described as a new strategic axis between Japan and India, two countries that are worried about Chinas growing influence in the region with the Chinese ASAT test in 2007 definitely having been a wake-up call for both nations. India and Japan have inked a series of joint collaborations in many fields. In September 2017, both governments signed a series of Memorandums of Understanding covering 10 areas of cooperation including the space sector with a MoU having been signed between the Indian Space Research Organisation (ISRO) and Japan Aerospace Exploration Agency (JAXA) covering the fields of satellite navigation and planetary exploration. Both space powers share similar goals in the Indo-Pacific region and are actively engaged in providing capacity building assistance to *allied* countries in the region in a bid to expand their regional influence and hinder the Chinese expansion. ISRO for instance launched the *South Asia Satellite* into geostationary orbit in 2017 as part of Narendra Modis *Neighbourhood first policy.* The satellite provides both broadcasting services as well meteorological capacities being equipped with remote sensing technology which enables collection of real-time weather data to the region and has been provided free of cost by the Indian government. Japan has analogously developed similar programmes providing capacity building to countries in the region. Vietnam benefitted extensively from the cooperation with Japan which has helped the country mature knowledge in the fields of satellite development and manufacturing.

China on the other hand is boosting its cooperation with Pakistan as well as other countries in the region. Chinese cooperation with Pakistan in the space field spans already over two decades. Pakistan's first satellite, Badr-1, was launched by China in 1990 and both countries subsequently signed an agreement to enhance cooperation for peaceful applications of space technologies. As of today, China continues to provide both financial loans as well as technical assistance to the Pakistani space industry which has also announced the desire for a Pakistani to be the first foreign astronaut on a Chinese spacecraft.

Space and Militarisation Issues

In 1947, the United States government has enacted its National Security Act of 1947 which, a part of having been a major reorganization of the US military organization following the World War II, has also separated the aerial forces of the US Army into an independent service branch, the US Air Force. More than 70 years later, in 2018, President Donald Trump has signed an executive order envisaging the formation of a new, sixth force of the US Armed Forces: the US Space Force. Addressing the National Space Council following the order of Trump, Vice-President Mike Pence, the Council´s chair claimed that " *time has come to write the next great chapter in the history of the Armed Forces of the United States*" and "*turn the page to an evolution of our armed forces to meet the challenges and opportunities on that limitless frontier."* (US Today, 2018)

That space is a potential dimension of conflict and that threats such as nuclear weapons might originate from space is no new fear especially in US public policy. The launch of the world's first artificial satellite Sputnik by the Soviet Union in 1957 did not only stun the entire world but clearly indicated that the arms between both super-powers was about to enter the fourth dimension. Despite the Sputnik being a simple artificial device with no offensive capabilities and of a short lifespan, it was understood that space technologies would come with an immense military value.

Shortly following the launch of Sputnik, the United Nations started dealing with the establishment of a framework which would guarantee the sole peaceful use of Outer Space and prevent the expansion of the Arms Race into Space. The Outer Space Treaty of 1967 set the first legislative framework aimed at preventing a weaponization of Outer Space. Article IV of the treaty states that states should "undertake not to place in orbit around the Earth any objects carrying nuclear weapons or any other kinds of weapons of mass destruction, install such weapons on celestial bodies, or station such weapons in outer space in any other manner". (OST, 1967) It has never been an secret that the Outer Space Treaty, just as all other international treaties, had left a series of loopholes. The treaty, for instance, does not regulate ICBMs which on their way to their target travel through Outer Space nor does it forbid weapons which could be launched from Earth and target objects in space.

So far, no nation has – at least officially – deployed military infrastructure with clearly dedicated offensive capabilities that could strike targets on Earth outside the Earth's Atmosphere. Despite the Outer Space not being weaponised, it is however often referred to as a weaponized environment whilst the relevance of space for military purposes is commonly accepted. (Saperstein, 2018) For example, satellite navigation, something that with little caution can be considered a public good, is known for its numerous civilian applications but is also well known that it traces its origins in a military context.

Space Militarization and Privatization

Patrick A. Salin has published an article back in 2001 claiming that the advancement of privatization of space ventures following the end of the Cold War has led to the continuation of the militarization of Outer Space. "We are slowly discovering that the militarization process of outer space seems to be a given, thanks to increasing competition within the space business environment. And, as privatization has accelerated during the last decade, we can clearly see an acceleration of the militarization process of outer space" (Salin, 2001)

He backs his claim with the fact that the big military aerospace contractors that have developed and constructed most military hardware for the Cold War did not vanish but continued to exist and gradually redirected their activities towards the production of civilian application for space. Most of these applications are dual-use goods that can serve for military purposes but are being commercialized in order to generate economic output financing the research and development. Apart of that it is important to stress that big industrial complexes, despite being private in most cases as is for instance in the USA, are still national players that are a key strategic interest every state will protect. Big states such as China and Russia aiming at expanding their economic activities beyond their internationally recognized borders are increasingly militarizing these spaces in order to guarantee and secure their access to specific resources.[12] A legal solution based on international law seems to be in no sight for disputes of this kind on Earth and yet is there no even vague solution for potential analogous situations in Outer Space which would be far more complex to solve by diplomatic means likely paving the way to further militarization of Outer Space.

LOOKING AHEAD: THE FUTURE OF SPACE POLICY

This chapter revealed that space activities and ambitions worldwide are constantly increasing and the club of nations having ongoing space programmes is getting ever bigger. It also revealed how much dynamic space has become recently: with the witnessing of non-space entrepreneurs entering the space domain, a diversifying number of space activities and, the growing national strategic value of space capabilities. Added on top of that, the ambitions for space exploration seem to be converging with the technology development and innovation necessary to materialise them. Looking at the future, this opens up a number of questions. How will our new ambitions in space be materialised? How will the global governance of space be managed in the future, when there are more players and diverse – at times conflicting – interests to be managed and reconciled?

Are the current governance frameworks and mechanisms in place as described throughout this chapter sufficient in dealing with current policy issues such as the weaponization of space and the crowding of the near-Earth environment? Guaranteeing a merely peaceful use of Outer Space in a currently multipolar global setting seems difficult as we lack global mechanisms that are able converge national space efforts of the key players towards a peaceful use.

This final section presents some initiatives that have the potential to bring the major key space players on one table. The activities proposed are in line with certain legacies of the past, such as the Apollo–Soyuz Test Project (ASTP), and thus may serve as pilot missions that will further enhance an inclusive global space cooperation with an exclusively peaceful approach.

Venturing Into New Frontiers

The Moon as a destination for further solar system exploration is present in almost all space policies of the major spacefaring nations. Most recently, with President Trump having signed the *Space Policy Directive #1,* directing NASA to land astronauts on the Moon instead of asteroids, it seems that the

Moon has once again become a major topic for US Space Policy providing space for cooperation with other agencies such as ESA, Roscosmos, JAXA and even the Chinese National Space Administration.

The Moon Village

The Moon Village idea as envisaged by ESA Director-General Johann-Dietrich Woerner was originally presented at the International Astronautical Congress (IAC) of October 2016 in Jerusalem. Its primary goal is the establishment of a sustainable human presence and activity on the lunar surface. ESA's Moon Village as such is not a concretely proposed ESA programme, but should instead be seen as a proposed concept for a joint global endeavour to establish a permanently inhabited settlement on the Moon. The Moon Village is intended to serve multiple fields including planetary sciences and human spaceflight but will, most importantly, constitute a paradigm shift in the context of Space 4.0 as it will *"create an environment where both international cooperation and the commercialisation of space can thrive"*. (Ops Journal, 2018)

The Moon Village will thus not only be a project rallying for contributions from both the governmental as well as private sector boosting scientific and technological cooperation. It will also act as a potential precursor mission for financially maintainable exploration of the solar system. Most important, however, is the potential role of the village as a role model that will substitute the ISS with a new large-scale collaborative framework capable of acting as role models for even more sophisticated missions towards a Martian settlement. In early 2016 ESA Director-General Wörner paid a visit to China calling for "space without border" approach in which he also called on China to support the Moon Village given the extensive lunar exploitation legacy China has acquired in the past. (Wang, 2016)

By creating a joint endeavour as far-reaching as such a type of lunar settlement would be, a new paradigm shift would occur in the way how international space exploration projects are being architected and governed. It would inevitably lead to a much greater cohesion and assimilation of space policies creating a new innovative platform on the Moon for further robotic and human exploration of outer space. (Doboš, 2015)

Two main hurdles will need to be overcome in order to proceed. A first hurdle is represented by the immense funding which will be needed to realise the project. Governmental budgets for space exploration are limited and are usually the first subject to cost-cutting in times of austerity or savings. A viable and sustainable funding model would require the participation of private actors which would require guarantees in regards to the longevity of the project as well as the economic exploitation schemes that would apply.

If successful, the Moon Village would pave the way for national space policies to become much more cohesive and mutually reinforcing instead of an agglomerate of monolithic infrastructures which are far more expensive and less effective. As with many projects of this magnitude, these benefits can only be had if the immense hurdles are overcome in the first place. Considering the increasingly tense geopolitical context of the late 2010s, however, it seems that the dream of a Moon Village established in the spirit of global collaboration is becoming ever harder to achieve.

Governance and Cooperation: Space Without Borders?

If an unprecedentedly challenging space project like the Moon Village would prove to be too ambitious at this stage, could the slightly lower ambitions of a more inclusive global space governance and stronger cooperation schemes be more realistic? The US government exclusion policy of China is a potential danger to any far-reaching global cooperation target. Current US-China relations, especially under the current US Administration under President Donald Trump are, to a certain extent, comparable to the space relations between the USA and the Soviet Union during the Cold War.

The Apollo-Soyuz Test Project was conducted in 1975 and was the first joint US-USSR space mission which has seen the in-orbit docking of an US Apollo and a Soviet Soyuz module. It was the first international cooperation project between the two superpowers in the field of space exploration laying the first foundation towards the creation of the ISS. An Agreement denominated *Agreement Concerning Cooperation in the Exploration and Use of Outer Space for Peaceful Purposes* envisaging such a project had already been signed in 1972 by Moscow and Washington D.C and was the first attempt to expand the politics of détente into Outer Space.

Similar projects were not pursued following the 1980 invasion of Afghanistan by the Soviet Union until the breakup of the USSR in 1991 after which Russia, the legal successor of the Union and the US had started cooperating more intensely.

The proliferation in bilateral agreements between both countries was mainly catalysed by the Russian financial situation which, following the dissolution of the Soviet Union, was catastrophic. Russia was in urgent needs for fresh capital inflows whilst the US had an great interest in collaborating with Russian engineers. This collaboration was also deemed as highly positive for the US as this kind of engagement had prevented Russian engineers becoming potentially available to offer technical support to countries the US regarded as enemies. Intensive cooperation with the Russians also helped the US accomplish important foreign policy objectives such as the Russian adherence to the Missile Technology Control Regime (MTCR) regime as well as other measures of non-proliferation. Russia on the other hand profited from strong capital inflows as well as a transfer of technology which has partially helped revive the countries space industry. The early years of space cooperation between both countries culminated in 1993 when Russian was officially invited to join the ISS as a full partner.

Collaboration in itself was not entirely new, as the two power blocks had even been modestly working together during the Cold War in the fields of meteorology and science. *"Despite the continued space competition between the United States and U.S.S.R., Khrushchev sent Kennedy a letter raising the possibility of space cooperation on a modest level after John Glenn became the first American to orbit Earth on Feb. 20, 1962. That led to two rounds of discussions between NASA's Deputy Administrator Hugh Dryden and Soviet academician Blagonravov. An agreement led to the opening of cooperation in three areas: 1) the exchange of weather data from satellites and the eventual coordinated launching of meteorological satellites; 2) a joint effort to map the geomagnetic field of Earth; and 3) cooperation in the experimental relay of communications. This link became a primary forum for subsequent U.S.-U.S.S.R. interaction on space."* (Sagdeev, 2018)

The positive effects of the project were the inter-locking nature of cooperation that was the result of numerous exchanges between scientists of both superpowers that have generated strong transparency and confidence building measures. Today, we see a similar dynamic occurring between the US and China, and the question should be raised whether space could be – as it has been at times in the past – a bridge-builder between rivalling nations, even – or perhaps especially – at times of geopolitical turmoil.

The fact is that China – in all likelihood – will continue to expand its space capabilities regardless of other countries' intentions or actions, as its ambition to be recognised as a global power is expressed throughout its different policies and its stance on international relations. Because of this, China is on the path of becoming an autonomous proprietary of world-class space capabilities throughout across the board. By inviting China around the table of outer space affairs, programmes and governance, the slow pace of progress in space governance could be accelerated.

In 2007, the EU proposed the "draft International Code of Conduct for Outer Space Activities" intended to provide an international framework which could, potentially, be denouement of the current dual-use approach in space activities. The draft proposal aims to enhance "safety, security, and sustainability of all outer space activities" as well as form a "regime of transparency and confidence-building measures" but has received limited support on an international level. China and Russia have dubbed the draft as they would like to limit the code to civilian and commercial space activities whilst many developing nations have raised concerns as the Code of Conduct has been drafted outside an UN setting. Because of this, animosity for the draft code is at a low level a decade after its proposal, preventing the rise of new tools for addressing the many space governance challenges.

CONCLUSION

Space policies have constantly evolved and changed in scope and articulation ever since humankind entered and started conquering the outer space or '4th' environment. This chapter sketched a comprehensive overview of the various elements that constitute space policies and, it explained how these are typically formed and evolving. A crucial element in this respect is the growing utility of the space environment, which has increasingly diversified the stakes and interest in conducting space activities.

At first, only a handful of nations disposed over the financial and technological capacities to expand their activities into Outer Space. In that context space policies were mainly a national domain and "not romantic visions of space exploration but down-to-earth considerations of national security, international competition, and international-prestige-building within the scope of the Cold War". (Harding, 2016) This period roughly coincides in what we have thematised as Space 2.0 in the beginning of this chapter. Despite space policies having been an exclusive national domain in the late 1950s, the international community was fast to react on the potential outreach space activities would have on Earth and were fast in reacting. A series of international fora aiming to address activities in Outer Space has been set up. The United Nations Committee on the Peaceful Uses of Outer Space is undoubtedly the most relevant one responsible for having negotiated and drafted the five UN treaties and agreements relating to activities in outer space.

It was however not until the arrival of the 'Space 3.0' paradigm - the period following the End of the Cold War - that space endeavours were gradually launched as multilateral projects encompassing an ever-growing number of actors. The United States (re-)discovered the potential of space diplomacy in an attempt to approach the Russian Federation following the dissolution of the Soviet Union. Space cooperation as a foreign policy tool also developed into a key element of Chinese and Indian space policies which have effectively used it to expand their networks and their reach of influence.

Security, one of the major strategic interests of space policy has seen a revival in the last years. The announcement of President Trump that the US government will form a sixth force of its Armed Forces, the Space Force is just an elucidation that space continues to be a crucial part of military and security policy. The growing number of forms of threats in development aimed against space assets, be it ASAT technology that might destroy satellites or some form of electronic warfare disabling communication with satellites make it highly likely that outer space issues will be integrated into military doctrines of more and more countries.

Over the last decade, space technologies and applications have matured to a point where they serve a crucial role in supporting a whole range of societal functions such as: mobility & logistics, agriculture, environmental monitoring, telecommunication and many more. This trend, which includes the development of 'New Space' but is in fact more comprehensive – is referred to as 'Space 4.0'. Enabled by the digital revolution and by a changing context of space activities, this new era offers a promising, yet disrupting outlook for the future.

New technologies will yet again empower to surpass historic frontiers. This time, humanity will be able to go beyond the historic confinement of the Earth-Moon system, and venture to places no humans have yet laid eyes on, such as Mars. At the same time, vast improvement and miniaturization of space technologies will allow us to increase the understanding of our planet and our place in the universe. On the economic side, new markets might be opened up - such as asteroid mining - as technology gives us a firmer grip on our near-Earth surroundings.

Yet as it often goes, new opportunities are accompanied by growing or even unseen challenges. Will we be able to maintain the peaceful conduct of space activities in a world where terrestrial geopolitical tensions are rising again? Will be able to achieve new milestones in a spirit of international cooperation, especially when the challenges (manned Mars missions) or competition dynamics (asteroid resource mining) have to be overcome? Will we be able to agree on the measures that are necessary to avoid a dangerous excess of space debris – something that could jeopardize nearly all our future space endeavours?

Looking ahead into the future remains a tricky exercise, but it is clear that in order to make humanity's future ambitions and presence in space a success, space policy actors have a huge task ahead of them. In this respect, the historic development trends in space policy can be further extrapolated to make some estimated guesses regarding the next steps that need to be undertaken.

First and foremost, space policies will become more important in their communication function. With the huge interest in space activities, and many new avenues to be explored, many countries will have to make choices and set priorities. This is reinforced by the fact that space policies will have to reflect ever more how countries will relate to other countries in pursuing their space interest, in terms of sustainability, international cooperation and space governance.

Secondly, it will become nearly unavoidable to set up newly agreed governance schemes to deal with fast technological progress and issues such as space debris. These new schemes can be brought about in various ways. Since the road of hard law creation mechanisms seems to have been deserted in the space domain, efforts will probably be geared to consensus finding and the agreement of new guidelines to be followed by space nations in conducting their activities. Creativity will be more crucial than in the past, as the concept of 'launching state' - as stipulated in the legal regime for outer space - is increasingly challenged in a space environment where private companies and entrepreneurs operating on an international basis and, in cooperation with different nations.

In the end, the overall direction is becoming increasingly clear as it unfolds. After having seized stewardship over Earth's land, oceans and atmosphere, we will also conquer the 4th environment: space. At least this will occur for the local area in our solar system. The way in which we take this next step depends on how we project ourselves as a species in this new environment, which ultimately reveals a great deal about ourselves. One might say that space policies reflect a great deal of the human mind and condition in this most recent chapter of our history.

REFERENCES

Allen, J. R., & Husain, A. (2017). The Next Space Race Is Artificial Intelligence. *Foreign Policy*, *3*. Retrieved from https://foreignpolicy.com/2017/11/03/the-next-space-race-is-artificial-intelligence-and-america-is-losing-to-china/

Bi, J. (2014). *China's Space Policy: Prospects for Collaboration between the EU and China*. Europe China Research and Advice Network.

Brown, P. J. (2009, August 11). South Korea's first rocket ready - at last. *Asia Times*. Retrieved from http://www.atimes.com/atimes/Korea/KH11Dg01.html

Chen, Y. (1991). China's space policy—A historical review. *Space Policy*, *7*(2), 116–128. doi:10.1016/0265-9646(91)90022-A

Countdown to Space Force continues as Trump administration details trajectory of plan. (n.d.). Retrieved from https://eu.usatoday.com/story/news/politics/2018/10/23/space-force-trump-administration-details-trajectory-pentagon-plan/1739251002/

Doboš, B. (2015). Geopolitics of the Moon: A European Perspective. *Astropolitics*, *13*(1), 78–87. doi:10.1080/14777622.2015.1012005

EU space policy summary. (n.d.). Retrieved from http://ec.europa.eu/growth/sectors/space_en

Galimov, E. (2015). Intentions and Failures. Fundamental Space Investigations in Russia of the Last Twenty Years. Twenty Years of Fruitless Efforts. *Geochemistry International*, *54*(13), 1151–1248. doi:10.1134/S0016702915130029

GEO Strategic Plan 2016-2025: Implementing GEOSS. (n.d.). Retrieved from https://www.earthobservations.org/documents/GEO_Strategic_Plan_2016_2025_Implementing_GEOSS.pdf

GPS and the World's First "Space War". (n.d.). Retrieved from https://www.scientificamerican.com/article/gps-and-the-world-s-first-space-war/

Gruss, M. (2016). China Turned to Debris-free ASAT Tests Following 2007 Outcry. *Spacenews*. Retrieved from http://spacenews.com/u-s-official-china-turned-to-debris-free-asat-tests-following-2007-outcry/

Harding, R. C. (2016). Introduction: Space policy in developing countries. *Space Policy*, *37*(Part 1).

Harkins, H. (2017). *Counter-Space Defence Co-Orbital Satellite Fighter - The Soviet Istrebitel Sputnik Anti-Satellite Complex*. Centurion Publishing.

ISRO. (n.d.). *APRSAF-24 held in Bengaluru*. Retrieved from https://www.isro.gov.in/aprsaf-24-held-bengaluru

Krige, J. (2007). NASA as an Instrument of U.S. Foreign Policy. In S. J. Dick & R. D. Launius (Eds.), *Societal Impact of Spaceflight* (pp. 2007–4801). Washington, DC: NASA.

Lahcen, A. (2013). *EUMETSAT-NOAA Collaboration in Meteorology from Space. Review of a Long-standing Trans-Atlantic Partnership*. ESPI Report 46.

Luzin, P., (2013). Outer Space as Russia's Soft-Power Tool. *Security Index: A Russian Journal on International Security, 19*(1), 25-29. Doi:10.1080/19934270.2013.757117

Manriquez, M. (2018). Japan's Space Law Revision: the Next Step Toward Re-Militarization? *NTI.* Retrieved from http://www.nti.org/analysis/articles/japans-space-law-revision/

Martinez, P. (2012). Is there a need for an African space agency? *Space Policy*, *28*(3), 142–145. doi:10.1016/j.spacepol.2012.06.011

Nagai, Y. (2015). Policy Analysis: Space Programmes of Japan And India. *66th International Astronautical Congress*, 1–14.

National Space Policy. (n.d.). Retrieved from https://space.gov.ae//Documents/PublicationPDFFiles/National_Space_Policy_English.pdf

National Space Policy of the United States of America. (2010). Retrieved from https://www.nasa.gov/sites/default/files/national_space_policy_6-28-10.pdf

New Space Policy Directive Calls for Human Expansion Across Solar System. (n.d.). Retrieved from https://www.nasa.gov/press-release/new-space-policy-directive-calls-for-human-expansion-across-solar-system

Rajagopalan, R. P. (2018, January 20). What's Next for India's Space Program? *The Diplomat*. Retrieved from https://thediplomat.com/2018/01/whats-next-for-indias-space-program/

Sagdeev, R. (n.d.). *United States-Soviet Space Cooperation during the Cold War*. Retrieved from https://www.nasa.gov/50th/50th_magazine/coldWarCoOp.html

Salin, P.A. (n.d.). *Privatization and militarization in the space business environment*. Academic Press.

Saperstein, A. M. (n.d.). *"Weaponization" vs. "militarization" of space*. Retrieved from https://www.aps.org/units/fps/newsletters/2002/july/saperstein.pdf

Sariak, G. (2017). Between a Rocket and a Hard Place: Military Space Technology and Stability in International Relations. *Astropolitics*, *15*(1), 51–64. doi:10.1080/14777622.2017.1288509

Sarli, B., Marco Cabero, M., Lopez, A., Cardoso, J., & Jimenez, D. (2015). *South American Space Era*. 66th International Astronautical Congress - IAC 2015, Jerusalem, Israel.

Signaling strength: Russia's real Syria success is electronic warfare against the US. (n.d.). Retrieved from https://thedefensepost.com/2018/05/01/russia-syria-electronic-warfare/

Sputniknews. (n.d.). *Putin stresses importance of new Far East space center*. Retrieved from https://sputniknews.com/russia/20100828160373755/

Statement from Vice President Mike Pence on the President's Signing of Space Policy Directive. White House, May 24, 2018. (n.d.). Retrieved from https://www.whitehouse.gov/briefings-statements/statement-vice-president-mike-pence-presidents-signing-space-policy-directive-2/

The history of American space policy. (n.d.). Retrieved from https://history.nasa.gov/spdocs.html

Treaty on Principles Governing the Activities of States in the Exploration and Use of Outer Space, including the Moon and Other Celestial Bodies. (n.d.). Retrieved from http://www.unoosa.org/pdf/gares/ARES_21_2222E.pdf

UNOOSA. (n.d.). Retrieved from http://www.unoosa.org/pdf/pres/lsc2009/pres-09.pdf

Wang, J. (2016, April 14). ESA head calls for 'space without borders' during first China visit. *GBTimes*. Retrieved from https://gbtimes.com/esa-head-calls-space-without-borders-during-first-china-visit

Wright, D., Grego, L., & Gronlund, L. (2005). *The Physics of Space Security – A reference manual.* American Academy of Arts and Sciences.

Zelnio, R. (2005, December 5). A Model for the International Development of the Moon. *The Space Review*.

Zhao, Y. (2016). The Role of bilateral and multilateral agreements in international space cooperation. *Space Policy, 36.*

ENDNOTES

1 The used delineation is based upon the Cambridge Dictionary definition. One can, however, find many similar or nearly identical definitions of policy in various sources.

2 It should be recalled that this was indeed an era of great technological achievements in space as well as on Earth, and that this time was characterised by stable and vast economic growth, which further fuelled optimism regarding the future.

3 Here it should be noted that in the past, space policies were less comprehensive than today, and were less strategically oriented but more focussing on the programmatic aspects of space programmes.

4 At the time of the Space Race, the Soviet-Union and the United States were the main pioneers and still are very much among the top of space nations today.

5 In the case of launchers, most manufacturers are private companies, but the market demand has traditionally been very institutional in that public agencies typically procure launch services from private space industry.

6 The term 'market failure' in economics refers to situations where the dynamics of the market are not sufficient to support an optimal level of provision of goods or services for society or the market. In these cases, governmental intervention in the market is typically considered as justifiable or even necessary.

7 Many consider this fact to be one of the main reasons why Russia could not hold up the race with the United States as it space program was far less coordinated that the US one under the leadership of the NASA.

8 This article was written before the British withdrawal from the EU, often coined 'Brexit', was officially concluded.

9 Excerpt from Article II, Purpose, Convention of establishment of a European Space Agency, SP-1271(E), 2003.

10 The satellite obtained funding from a competitive grant from the United Nations Office for Outer Space Affairs (UNOOSA) and received technical assistance by the Japanese Space Agency JAXA.

11 The permanent members of the UN Security Council are China, France, Russia, the United Kingdom and, the United States.

12 In this context we refer in particular to the Russian territorial claims in the Arctic as well as China's sovereign claims in the South China Sea.

Compilation of References

1992. Rio Declaration on Environment and Development, UN Doc. A/CONF.151/26 (vol. I)

51 U.S. Code § 50904 and 14 CFR § 415.3 (a).

Abney, M. B., Mansell, J., Rabenberg, E., Stanley, M., Edmunson, J., Alleman, E., . . . Dumez, S. (2014). Series-Bosch Technology For Oxygen Recovery During Lunar or Martian Surface Missions. *Proceedings of the 44th International Conference on Environmental Systems.*

Acuña, M. H., Connerney, J. E. P., Wasilewski, P., Lin, R. P., Mitchell, D., Anderson, K. A., ... Ness, N. F. (2001). Magnetic field of Mars: Summary of results from the aerobraking and mapping orbits. *Journal of Geophysical Research, 106*(E10), 23403–23417. doi:10.1029/2000JE001404

Africano, J. L., Stansbery, E. G., & Kervin, P. W. (2004). The optical orbital debris measurement program at NASA and AMOS. *Advances in Space Research, 34*(5), 892–900. doi:10.1016/j.asr.2003.02.022

Agreement among the Government of Canada, Governments of Member States of the European Space Agency, the Government of Japan, the Government of the Russian Federation, and the Government of the United States of America concerning Cooperation on the Civil International Space Station (hereafter Intergovernmental Agreement), Washington, done 29 January 1998, entered into force 27 March 2001; Space Law e Basic Legal Documents, D.II.4.

Agreement among the Government of Canada, Governments of Member States of the European Space Agency, the Government of Japan, the Government of the Russian Federation, and the Government of the United States of America concerning Cooperation on the Civil International Space Station, Washington, 29 January 1998, entered into force 27 March 2001 (1998 IGA-Agreement)

Agreement among the Government of the United States of America, Governments of Member States of the European Space Agency, the Government of Japan, and the Government of Canada on Cooperation in the Detailed Design, Development, Operation, and Utilization of the Permanently Manned Civil Space Station, Washington, done 29 September 1988, entered into force between the Government of Japan and the Government of the United States 30 January 1992, (1988 IGA-Agreement)

Agreement Governing the Activities of States on the Moon and Other Celestial Bodies, 5 December 1979, 1363 *UNTS* 3 / 18 *ILM* 1434.

Agreement Governing the Activities of States on the Moon and Other Celestial Bodies, entered into force 11 July 1984, 1363 UNTS 21, (Moon Agreement, 1984)

Agreement Governing the Activities of States on the Moon and Other Celestial Bodies, entered into force July 11, 1984, 1363 U.N.T.S. 3.

Agreement on the Rescue of Astronauts, the Return of Astronauts and the Return of Objects Launched into Outer Space, 22 April 1968, 19 UST 7570, 672 UNTS 119.

Agreement on the Rescue of Astronauts, the Return of Astronauts, and the Return of Objects Launched into Outer Space, entered into force on 3 December 1968, 672 UNTS 119, (Rescue and Return Agreement, 1968)

Aguirre. (2013). *Introduction to Space Systems: Design and Synthesis*. New York: Springer.

Ailor, W. (2015). Space Traffic Management. In Handbook of Space Security. New York: Springer Science+Business Media. doi:10.1007/978-1-4614-2029-3_58

Aimé, C. (2000). *Discourse on Colonialism*. Monthly Review Press.

Alain du Parquet, C., & the Specific Clauses of Launch Services Agreements in Ingo Baumann and Lesley Jane Smith (Ed.). Contracting for Space: Contract Practice in the European Space Sector (Farnham; UK: Ashgate Publishing, 2011) ['Contracting for Space'] at 385

Alain du Parquet, C., Contracting for Space, supra note 11 at 390

Alan, C. C. (1997). *Oxygen Extraction from Lunar Samples*. Retrieved from https://curator.jsc.nasa.gov/lunar/lnews/lnmar97/oxygen.htm

Alessandro, A. (2016). *NASA's Restore-L Mission to Refuel Landsat 7*. Demonstrate Crosscutting Technologies. Retrieved April 30, 2018, from https://www.nasa.gov/feature/nasa-s-restore-l-mission-to-refuel-landsat-7-demonstrate-crosscutting-technologies

Allen, C. C., Morris, R. V., & McKay, D. S. (1996). Oxygen extraction from lunar soils and pyroclastic glass. *Journal of Geophysical Research*, *101*(E11), 26085–26095. doi:10.1029/96JE02726

Allen, J. R., & Husain, A. (2017). The Next Space Race Is Artificial Intelligence. *Foreign Policy*, *3*. Retrieved from https://foreignpolicy.com/2017/11/03/the-next-space-race-is-artificial-intelligence-and-america-is-losing-to-china/

Allison, A. L. (2014). The ITU and managing satellite orbital and spectrum resources in the 21st century. *SpringerBriefs in Space Development*.

Anderson, P. V., & Schaub, H. (2014). Local debris congestion in the geosynchronous environment with population augmentation. *Acta Astronautica*, *94*(2), 619–628. doi:10.1016/j.actaastro.2013.08.023

Andrews, E. (1990). *Tiny Tonga seeks Satellite Empire in Space*. Retrieved from https://www.nytimes.com/1990/08/28/business/tiny-tonga-seeks-satellite-empire-in-space.html

Anglo-Norwegian Fisheries (U.K v. Nor.), 1951 LC.J. 116 (Dec. 18).

AnsariX. Prize (2004). Retrieved from https://ansari.xprize.org

Antarctic Treaty of December 1, 1959, entered into force June 23, 1961, 402 U.N.T.S. 71.

Aoki, S. (2013). Efficient and equitable use of orbit by satellite systems: "Paper satellite" issues revisited. In *Proceedings of the International Astronautical Congress, IAC* (Vol. 14, pp. 11149-11155). IAF.

Aoki, S. (2009). Space Traffic Management for the Prevention of Weaponization of Outer Space. In *Proceedings of the International Institute of Space Law*. American Institute of Aeronautics and Astronautics.

Ariane 5 User's Manual Issue 5. (2011). *Process for launch window definition, s.2.7.2*. Available at: http://www.arianespace.com/wp-content/uploads/2015/09/Ariane5_users_manual_Issue5_July2011.pdf

Arianespace. (2011). *Ariane 5 User's Manual* (5th ed.). Author.

Arianespace. (2012). *Soyuz User's Manual.* Author.

Arianespace. (2014). *Vega User's Manual. Issue 4, Revision 0.* Author.

ASD-Eurospace. (2012). *Space Trends: Global Space Activity Overview 1986-2011* (1st ed.). Paris, France: Eurospace.

Autorité de Régulation des communications électroniques et des Postes (ARCEP). (2018). *4G.* Retrieved from https://www.arcep.fr/index.php?id=9971

Badescu, V. (Ed.). (2013). *Asteroids: Prospective energy and material resources.* Heidelberg, Germany: Springer. doi:10.1007/978-3-642-39244-3

Baker, H. A. (1989). *Space debris: legal and policy implications.* Nijhoff.

Balogh, W. (2012). Institutional Aspects. In *Outer Space in Society, Politics and Law.* Wien: Springer Verlag.

Balsano, A. M., & Wheeler, J. (2006). The IGA and ESA, Protecting Intellectual Property Rights in the Context of ISS Activities. In *The International Space Station: Commercial Utilisation form a European Legal Perspective.* Leiden: Martinus Nijhoff Publishers. doi:10.1163/ej.9789004152564.i-267.41

Bank of America Merill Lynch Thematic Research: "To Infinity and Beyond – Global Space Primer". October 30, 2017.

Barlow, N. (2008). *Mars: An introduction to its interior, surface and atmosphere.* Cambridge, UK: University Press Cambridge. doi:10.1017/CBO9780511536069

Bat Conservation Trust. (2018). *Echolocation.* Retrieved form http://www.bats.org.uk/pages/echolocation.html

Bender, R. (1998). *Launching and Operating Satellites: Legal Issues.* Kluwer Law Int.

Benko, M., & Schrogl, K. U. (2008). UN Committee on the Peaceful Uses of Outer Space: Adoption of the Resolution on Enhancing Registration Practice and of the UNCOPUOS Space Debris Mitigation Guidelines. *ZLW, 57*, 335.

Bertenyi, B. (n.d.). *3GPP system standards heading into the 5G era.* Retrieved from http://www.3gpp.org/news-events/3gpp-news/1614-sa_5g

Biesbroeck, R., Soares, T., Hüsing, J., & Innocenti, L. (2013). The e.Deorbit CDF Study: A Design Study for the Safe Removal of a Large Space Debris. In L. Ouwehand (Ed.), *Proceedings of the 6th European Conference on Space Debris.* Darmstadt, Germany: European Space Agency.

Bi, J. (2014). *China's Space Policy: Prospects for Collaboration between the EU and China.* Europe China Research and Advice Network.

Blackerby, C., Okamoto, A., Fujimoto, K., Okada, N., Forshaw, J. L., & Auburn, J. (2018, September 14). ELSA-d: An In-Orbit End-of-Life Demonstration Mission. In *Proceedings of the 69th International Astronautical Congress.* Bremen, Germany: International Astronautical Federation.

Blount, P. J. (2009). Limits on Space Weapons: Incorporating the Law of War Into the Corpus Juris Spatialis. *Proceedings of the 51st Colloquium on the Law of Outer Space.* Available at http://papers.ssrn.com/abstract=1393321

Blount, P. J. (2018). Innovating the Law: Fifty Years of the Outer Space Treaty. In M. Hofmann & P.J. Blount, (Eds.), Innovation in Outer Space: International and African Legal Perspectives (pp. 31-52). Oxford, UK: Nomos.

Blount, P.J. (2011). Developments in Space Security and Their Legal Implications. *Law/Technology, 44*(2), 18-39.

Blue Origin. (2015). Retrieved from https://www.blueorigin.com/

Bobe, L. S., Samsonov, N. M., Soloukhin, V. A., Farafonov, N. S., Novikov, V. M., Andreichuk, P. O., ... Skuratov, V. M. (2005). Water Supply of the Crew of a Space Station Through Water Recovery and Water Delivery: SRV-K and SPK-U System Operation on ISS. *Proceedings of the 35th International Conference on Environmental Systems.* 10.4271/2005-01-2806

Bockstahler, K., Matthias, C., Lucas, J., Witt, J., Hovland, S., & Laurini, D. (2015). Development Status of the Advanced Closed Loop System ACLS. *Proceedings of the 45th International Conference on Environmental Systems.*

Böckstiegel, K. H. (1981). Legal implications of commercial space activities. *Proc. Coll. L. Outer Space*, *24*, 1–17.

Bolden, C. (2014). Obama Administration Extends International Space Station Until at Least 2024. *NASA Blogs*. Available at https://blogs.nasa.gov/bolden/2014/01/08/obama-administration-extends-international-space-station-until-at-least-2024/

Bombardelli, C. (2012). A Plan to De-orbit Envisat. In *Proceedings of the 2nd European Workshop on Active Debris Removal*. Paris, France: Centre national d'études spatiales.

Bonnal, C., Ruault, J.-M., & Desjean, M.-C. (2013). Active debris removal: Recent progress and current trends. *Acta Astronautica*, *85*, 51–60. doi:10.1016/j.actaastro.2012.11.009

Bora, K. (2014). Russia Threatens to Ban US from using International Space Station, Bars Rocket Engines for military Use over Ukraine Crisis. *International Business Times*. Available at http://www.ibtimes.com/russia-threatens-ban-us-using-international-space-station-bars-rocket-engines-military-use-1584065

Bourbonniére, M. (2005). National-Security Law in Outer Space: The Interface of Exploration and Security. *Journal of Air Law and Commerce*, *70*, 3–62.

Bourbonnière, M., & Lee, R. J. (2007). Legality of the Deployment of Conventional Weapons in Earth Orbit: Balancing Space Law and the Law of Armed Conflict. *European Journal of International Law*, *18*(5), 873–901. doi:10.1093/ejil/chm051

Bousquet, M. (2018). ISAE-SUPAERO lecture: Introduction to satellite communication systems. Toulouse, France: ISAE-SUPAERO.

Braun, J. (2015). *Influence Analysis for Liquid Satellite Attitude Control Thruster Design* (Bachelor's thesis). Technische Universität München, Institute of Flight Propulsion.

Braun, R. D., & Manning, R. M. (2007). Mars exploration entry, descent, and landing challenges. *Journal of Spacecraft and Rockets*, *44*(2), 310–323. doi:10.2514/1.25116

Braun, V., Lüpken, A., Flegel, S., Gelhaus, J., Möckel, M., Kebschull, C., ... Vörsmann, P. (2013). Active debris removal of multiple priority targets. *Advances in Space Research*, *51*(9), 1638–1648. doi:10.1016/j.asr.2012.12.003

Broad, W. J. (2010). China Explores a Rich Frontier, Two Miles Deep. *New York Times*, p. A1.

Brodkin, J. (2017, March). With latency as low as 25ms, SpaceX to launch broadband satellites in 2019. *Ars Technica*. Retrieved from https://arstechnica.com/information-technology/2017/05/spacexs-falcon-9-rocket-will-launch-thousands-of-broadband-satellites/

Brown, P. J. (2009, August 11). South Korea's first rocket ready - at last. *Asia Times*. Retrieved from http://www.atimes.com/atimes/Korea/KH11Dg01.html

Brownlie, I. (1990). *Principles of Public International Law*. Oxford University Press.

Buhrig, A., Cox, S., Alexander, M., Krenitsky, M., & Huebner, C. (1996). Sea Launch Program Overview. In *16th International Communications Satellite Systems Conference* (pp. 58-67). Washington, DC: AIAA.

Bussey, B., & Davis, R. (2015). *First landing site/ exploration zone workshop for human missions to the surface of Mars: Second announcement.* Retrieved from https://www.nasa.gov/sites/default/files/atoms/files/hls2-2nd-announcement-7-8.pdf

Butler, L. B., Faszcza, J. J., & Thomas, G. A. (1997). Development Status of the Metal Oxide Regenerable CO2 Removal System for the NASA EMU. *Proceedings of the 27th International Conference on Environmental Systems.* 10.4271/972505

Canada Island Visit Angers Danes. (2005). BBC News. http://news.bbc.co.uk/2/hi/europe/4715245.stm

Carpanelli, E., & Cohen, B. (2014). Interpreting "Damage Caused by Space Objects" under the 1972 Liability Convention. In *Proceedings of the International Institute of Space Law 2013.* The Hague: Eleven International Publishing.

Carpenter, J., & Fisackerly, R. (2017). PROSPECT: ESA's package for resource observation and in situ prospecting for exploration, commercial exploitation and transportation. *Proceedings of the 48th Lunar and Planetary Science Conference.*

Carrasquillo, R. L., Cloud, D., & Kundrotas, R. (2004). Status of the Node 3 Regenerative ECLSS Water Recovery and Oxygen Generation Systems. *Proceedings of the 34th International Conference on Environmental Systems.* 10.4271/2004-01-2384

Carter, L., Tobias, B., & Orozco, N. (2012). Status of ISS Water Management and Recovery. *Proceedings of the 42nd International Conference on Environmental Systems.* 10.2514/6.2012-3594

Cassidy, J. L.; Johnson, R. I.; Leveye, J. C.; Miller, F. E. (1964). *The Mercury-Redstone Project.* NASA.

CBC News. (2015). US space-mining law seen leading to possible treaty violations. *CBC News.* Retrieved on 30 April 2018 from http://www.cbc.ca/news/technology/space-mining-us-treaty-1.3339104

Center for the Advancement of Science in Space. (2015). *History and Timeline of the ISS.* Retrieved January 12, 2015, from http://www.iss-casis.org/About/ISSTimeline.aspx

Cesaretti, G., Dini, E., de Kestelier, X., Colla, V., & Pambaguian, L. (2014). Building components for an outpost on the Lunar soil by means of a novel 3D printing technology. *Acta Astronautica, 93*, 430–450. doi:10.1016/j.actaastro.2013.07.034

Chandler, D. (1991, May). Weightlessness and Microgravity. *The Physics Teacher, 312–13.* doi:10.1119/1.2343327

Charter of the United Nations, entered into force 24 October 1945, 1 UNTS XVI (UN-Charter)

Charter of the United Nations, entered into force Oct. 24, 1945, U.N.T.S. XVI.

Charter of the United Nations. June 26, 1945.

Check out this incredible Mars City the UAE is building for training purposes. Last accessed May 8, 2018,

Cheng, B. (1997). *Studies in International Space Law.* Oxford, UK: Clarendon Press. doi:10.1093/acprof:oso/9780198257301.001.0001

Chen, Y. (1991). China's space policy—A historical review. *Space Policy, 7*(2), 116–128. doi:10.1016/0265-9646(91)90022-A

Chowdhury, H. (2018, March 29). Mega-constellations of satellites increase space junk risk. *Financial Times.* Retrieved from http://www.ft.com

Christensen, P. R. (2003). Formation of recent martian gullies through melting of extensive water-rich snow deposits. *Nature*, *422*(6927), 45–47. doi:10.1038/nature01436 PMID:12594459

Christensen, P. R. (2006). Water at the Poles and in Permafrost Regions of Mars. *Elements*, *2*(3), 151–155. doi:10.2113/gselements.2.3.151

Christiansen, E. L., Crews, J. L., Williamsen, J. E., Robinson, J. H., & Nolen, A. M. (1995). Enhanced meteoroid and orbital debris shielding. *International Journal of Impact Engineering*, *17*(1–3), 217–228. doi:10.1016/0734-743X(95)99848-L

Christol, C. Q. (1992). Legal Aspects of Aerospace Planes. In *International Conference on the Law, Policy, and Commerce of International Air Transport and Space Activities. The Highways of air and outer space over Asia.* Dordrecht: Martinus Nijhoff Publishers.

Christol, C. (1982). *The Modern International Law of Outer Space*. New York: Pergamon Press.

Christol, C. Q. (1984). Article 2 of the 1967 Principles Treaty revisited. *Annals of Air and Space Law*, *9*, 217–264.

Chullen, C., Campbell, C., Papale, W. G., Hawes, K., & Wichowski, R. (2015). Rapid Cycle Amine 3.0 System Development. *Proceedings of the 45th International Conference on Environmental Systems.*

Clark, S. (2015, July). OneWeb launch deal called largest commercial rocket buy in history. *Spaceflight Now.* Retrieved from https://spaceflightnow.com/2015/07/01/oneweb-launch-deal-called-largest-commercial-rocket-buy-in-history/

Clark, D. L., Keller, B. W., & Kirkland, J. A. (2009). Field Test Results of the PILOT Hydrogen Reduction Reactor. *AIAA SPACE Conference & Exposition.*

Cobley, P. (n.d.). *Communication.* Retrieved from https://semioticon.com/seo/C/communication.html

Cocks, F. H. (2010). 3He in permanently shadowed lunar polar surfaces. *Icarus*, *206*(2), 778–779. doi:10.1016/j.icarus.2009.12.032

Codding, G. A., & Rutkowski, A. M. (1982). *The International Telecommunication Union in a changing world.* Artech House on Demand.

Cogliati-Bantz, V. (2015). *Means of Transportation and Registration of Nationality: Transportation registered by International Organizations.* Abingdon, UK: Routledge.

Cohen, M. R. (1927). Property and sovereignty. *Cornell L.Q.*, *13*, 8–30.

Colaprete, A., Schultz, P., Heldmann, J., Wooden, D., Shirley, M., Ennico, K., ... Sollitt, L. (2010). Detection of water in the LCROSS ejecta plume. *Science*, *330*(6003), 463–468. doi:10.1126cience.1186986 PMID:20966242

Collins, P., Iwasaki, Y., Kanayama, H., & Ohnuki, M. (1994). Commercial Implications of Market Research on Space Tourism. *The Journal of Space Technology and Science.*, *10*(2), 3–11.

Colony Law and Legal Definition. (n.d.). Last accessed January 16, 2018. http://definitions.uslegal.com/c/colony/

Constitution of the International Telecommunication Union (ITU Constitution), ATS (1994) 28; BTS 24 (1996)

Contant-Jorgenson, C., Lala, P., & Schrogl, K. U. (2006). *Cosmic Study on Space Traffic Management.* Paris: International Academy of Astronautics. doi:10.1016/j.spacepol.2006.08.004

Convention Establishing the World Intellectual Property Organization. done at Stockholm on 14 July 1967, as amended September 1979, 828 UNTS 3, (WIPO Convention)

Convention on International Civil Aviation (Chicago Convention), Chicago, 7 December 1944; 15 UNTS 295; TIAS 1591; 61 Stat. 1180; Cmd. 6614; UKTS 1953 No. 8; ATS 1957 No. 5; ICAO Doc. 7300; Annex 2, Rules of Air, Chapter I Definitions.

Convention on International Liability for Damage Caused by Space Objects, 1972, 961 UN.T.S. 187.

Convention on International Liability for Damage Caused by Space Objects, 29 March 1972, 961 *UNTS* 187.

Convention on International Liability for Damage Caused by Space Objects, 961 UNTS 187 (1971): As of 2016, the Liability Convention has been ratified by 92 States (including all major space-faring nations)

Convention on International Liability for Damage Caused by Space Objects, entered into force 1 September 1972, 961 UNTS 187, (Liability Convention, 1972)

Convention on Registration of Objects Launched into Outer Space, entered into force 15 September 1976, 1023 UNTS 15, (Registration Convention, 1976)

Convention on the Registration of Objects Launched into Outer Space, Mar. 29 1972, 1023 U.N.T.S. 15

COPUOS History. (n.d.). *The website of United Nations Office for Outer Space Affairs*. Available at: http://www.unoosa.org/oosa/en/ourwork/copuos/history.html

COSPAR. (n.d.). Retrieved from https://cosparhq.cnes.fr/about

Countdown to Space Force continues as Trump administration details trajectory of plan. (n.d.). Retrieved from https://eu.usatoday.com/story/news/politics/2018/10/23/space-force-trump-administration-details-trajectory-pentagon-plan/1739251002/

Crawford, J. (n.d.). *Responsibility to the International Community as a Whole*. Retrieved from http://lcil.law.cam.ac.ukISnyderlectOO(f).doc

Crawford, I. A. (2015). Lunar resources. *Progress in Physical Geography*, *39*(2), 137–167. doi:10.1177/0309133314567585

D'Errico, P. (2001). Native Americans in America: A Theoretical and Historical Overview. In F. E. Hoxie, P. C. Mancall, & J. H. Merrell (Eds.), *American Nations: Encounters In Indian Country, 1850 To The Present*. New York: Routledge.

DARPA. (2016, Mar. 25). *Program Aims to Facilitate Robotic Servicing of Geosynchronous Satellites*. Retrieved from: http://www.darpa.mil/news-events/2016-03-25

Davenas, A. (1993). *Solid Rocket Propulsion Technology*. Pergamon Press.

Davenport, C. (2016). *Companies flood Earth's orbit with satellites, but no one's directing traffic*. Retrieved from: https://www.washingtonpost.com/business/economy/companies-flood-earths-orbit-with-satellites/2016/07/22/9db2e522-49ee-11e6-90a8-fb84201e0645_story.html?utm_term=.5066ccce7f07

David, L. (2017). *Private Space Station Coming Soon?* Available at https://www.scientificamerican.com/article/private-space-station-coming-soon-company-aiming-for-2020-launch/

Davis, J. (2018). Goodbye, ISS. Hello, private space stations? *The Planetary Society*. Available at http://www.planetary.org/blogs/jason-davis/2018/20180220-bye-iss-hello-private-stations.html

De Man, P. (2013). The relevance of ITU regulations for clarifying the space debris concept and strengthening guidelines on the removal of satellites at the end of their functional life. *Annals of Air and Space Law*, *38*, 203-236.

De Man, P. (2017). Luxembourg law on space resources rests on contentious relationship with international law. *The Space Review*. Retrieved on 30 April 2018 from http://thespacereview.com/article/3355/1

De Man, P. (2015). *Exclusive use in an inclusive environment: the meaning of the non-appropriation principle for space resource exploitation*. Berlin: Springer.

de Peuter, W., Visentin, G., Fehse, W., Elfving, A., Brown, D., & Ashford, E. (1994). Satellite Servicing in GEO by Robotic Servicing Vehicle. *ESA Bulletin*, (78): 33–39.

De Selding, P. B. (2016). *Enough satellites to darken the skies*. Retrieved from: http://www.spacenewsmag.com/the-bottom-line/enough-satellites-to-darken-the-skies/

Decision City of Sherrill vs. Oneida Indian Nation of N.Y.

Declaration of the United Nations Conference on the Human Environment, U.N. Doc. A/Conf.48/14/Rev. 1, Princ. 21 (1973)

Delzeit, A. N., & Beal, R. F. (1996). The Vulnerability of the Pacific Rim Orbital Spectrum Under International Space Law. *NY Int'l L. Rev.*, *9*, 69.

Dembling, P., & Arons, D. (1967). The Evolution of the Outer Space Treaty. *Journal of Air Law and Commerce*, *33*, 419–456.

Dempsey, P. S. (2015). *The Emergence of National Space Law*. Available at SSRN: http://ssrn.com/abstract=2692639

Department of Defense. (2008, Feb. 14). Transcript: Department of Defense News Briefing with Deputy National Security Advisor Jeffrey, Gen. Cartwright and NASA Administrator Griffin. In P. J. Blount & J. I. Gabrynowicz (Eds.), *USA-193: Selected Documents (51). University*. National Center for Remote Sensing, Air, and Space Law.

Detter, I. (1996), The effect of Resolutions of International Organizations. In Theory of International Law at the Threshold of the 21st Century– Essays in Honor of Krzysztof Skubiszewski. The Hague: Academic Press.

Dicky, V., Khutorovsky, Z., Kuricshah, A., Menshicov, A., Nazirov, R., Pitsyck, A., ... Yurasov, V. (1993). The Russian space surveillance system and some aspects of spaceflight safety. *Advances in Space Research*, *13*(8), 21–31. doi:10.1016/0273-1177(93)90564-R

Dimopolous, A. (2015). An Institutional Perspective II: The Role of the CJEU in the Unitary (EU) Patent System. In *The Unitary EU Patent System*. London: Hart/Bloomsbury.

Dinerman, T. (2008). *Business on the ISS beyond space tourism*. The Space Review.

Dirk Krebs, G. (2017). *Syncom 1, 2, 3*. Retrieved from http://space.skyrocket.de/doc_sdat/syncom-1.htm

Doboš, B. (2015). Geopolitics of the Moon: A European Perspective. *Astropolitics*, *13*(1), 78–87. doi:10.1080/14777622.2015.1012005

Doc, U. S. Space Policy Directive-3, National Space Traffic Management Policy, issued on 10 June 2018. (n.d.). Available at: https://www.whitehouse.gov/presidential-actions/space-policy-directive-3-national-space-traffic-management-policy/

Doctrine of Discovery". HR/5088. United Nations Economic and Social Council. May 8, 2012.

Does Elon Musk's Plan Violate The Outer Space Treaty - Planetary Protection For Mars After Human Crashes. (n.d.). Retrieved from http://www.science20.com/robert_inventor/does_elon_musks_plan_violate_the_outer_space_treaty_planetary_protection_for_mars_after_human_crashes-180496

Dolado-Perez, J. C., Pardini, C., & Anselmo, L. (2015). Review of uncertainty sources affecting the long-term predictions of space debris evolutionary models. *Acta Astronautica*, *113*, 51–65. doi:10.1016/j.actaastro.2015.03.033

Dorizon, S., Ciarletti, V., Plettemeier, D., & Benedix, W. S. (2016). Performance validation of the ExoMars 2018 WISDOM GPR in ice caves, Austria. *Planetary and Space Science*, *120*, 1–14. doi:10.1016/j.pss.2015.10.008

Dorsey, J. T., Chauncey Wu, K., & Smith, R. (2008). *Structural definition and mass estimation of lunar surface habitats for the lunar architecture team phase 2 (LAT-2) study.* Retrieved from https://ntrs.nasa.gov/search.jsp?R=20080013395

Do, S., Owens, A., Ho, K., Schreiner, S., & de Weck, O. (2016). An independent assessment of the technical feasibility of the Mars One mission plan – Updated analysis. *Acta Astronautica, 120*, 192–228. doi:10.1016/j.actaastro.2015.11.025

Drake, B. G. (2009). *Human Exploration of Mars Design Reference Architecture 5.0* (No. NASA/SP-2009.566).

Drake, B. G., & Watts, K. D. (2009). *Human exploration of Mars design reference architecture 5.0.* Retrieved form https://www.nasa.gov/sites/default/files/files/NASA-SP-2009-566-ADD2.pdf

Duhaime's Law Dictionary. (n.d.). Retrieved from http://www.duhaime.org/LegalDictionary/E/ErgaOmnes.aspx

Dundas, C. M., Bramson, A. M., Ojha, L., Wray, J. J., Mellon, M. T., Byrne, S., ... Holt, J. W. (2018). Exposed subsurface ice sheets in the Martian mid-latitudes. *Science, 359*(6372), 199–201. doi:10.1126cience.aao1619 PMID:29326269

Easton, R. L., & Votaw, M. J. (1959). Vanguard I IGY Satellite (1958 Beta). *The Review of Scientific Instruments, 30*(2), 70–75. doi:10.1063/1.1716492

Ebbesen, S. D., & Mogensen, M. (2009). Electrolysis of carbon dioxide in Solid Oxide Electrolysis Cells. *Journal of Power Sources, 193*(1), 349–358. doi:10.1016/j.jpowsour.2009.02.093

Eberspeaker, P., & Pierce, D. (2013, September 21). *An Overview of the NASA Sounding Rockets and Balloon Programs.* NASA.

Ebisuzaki, T., Quinn, M. N., Wada, S., Piotrowski, L. W., Takizawa, Y., Casolino, M., ... Mourou, G. (2015). Demonstration designs for the remediation of space debris from the International Space Station. *Acta Astronautica, 112*, 102–113. doi:10.1016/j.actaastro.2015.03.004

Eckart, P. (1996). *Spaceflight Support and Biospherics.* Microcosm Press, Kluwer Academic Publishers Incorporated.

Einstein, A. (1916). Die Grundlage der allgemeinen Relativitätstheorie. *Annalen der Physik, 49*, 769–822. doi:10.1002/andp.19163540702

Elleman, M. (2013). Prelude to an ICBM? Putting North Korea's Unha-3 Launch into Context. *Arms Control Today, 43*(2), 8–13.

Ellery, A., Lowing, P., Wanjara, P., Kirby, M., Mellor, I., & Doughty, G. R. (2017). FCC Cambridge Process with Metal 3D Printing as Universal In-Situ Resource Utilisation. *Proceedings of the 14th Symposium on Advanced Space Technologies in Robotics and Automation (ASTRA).*

Erwin, S., & Henry, C. (2018, January 4). Orbital ATK lands second Intelsat satellite servicing deal. *SpaceNews.* Retrieved from http://www.spacenews.com

ESA and Arianespace sign a Frame Contract for the Procurement of Launch Services. (2009, June 15). ESA Press Release. Available at: http://www.esa.int/Our_Activities/Launchers/ESA_and_Arianespace_sign_a_Frame_Contract_for_the_procurement_of_launch_services

ESA. (n.d.a). *International Space Station Legal Framework.* Available at http://www.esa.int/Our_Activities/Human_Spaceflight/International_Space_Station/International_Space_Station_legal_framework

ESA. (n.d.b). *International Space Station, Barter Agreements.* Available at http://m.esa.int/Our_Activities/Human_Spaceflight/International_Space_Station/ESA_s_International_Space_Station_barter_agreements

ESA. (n.d.c). *Patents and Space-Related Inventions*. Available at http://m.esa.int/About_Us/Law_at_ESA/Intellectual_Property_Rights/Patents_and_space-related_inventions

EU space policy summary. (n.d.). Retrieved from http://ec.europa.eu/growth/sectors/space_en

EUROCONTROL. (n.d.). *What is air traffic management?* Available from: http://www.eurocontrol.int/articles/what-air-traffic-management

European Space Agency. (1993). *Meteosat Second Generation. The Satellite Development*. ESA BR-153.

European Space Agency. (2010). *Environment Control and Life Support System*. ECLSS.

European Space Agency. (2013). *Cupola*. Retrieved January 12, 2015, from http://www.esa.int/Our_Activities/Human_Spaceflight/International_Space_Station/Cupola

European Space Agency. (2017, January). *Space Debris by Numbers*. Retrieved from https://www.esa.int/Our_Activities/Operations/Space_Debris/Space_debris_by_the_numbers

FAA Regulation 14 C.F.R. § 415 (1998)

Falcon 9 Launch Services Agreement entered into between SpaceX and Orbcomm in 2012, Exhibit A 'Defined Terms' at p.24. (n.d.). Available at: http://www.sec.gov/Archives/edgar/data/1361983/000119312513112208/d468141dex102.htm

Fanon, F. (1967). *The Wretched of the Earth*. London: Penguin.

Farand, A. (2005). *Astronauts' behaviour onboard the International Space Station: regulatory framework*. available at http://portal.unesco.org/shs/en/files/8480/11224719601Farand.pdf/Farand.pdf

Farand, A. (1998). The Space Station Cooperation Framework. *ESA Bulletin*, 94.

Fasan, E. (1984). Large space structures and celestial bodies. *Proc. Coll. L. Outer Space*, *27*, 243–246.

Fasan, E. (1987). Celestial bodies and the exploitative use of outer space. *Annals of Air and Space Law*, *12*, 227–234.

Fasan, E. (1998). Asteroids and other celestial bodies - some legal differences. *J. Space L.*, *26*, 33–40.

Fateri, M., & Gebhardt, A. (2015). Process Parameters Development of Selective Laser Melting of Lunar Regolith for On-Site Manufacturing Applications. *International Journal of Applied Ceramic Technology*, *12*(1), 46–52. doi:10.1111/ijac.12326

Federation of American Scientists. (n.d.). *Prevention of an Arms Race in Outer Space*. Retrieved from https://fas.org/programs/ssp/nukes/ArmsControl_NEW/nonproliferation/NFZ/NP-NFZ-PAROS.html

Feldman, W. C. (2004). Global distribution of near-surface hydrogen on Mars. *Journal of Geophysical Research*, *109*(E9), E09006. doi:10.1029/2003JE002160

Flegel, S., Krisko, P., Gelhaus, J., Wiedemann, C., Möckel, M., Vörsmann, P., … Matney, M. (2010). Modeling the Space Debris Environment with MASTER2009 and ORDEM2010. In *Proceedings of the 38th COSPAR Sceintific Assembly*. Bremen, Germany: COSPAR.

Forshaw, J. L., Aglietti, G. S., Navarathinam, N., Kadhem, H., Salmon, T., Pisseloup, A., ... Steyn, W. H. (2016). RemoveDEBRIS: An in-orbit active debris removal demonstration mission. *Acta Astronautica*, *127*, 448–463. doi:10.1016/j.actaastro.2016.06.018

Fortescue, B., Swinerd, G., & Stark, J. (2011). *Spacecraft Systems Engineering*. John Wiley & Sons Ltd. doi:10.1002/9781119971009

Forward, R. L., Hoyt, R. P., & Uphoff, C. W. (2000). Terminator Tether™: A Spacecraft Deorbit Device. *Journal of Spacecraft and Rockets*, *37*(2), 187–196. doi:10.2514/2.3565

Foster, J. L., Benbrook, J. R., & Stansbery, E. G. (2005). Detection of small radar cross-section orbital debris with the Haystack radar. *Advances in Space Research*, *35*(7), 1210–1213. doi:10.1016/j.asr.2005.05.041

Foust, J. (2014). Hearing raises questions about asteroid mining bill. *Space News*. Retrieved on 30 April 2018 from http://spacenews.com/41825hearing-raises-questions-about-asteroid-mining-bill/

Foust, J. (2016, June 16). *Space News - Blue origin flies New Shepard again*. Retrieved from http://spacenews.com/blue-origin-flies-new-shepard-again/

Foust, J. (2018a, January 17). Effective Space signs first contract for satellite life extension services. *SpaceNews*. Retrieved from http://www.spacenews.com

Foust, J. (2018b, March 14). Orbital ATK unveils new version of satellite servicing vehicle. *SpaceNews*. Retrieved from http://www.spacenews.com

Francis, R., Estlin, T., Doran, G., Johnstone, S., Gaines, D., Verma, V., ... Bornstein, B. (2017). AEGIS autonomous targeting for ChemCam on Mars Science Laboratory: Deployment and results of initial science team use. *Science Robotics*, *2*(7), 1–12. doi:10.1126cirobotics.aan4582

Freeland, S. (2010). Fly Me to the Moon: How Will International Law Cope with Commercial Space Tourism? *Melbourne Journal of International Law*, 11, 90.

Freeland, S. (2010). Matching detail with practice: The essential elements of National Space Legislation. *Proceedings of the International Institute of Space Law*, *540*, 541.

Freeland, S., & Nakarada Pecujlic, A. (Forthcoming). How do you like your Regulation – Hard or Soft? *The Antarctic Treaty and the Outer Space Treaty Compared.*

French Space Operations Act, No. 2008-518 (2008), Article 1, translation from Philippe Clerc 'Consequences of the French Space Law on Space Operations (FSOA) on CNES' Mission' in Contracting for Space, supra note 11

Frost, R. (2014). *Here's What Happens If You Commit A Crime in Outer Space*. Available at http://www.businessinsider.com/heres-what-happens-if-you-commit-a-crime-in-outer-space-2014-1?IR=T

Frowein, J. A. (2004). The Internal and External Effects of Resolutions by International Organizations. *Völkerrecht—Menschenrechte—Verfassungsfragen Deutschlands und Europas*, 81-94.

Futron Corporation. (2004). *Design Reliability Comparison for SpaceX Falcon Vehicles*. Research Report sponsored by SpaceX.

Gabčíkovo-Nagymaros Project (Separate Opinion Weeramantry), p. 118

Gabrynowicz, J. I. (2010). One half century and counting: The evolution of US national space law and three long-term emerging issues. *Harv. L. & Pol'y Rev*, *4*, 405.

Gabrynowicz, J. I. (2016). Review of the book: The Meaning of Liberty beyond Earth. *Quest.*, *23*(1).

Gál, G. (1996). Acquisition of property rights in the legal regime of celestial bodies. *Proc. Coll. L. Outer Space*, *39*, 45–49.

Galimov, E. (2015). Intentions and Failures. Fundamental Space Investigations in Russia of the Last Twenty Years. Twenty Years of Fruitless Efforts. *Geochemistry International*, *54*(13), 1151–1248. doi:10.1134/S0016702915130029

Garcia, M. (2016). *About the Space Station: Facts and Figures*. Retrieved January 1, 2016, from https://www.nasa.gov/mission_pages/station/main/onthestation/facts_and_figures.html

Garcia, M. (2018). *Visiting Vehicle Launches, Arrivals and Departures*. Retrieved from https://www.nasa.gov/feature/visiting-vehicle-launches-arrivals-and-departures

General Assembly Resolution 1514 (XV) of December 14, 1960.

General Assembly Resolution 65/119, A/RES/65/119, December 10, 2010.

GEO Strategic Plan 2016-2025: Implementing GEOSS. (n.d.). Retrieved from https://www.earthobservations.org/documents/GEO_Strategic_Plan_2016_2025_Implementing_GEOSS.pdf

Georgakas, S. (2012). *Contamination Risk During EVA. Space Safety Magazine*.

Gibson, M. A., & Knudsen, C. W. (1985). Lunar Oxygen Production from Ilmenite. In W. W. Mendel (Ed.), Lunar Bases and Space Activities of the 21st Century (pp. 543–550). Academic Press.

Glavin, D. P., Malespin, C., ten Kate, I. L., Getty, S. A., Holmes, V. E., Mumm, E., . . . Mahaffy, P. R. (2012). Volatile Analysis by Pyrolysis of Regolith for planetary resource exploration. In *2012 IEEE Aerospace Conference: Big Sky, Montana*, March 3 - 10, 2012 (pp. 1–11). Piscataway, NJ: IEEE. 10.1109/AERO.2012.6187065

Global Solutions for the Aerospace Market. (n.d.). Retrieved from http://magellan.aero/product/rockets/

Globalstar. (2000). *Description of Globalstar System*. Retrieved from https://gsproductsupport.files.wordpress.com/2009/04/description-of-the-globalstar-system-gs-tr-94-0001-rev-e-2000-12-07.pdf

Goedhuis, D. (1981). Some recent trends in the interpretation and the implementation of the rules of international space law. *Colum. J. Transnat'l L.*, *19*, 213–234.

Gorove, K.M. (2000). Delimitation of Outer Space and the Aerospace Object – Where is the Law? *Journal of Space Law*, *28*, 11-16.

Gorove, S. (1984). International space law in perspective - some major issues, trends and alternatives. *Rec. Cours*, *3*(181), 3-12.

Gorove, S. (1990). Environmental risks arising from space activities: focus on the Liability Convention. In *Environmental aspects of activities in outer space: state of the law and measures of protection*. Carl Heymanns.

Gorove, S. (1993). Toward a clarification of the term «space object» - an international legal and policy imperative? *J. Space L.*, *21*, 11–26.

GPS and the World's First "Space War". (n.d.). Retrieved from https://www.scientificamerican.com/article/gps-and-the-world-s-first-space-war/

Graham, A. R., & Kingston, J. (2015). Assessment of the commercial viability of selected options for on-orbit servicing (OOS). *Acta Astronautica*, *117*, 38–48. doi:10.1016/j.actaastro.2015.07.023

Grego, L. (2012). *A History of Anti-satellite Programs, 4*. Retrieved from: http://www.ucsusa.org/sites/default/files/legacy/assets/documents/nwgs/a-history-of-ASAT-programs_lo-res.pdf

Groemer, G., Soucek, A., Frischauf, N., Stumptner, W., Ragonig, C., Sams, S., ... Sivenesan, C. (2014). The MARS2013 Mars Analog Mission. *Astrobiology*, *14*(5), 360–376. doi:10.1089/ast.2013.1062 PMID:24823799

Grush, L. (2018). *The Verge*. Retrieved from SpaceX just launched two of its space internet satellites — the first of nearly 12,000: https://www.theverge.com/2018/2/15/17016208/spacex-falcon-9-launch-starlink-microsat-2a-2b-paz-watch-live

Gruss, M. (2016). China Turned to Debris-free ASAT Tests Following 2007 Outcry. *Spacenews*. Retrieved from http://spacenews.com/u-s-official-china-turned-to-debris-free-asat-tests-following-2007-outcry/

Haidn, O. J. (2008). *Advanced Rocket Engines*. Advances on Populsion Technology for High-Speed Aircraft.

Haidn, O. J. (2014). *Space Propulsion I, Lecture Script*. Garching: Technische Universität München, Institute of Turbomachinery and Flight Propulsion.

Harding, R. C. (2016). Introduction: Space policy in developing countries. *Space Policy*, *37*(Part 1).

Harkins, H. (2017). *Counter-Space Defence Co-Orbital Satellite Fighter - The Soviet Istrebitel Sputnik Anti-Satellite Complex*. Centurion Publishing.

Harrison, A. A. (2001). *Spacefaring The Human Dimension*. Berkeley, CA: University of California Press.

Hayne, P. O., Hendrix, A., Sefton-Nash, E., Siegler, M., Lucey, P. G., Retherford, K. D., ... Paige, D. (2015). Evidence for exposed water ice in the Moon's south polar regions from Lunar Reconnaissance Orbiter ultraviolet albedo and temperature measurements. *Icarus*, *255*, 58–69. doi:10.1016/j.icarus.2015.03.032

Hearsey, C. M. (2015). *The Evolution of Outer Space Law: An Economic Analysis of Rule Formation* (Masters Thesis). Available from ProQuest Dissertations and Theses database. (UMI No. 10003525)

Henry, C. (2017, March). Blue Origin signs OneWeb as second customer for New Glenn reusable rocket. *Space News*. Retrieved from http://spacenews.com/blue-origin-gets-oneweb-as-second-new-glenn-customer/

Henry, C. (2017, October). SpaceX, OneWeb detail constellation plans to Congress. *Space News*. Retrieved from http://spacenews.com/spacex-oneweb-detail-constellation-plans-to-congress/

Henry, C. (2017a, June 29). MDA restarts satellite servicing business with SES as first customer. *SpaceNews*. Retrieved from http://www.spacenews.com

Henry, C. (2017b, September 28). Airbus to challenge SSL, Orbital ATK with new space tug business. *SpaceNews*. Retrieved from http://www.spacenews.com

Henry, C. (2018, April). Startup with Softbank, Airbus investment planning video constellation with several hundred satellites. *Space News*. Retrieved from http://spacenews.com/startup-with-softbank-airbus-investment-planning-video-constellation-with-several-hundred-satellites/

Hertzfeld, H. (2009). Current and Future Issues In International Space Law. *ILSA Journal of International & Comparative Law*, *15*(2), 327.

Hettrich, S., Dinkelaker, A. N., Sejkora, N., Pfeil, I., Scornet, Q., Moser, L., ... Luger, U. (2014, May). Efficiency analysis of the MARS2013 planning strategy. *Astrobiology*, *14*(5), 377–390. doi:10.1089/ast.2013.1076 PMID:24766437

Hibbitts, C., Grieves, G., Poston, M. J., Dyar, M. D., Alexandrov, B., Johnson, M., & Orlando, T. M. (2011). Thermal stability of water and hydroxyl on the surface of the Moon from temperature-programmed desorption measurements of lunar analog materials. *Icarus*, *213*(1), 64–72. doi:10.1016/j.icarus.2011.02.015

Hobe, S. (2015). Rights and Obligations in the International Commons: The Case of Outer Space. In *IISL Proceedings*. Eleven Publisher.

Hobe, S. (Ed.). (2009). *Cologne Commentary on Space Law: Volume 1. Outer Space Treaty*. Koln, Germany: Carl Heymanns Verlag.

Hobe, S., & Cloppenburg, J. (2004). Towards a New Aerospace Convention? Selected Legal Issues of "Space Tourism". *Proceedings of the Forty-Seventh Colloquium on the Law of Outer Space*, 377.

Hobe, S., & Tronchetti, F. (2013). [Moon Agreement:] Article 6 (scientific investigations/samples/minerals). In *Cologne commentary on space law, Volume II: Rescue Agreement, Liability Convention, Registration Convention, Moon Agreement* (pp. 369-371). Cologne: Carl Heymanns.

Hobe, S. (1992). *Die rechtlichen Rahmenbedingungen der wirtschaftlichen Nutzung des Weltraums*. Berlin: Duncker und Humblot.

Hobe, S., Goh, G., & Neumann, J. (Eds.). (2007). Space Tourism Activities-Emerging Challenges to Air and Space Law. *J. Space L.*, *33*, 366.

Hoffman, J. A., Rapp, D., & Hecht, M. The Mars Oxygen ISRU Experiment (MOXIE) on the Mars 2020 Rover. AIAA SPACE 2014 Conference and Exposition. doi:10.2514/6.2015-4561

Hohfeld, W. N. (1913). Some fundamental legal conceptions as applied in judicial reasoning. *The Yale Law Journal*, *23*(1), 16–59. doi:10.2307/785533

Hohfeld, W. N. (1917). Fundamental legal conceptions as applied in judicial reasoning. *The Yale Law Journal*, *26*(8), 710–770. doi:10.2307/786270

Hollingham, R. (2014, September 8). *V2: The Nazi Rocket that Launched the Space Age*. Retrieved from BBC: http://www.bbc.com/future/story/20140905-the-nazis-space-age-rocket

Honoré, A. M. (1961). Ownership. In Oxford essays in jurisprudence: a collaborative work (pp. 107-147). Oxford, UK: Clarendon.

Horneck, G., Facius, R., Reichert, M., Rettberg, P., Seboldt, W., Manzey, D., ... Gerzer, R. (2003). Humex, a study on the survivability and adaptation of humans to long-duration exploratory missions, part I: Lunar missions. *Advances in Space Research*, *31*(11), 2389–2401. doi:10.1016/S0273-1177(03)00568-4 PMID:14696589

IEEE. (2002). *IEEE Std 521-2002 (Revision of IEEE Std 521-1984)*. IEEE Standard Letter Designations for Radar-Frequency Bands.

IISL-ECSL Symposium Transfer of Ownership of Space Objects: Issues of Responsibility, Liability and Registration held on 19 March 2012 at UN COPUOS Legal Subcommittee (Vienna). (n.d.). Available at: http://www.unoosa.org/oosa/en/ourwork/copuos/lsc/2012/symposium.html

Imburgia, J. S. (2011). Space debris and its threat to national security: A proposal for a binding international agreement to clean up the junk. *Vand. J. Transnat'l L.*, *44*, 589.

Innocenti, L., Soares, T., & Delaval, J. (2013). ESA Clean Space Initiative. In L. Ouwehand (Ed.), *Proceedings of the 6th European Conference on Space Debris*. Nordwijk, The Netherlands: European Space Agency.

Inter-Agency Space Debris Coordination Committee. (2007). *IADC Debris Mitigation Guidelines*. Retrieved from: http://www.unoosa.org/documents/pdf/spacelaw/sd/IADC-2002-01-IADC-Space_Debris-Guidelines-Revision1.pdf

Intergovernmental Agreement (IGA) on Space Station Cooperation (1998).

Interim Agreement Between the National Aeronautics and Space Administration of the United States of America and Russian Space Agency for the Conduct of Activities Leading to Russian Partnership in Detailed Design. (1994). *Development, Operation and Utilization on the Permanently Manned Civil Space Station, done at Washington, 23 June, 1994*. Interim Agreement.

International Civil Aviation Organization. Convention on Civil Aviation, 7 December 1944, 15 *UNTS*. 295. (n.d.). Available at: http://www.refworld.org/docid/3ddca0dd4.html

International Docking Standard. (n.d.). Retrieved from: http://www.internationaldockingstandard.com/

International Energy Agency. (2014). *Key World Energy Statistics 2014*. IEA.

International Space Elevator Consortium (ISEC). (n.d.). online At: http://www.isec.org

International Space Exploration Coordination Group (ISECG). (2018). *The Global Exploration Roadmap*. Author.

International Space Exploration Coordination Group (ISECG). (2018, February 2). *The Global Exploration Roadmap*. Retrieved from https://www.globalspaceexploration.org/wordpress/?cat=3

International Space Station (ISS) EVA Suit Water Intrusion High Visibility Close Call. (2013).

International Space Station. (2008). *Environmental Control and Life Support System*. National Aeronautic and Space Administration.

International Telecommunication Union. (2016). *ITU Radio Regulations*. Retrieved from https://www.itu.int/pub/R-REG-RR/en

International Telecommunication Union. (2016). *Radio Regulations*. ITU.

Internet World Stats. (2017, December). *Internet Users in the World by Regions*. Retrieved from https://www.internetworldstats.com/stats.htm

ISRO. (n.d.). *APRSAF-24 held in Bengaluru*. Retrieved from https://www.isro.gov.in/aprsaf-24-held-bengaluru

ISS US National Laboratory. (2010). *Reference Model for the International Space Station*. Available at https://www.nasa.gov/pdf/487816main_10_09_22%20ISS%20National%20Lab%20ProOrbis%20Reference%20Model.pdf

Jacobson, A., & Hume, A. (2016). *NASA Communications Network to Double Space Station Data Rates*. Retrieved from https://www.nasa.gov/feature/goddard/2016/nasa-communications-network-to-double-space-station-data-rates

Jakhu, R. (2006). Legal Issues Relating to the Global Public Interest in Outer Space. *Journal of Space Law, 32*, 31.

Jakhu, R. (2007). Legal issues of satellite telecommunications, the geostationary orbit, and space debris. *Astropolitics, 5*(2), 173-208.

Jakhu, R. S., Sgobba, T., & Dempsey, P. S. (Eds.). (2011). The Need for an Integrated Regulatory Regime for Aviation and Space: ICAO for Space? Studies in Space Policy Vol. 7. Wien: Springer. doi:10.1007/978-3-7091-0718-8

Jakhu, R. (2007). Legal issues of satellite telecommunications, the geostationary orbit, and space debris. *Astropolitics, 5*(2), 173–208.

Jakhu, R. S. (1983). The evolution of the ITU's regulatory regime governing space radio-communication services and the geostationary satellite orbit. *Annals of Air and Space Law, 3*, 381–406.

Jakhu, R. S. (1983). *The legal regime of the geostationary orbit*. Montréal: McGill University.

Jakhu, R. S., Pelton, J., & Nyampong, Y. (2017). *Space Mining and Its Regulation*. Berlin: Springer. doi:10.1007/978-3-319-39246-2

Jayakar, K. (1998). Globalization and the Legitimacy of International Telecommunications Standard-Setting Organizations. *Indiana Journal of Global Legal Studies, 5*, 711–738.

Jenkins, D. R. (2007). X-15: Extending the Frontiers of Flight. NASA.

Johnson vs. M'Intosh (1823). 9 21 U.S. 543 (8 Wheat.). The case involved land purchases made by British citizens in 1773 and 1775.

Johnson, N. L. (1993). US space surveillance. *Advances in Space Research*, *13*(8), 5–20. doi:10.1016/0273-1177(93)90563-Q

Johnson, N. L., & Loftus, J. P. (1999). Reducing orbital debris- Standards and practices. *Launchspace*, *4*(2), 24–26.

Jones, D. (2004). *Reusable Rocket Propulsion for Space Tourism Vehicles. Joint Propulsion Conference and Exhib.* Fort Lauderdale, FL: AIAA.

Judge Max Huber in the Island of Palmas Case. Netherlands vs. US (1928). Permanent Court of Arbitration. 2 RIAA 829 at 845-6.

Jurist, J. M. (2008, October 13). *The Commercial Suborbital Sounding Rocket Market: A Role for RLVs?* Retrieved from The Space Review: http://www.thespacereview.com/article/1228/1

Kaiser, C., Sjöberg, F., Delcura, J. M., & Eilertsen, B. (2008). SMART-OLEV—An orbital life extension vehicle for servicing commercial spacecrafts in GEO. *Acta Astronautica*, *63*(1–4), 400–410. doi:10.1016/j.actaastro.2007.12.053

Kaplan, L. S. (1988). Space-Specific Remedies for Torts in Outer Space: What Path Will US Law Follow. *International Law*, *22*, 1145–1156.

Karmali, F., & Shelhamer, M. (2008). The dynamics of parabolic flight: Flight characteristics and passenger percepts. *Acta Astronautica*, *63*(5-6), 594–602. doi:10.1016/j.actaastro.2008.04.009 PMID:19727328

Kauderer, A. (2013). *Canadarm2 and the Mobile Servicing System*. Retrieved January 12, 2015, from http://www.nasa.gov/mission_pages/station/structure/elements/mss.html

Kerrest, A., & Smith, L. (2009). Article VII. In S. Hobe, B. Schmidt-Tedd, & K. U. Schrogl (Eds.), *Cologne Commentary on Space Law: Outer Space Treaty* (pp. 126–145). Carl Heymanns Verlag.

Kessler, D. J., Johnson, N. L., Liou, J.-C., & Matney, M. (2010). The Kessler Syndrome: Implications to Future Space Operations. In S. C. McQuerry (Ed.), *Proceedings of the 33rd Annual AAS Rocky Mountain Guidance and Control Conference*. Breckenridge, CO: Univelt.

Kessler, D. J., & Cour-Palais, B. G. (1978). Collision frequency of artificial satellites: The creation of a debris belt. *Journal of Geophysical Research: Space Physics*, *83*(A6), 2637–2646. doi:10.1029/JA083iA06p02637

Kitamura, S., Hayakawa, Y., & Kawamoto, S. (2014). A reorbiter for large GEO debris objects using ion beam irradiation. *Acta Astronautica*, *94*(2), 725–735. doi:10.1016/j.actaastro.2013.07.037

Kleinhenz, J., Sacksteder, K., & Nayagam, V. (2008). Lunar Resource Utilization: Development of a Reactor for Volatile Extraction from Regolith. In *46th AIAA Aerospace Sciences Meeting and Exhibit*. Reston, VA: American Institute of Aeronautics and Astronautics. 10.2514/6.2008-1415

Klinkrad, H. (2006). *Space Debris: Models and Risk Analysis*. Berlin, Germany: Springer.

Klinkrad, H. (2015). Hazards of Orbital Debris. In J. N. Pelton & F. Allahdadi (Eds.), *Handbook of Cosmic Hazards and Planetary Defense*. Cham, Switzerland: Springer.

Klinkrad, H., Beltrami, P., Hauptmann, S., Martin, C., Sdunnus, H., Stokes, H., ... Wilkinson, J. (2004). The ESA space debris mitigation handbook 2002. *Advances in Space Research*, *34*(5), 1251–1259. doi:10.1016/j.asr.2003.01.018

Kniffen, McKinney, & Estey. (1990). Hybrid Rocket Development at the Ameriacan Rocket Company. In *26th Joint Propulsion Conference*. Orlando, FL: AIAA.

Knörnschild, M., Jung, M., Nagy, M., Metz, M., & Kalko, E. (2008). Bat echolocation calls facilitate social communication. *Proceedings of the Royal Society, Biological sciences*. Retrieved from http://rspb.royalsocietypublishing.org/content/279/1748/4827

Krag, H., Flohrer, T., & Lemmens, S. (2013). Consideration of Space Debris Mitigation Requirements in the Operation of LEO Missions. In *Space Operations: Experience, Mission Systems, and Advanced Concepts* (pp. 413–429). Reston, VA: American Institute of Aeronautics and Astronautics. doi:10.2514/5.9781624102080.0413.0430

Kramer, H. J. (1996). *Observation of the Earth and Its Environment: Survey of Missions and Sensors*. Berlin: Springer-Verlag. doi:10.1007/978-3-642-97678-0

Krige, J. (2007). NASA as an Instrument of U.S. Foreign Policy. In S. J. Dick & R. D. Launius (Eds.), *Societal Impact of Spaceflight* (pp. 2007–4801). Washington, DC: NASA.

Krisko, P. H., Flegel, S., Matney, M. J., Jarkey, D. R., & Braun, V. (2015). ORDEM 3.0 and MASTER-2009 modeled debris population comparison. *Acta Astronautica, 113*, 204–211.

Krisko, P. H. (2004). EVOLVE historical and projected orbital debris test environments. *Advances in Space Research, 34*(5), 975–980. doi:10.1016/j.asr.2003.02.025

Krisko, P. H. (2010). NASA's new orbital debris engineering model, ORDEM2010. In *Proceedings of the Fourth IAASS Conference*. Noordwijk, The Netherlands: European Space Agency.

Krolikowski, A., & David, E. (2013). Commercial On-Orbit Satellite Servicing: National and International Policy Considerations Raised by Industry Proposals. *New Space, 1*(1), 29–41. doi:10.1089pace.2013.0002

Kyle, E. (2018, November 21). *Space Launch Report*. Retrieved from http://www.spacelaunchreport.com/

Lafrance, A. (2015). The Internet in Space? Slow as Dial-Up. *Atlantic (Boston, Mass.)*.

Lahcen, A. (2013). *EUMETSAT-NOAA Collaboration in Meteorology from Space. Review of a Longstanding Trans-Atlantic Partnership*. ESPI Report 46.

Lamassoure, E. S. (2001). *A Framework to Account for Flexibility in Modeling the Value of On-Orbit Servicing for Space Systems*. Cambridge, MA: Massachusetts Institute of Technology.

Lambour, R., Rajan, N., Morgan, T., Kupiec, I., & Stansbery, E. (2004). Assessment of orbital debris size estimation from radar cross-section measurements. *Advances in Space Research, 34*(5), 1013–1020. doi:10.1016/j.asr.2003.02.043

Larsen, P. B. (2017). Small Satellite Legal Issues. *J. Air L. & Com, 82*, 275.

Larson, W. (2012). *Hydrogen Reduction of Regolith*. Retrieved from http://isru.nasa.gov/Hydrogen-Reduction-of-Regolith.html

Larson, W. J., & Pranke, L. K. (2000). *Human Spaceflight: Mission Analysis and Design* (W. J. Larson & L. K. Pranke, Eds.). McGraw-Hill Higher Education.

Larson, W. J., & Wertz, J. R. (1999). *Space Mission Analysis and Design*. Mcgraw-Hill.

Launch Vehicle Performance Website. (n.d.). Retrieved from http://elvperf.ksc.nasa.gov/Pages/Vehicles.aspx

Launius, R. D., & Jenkins, D. R. (2006). Is it finally Time for Space Tourism? *Astropolitics, 4*(3), 253–280. doi:10.1080/14777620601039693

Lee, K. A., Oryshchyn, L., Paz, A., Reddington, M., & Simon, T. M. (2013). The ROxygen Project: Outpost-Scale Lunar Oxygen Production System Development at Johnson Space Center. *Journal of Aerospace Engineering*, *26*(1), 67–73. doi:10.1061/(ASCE)AS.1943-5525.0000230

Lee, R. (2012). *Law and Regulation of Commercial Mining of Minerals in Outer Space*. Dordrecht: Springer. doi:10.1007/978-94-007-2039-8

Leinberg, G. (1989). Orbital space debris. *JL & Tech.*, *4*, 93.

Leitenberger, B. (2009). *Europäische Trägerraketen 1: von der Diamant zur Ariane 4*. Books on Demand.

Leone, D. (2014, August 11). NASA Ready to Put Robotic Refueling to the Test. *SpaceNews*. Retrieved from http://www.spacenews.com

Léveillé, R. J., & Datta, S. (2010). Lava tubes and basaltic caves as astrobiological targets on Earth and Mars: A review. *Planetary and Space Science*, *58*(4), 592–598. doi:10.1016/j.pss.2009.06.004

Lewis, J. A. (2013). Reconsidering Deterrence for Space and Cyberspace. In M. Krepon & J. Thompson (Eds.), Anti-Satellite Weapons, Deterrence and Sino-American Space Relations (pp. 61–80). Washington, DC: Stimson Center.

Lewis, H. G., Newland, R. J., Swinerd, G. G., & Saunders, A. (2010). A new analysis of debris mitigation and removal using networks. *Acta Astronautica*, *66*(1–2), 257–268. doi:10.1016/j.actaastro.2009.05.010

Lim, S., & Anand, M. (2015). In-Situ Resource Utilisation (ISRU) derived extraterrestria construction processes using sintering-based additive manufacturing techniques – focusing on a lunar sur- face environment. *European Lunar Symposium*.

Linder, A. (2009, December 9). Märkligt ljussken över Kiruna [Strange lightning over Kiruna]. *Norrländska Socialdemokraten*.

Liou, J.-C. (2011). An active debris removal parametric study for LEO environment remediation. *Advances in Space Research*, *47*(11), 1865–1876. doi:10.1016/j.asr.2011.02.003

Liou, J.-C., Hall, D., Krisko, P., & Opiela, J. (2004). LEGEND – a three-dimensional LEO-to-GEO debris evolutionary model. *Advances in Space Research*, *34*(5), 981–986. doi:10.1016/j.asr.2003.02.027

Liou, J.-C., & Johnson, N. L. (2008). Instability of the present LEO satellite populations. *Advances in Space Research*, *41*(7), 1046–1053. doi:10.1016/j.asr.2007.04.081

Liou, J.-C., Matney, M. J., Anz-Meador, P. D., Kessler, D. J., Jansen, M., & Theall, J. R. (2002). *The New NASA Orbital Debris Engineering Model ORDEM 3.0 (NASA/TP-2002-210780)*. Houston, TX: National Aeronautics and Space Administration.

Listner, M. J. (2018). *A Legal Look at Elon Musk's Plan to Colonize Mars*. Retrieved from http://www.thespacereview.com/article/3286/1

Logsdon, J. M. (Ed.). (1998). *Legislative Origins of the National Aeronautics and Space Act of 1958 : Proceedings of an Oral History Workshop : Conducted April 3, 1992*. Washington, DC: Government Printing Office.

Losekamm, M. J., Hacker, J., Sardesi, N., Peculjic, A. N., & Vigneron, A. (2015). Legal and Political Implications of Future On-Orbit Servicing Missions. Paper presented at the International Astronautical Congress, Jerusalem, Israel.

Loyd, S., Grimwood, J. M., & Alexander, C. C. (1989). 11-1 Suborbital Flights into Space. In This New Ocean: A History of Project Mercury. NASA Special Publication-4201 in the NASA History Series. NASA.

Lubos, P. (1982). Traffic Rules for Outer Space. In *Proceedings of the International Institute of Space Law*. American Institute of Aeronautics and Astronautics.

Luzin, P., (2013). Outer Space as Russia's Soft-Power Tool. *Security Index: A Russian Journal on International Security, 19*(1), 25-29. Doi:10.1080/19934270.2013.757117

Lyall, F. (1995). Posts and Telecommunications. In United Nations Legal Order (Vol. 2). Cambridge.

Lyall, F. (1996). Paralysis by phantom: problems of the ITU filling procedures. In *Proceedings of the Thirty-Ninth Colloquium on the Law of Outer Space* (p. 189). Academic Press.

Mahaffy, P. R., Webster, C. R., Atreya, S. K., Franz, H., Wong, M., Conrad, P. G., ... Trainer, M. (2013). Abundance and isotopic composition of gases in the martian atmosphere from the Curiosity rover. *Science, 341*(6143), 263–266. doi:10.1126cience.1237966 PMID:23869014

Mahoney, E. (2009). *Past and present: Field testing for the Moon.* Retrieved from https://www.nasa.gov/exploration/analogs/then-and-now.html

Maine, K., Devieux, C., & Swan, P. (1995). *Overview of IRIDIUM satellite network.* doi:10.1109/WESCON.1995.485428

Maini, A. K., & Agarwal, V. (2014). *Satellite Technology: Principles and Applications.* John Wiley & Sons Ltd. doi:10.1002/9781118636459

Maini, A. K., & Agrawal, V. (2007). *Compendium of Satellites and Satellite Launch Vehicles. In Satellite Technology: Principles and Applications*. John Wiley & Sons, Ltd.

Malik, T. (2014). *International Space Station gets Life Extension through 2024.* Space.com.

Manriquez, M. (2018). Japan's Space Law Revision: the Next Step Toward Re-Militarization? *NTI*. Retrieved from http://www.nti.org/analysis/articles/japans-space-law-revision/

Mantellato, R., Lorenzini, E. C., Sternberg, D., Roascio, D., Saenz-Otero, A., & Zachrau, H. J. (2017). Simulation of a tethered microgravity robot pair and validation on a planar air bearing. *Acta Astronautica, 138*, 579–589. doi:10.1016/j.actaastro.2016.12.029

Marder, E. (2008). *CPR for the OST: How China's Anti-Satellite Weapons Test Can Breathe New Life into Article IX of the Outer Space Treaty*. Center for Defense Information.

Mars Exploration Program Analysis Group. (2015). *Mars scientific goals, objectives, investigations, and priorities: 2015.* Retrieved from https://mepag.jpl.nasa.gov/reports/MEPAG%20Goals_Document_2015_v18_FINAL.pdf

Marshall, T. (1998). 15 Sign Pact for Space Station. *Washington Times.*

Martinez, P. (2012). Is there a need for an African space agency? *Space Policy, 28*(3), 142–145. doi:10.1016/j.spacepol.2012.06.011

Maskus, K. E. (2000). *Intellectual Property Rights in the Global Economy*. Washington, DC: Institute for International Economics.

Masson-Zwaan, T., & Richards, B. (2015, Dec. 8). International Perspectives on Space Resource Rights. *Space News*. Retrieved from: http://spacenews.com/op-ed-international-perspectives-on-space-resource-rights/

Mattei, U. (2000). *Basic Principles of Property Law: A Comparative Legal and Economic Introduction. Legal Studies No 93*. Westport, CT: Greenwood Publishing Group.

Maxwell, J. C. (1865, January). A Dynamical Theory of the Electromagnetic Field. *Philosophical Transactions of the Royal Society of London, 155*, 459–512. doi:10.1098/rstl.1865.0008

May, S. (2015). *Working in Space*. Retrieved January 12, 2015, from https://www.nasa.gov/audience/foreducators/stem-on-station/ditl_working

McLaughlin Green, C., & Lomask, M. (1970). *Vanguard - A History (SP-4202)*. Washington, DC: National Aeronautics and Space Administration.

McLeod, C., & Krekeler, M. (2017). Sources of Extraterrestrial Rare Earth Elements: To the Moon and Beyond. *Resources, 6*(3), 40. doi:10.3390/resources6030040

Mehpravar, A. (2014). *CubeSat Design Specification. Rev 13*. California Polytechnic State University.

Mehta, U., Aftosims, M., Bowles, J., & Pandya, S. (2016). Skylon Aerospace Plane and Its Aerodynamics and Plumes. *Journal of Spacecraft and Rockets, 53*(2), 340–353. doi:10.2514/1.A33408

Memorandum of Understanding between the National Aeronautics and Space Administration of the United States of America and the Canadian Space Agency concerning Cooperation on the Civil International Space Station, done at Washington D.C., Jan 29 1998. (n.d.). Available at http://web.archive.org/web/20091029094938/http://www.nasa.gov/mission_pages/station/structure/elements/nasa_csa.html

Memorandum of Understanding between the National Aeronautics and Space Administration of the United States of America and the Government of Japan concerning Cooperation on the Civil International Space Station, done at Washington, Feb 24 1998. (n.d.). Available at http://web.archive.org/web/20091029094939/http://www.nasa.gov/mission_pages/station/structure/elements/nasa_japan.html

Metz, M. (2012). DLR Perspective on Sustainable Use of Space. In *Proceedings of the CleanSpace Workshop on Active Debris Removal*. Darmstadt, Germany: European Space Agency.

Meurisse, A., Makaya, A., Willsch, C., & Sperl, M. (2018). Solar 3D printing of lunar regolith. *Acta Astronautica, 152*, 800–810. doi:10.1016/j.actaastro.2018.06.063

Meyer, P. (2014). *The Diplomacy of Space Security: Whither the International Code of Conduct?* Simons Papers in Security and Development. Retrieved from http://summit.sfu.ca/item/14921

Miller, J. M., Claybrook, A., Greenlund, S., Marquez, J. J., & Feigh, K. M. (2017). *Operational assessment of Apollo lunar surface extravehicular activity*. Retrieved from https://ntrs.nasa.gov/archive/nasa/casi.ntrs.nasa.gov/20170007261.pdf

Miller, R. J. (2007). Finders Keepers in the Arctic? *L.A. Times*, p. A19.

Miller, J., Taylor, L., Zeitlin, C., Heilbronn, L., Guetersloh, S., DiGiuseppe, M., ... Murakami, T. (2009). Lunar soil as shielding against space radiation. *Radiation Measurements, 44*(2), 163–167. doi:10.1016/j.radmeas.2009.01.010

Miller, R. J. (2010). Christianity, American Indians, and the Doctrine of Discovery. In A. Yong & B. Brown Zikmund (Eds.), *Remembering Jamestown: Hard Questions About Christian Mission* (p. 53). Pickwick.

Miller, R. J. (2011). *American Indians, Manifest Destiny, and the Doctrine of Discovery. Wyo. L. Rev.*

Miller, R. J., LeSage, L., & Lopez Escarcena, S. (2011). The International Law of Discovery, Indigenous Peoples, and Chile. *Nebraska Law Review, 89*(4), 6.

Miller, R. J., Ruru, J., Behrendt, L., & Lindberg, T. (2010). *Discovering Indigenous Lands: The Doctrine of Discovery in the English Colonies*. Oxford University Press. doi:10.1093/acprof:oso/9780199579815.001.0001

Mineiro, M. C. (2008). The United States and the Legality of Outer Space Weaponization: A Proposal for Greater Transparency and a Dispute Resolution Mechanism. *Annals of Air and Space Law, 2008*, 441–466.

Mirmina, S. A. (2004). The Regulation of Orbital Debris through National Measures. *Air and Space Law, 29*(2), 137–146.

Moltz, J. C. (2008). *The Politics of Space Security: Strategic Restraint and the Pursuit of National Interests*. Stanford, CA: Stanford University Press.

Monchieri, E., Hovland, S., Masi, M., Sliepcevich, A., Lavagna, M., Hoppenbrouwers, T., & Venditti, F. (2010). ESA Lunar In-Situ Resource Utilisation (ISRU) Concept Design and Breadboarding Activities. *61st International Astronautical Congress*.

Montes, C., Broussard, K., Gongre, M., Simicevic, N., Mejia, J., Tham, J., ... Davis, G. (2015). Evaluation of lunar regolith geopolymer binder as a radioactive shielding material for space exploration applications. *Advances in Space Research, 56*(6), 1212–1221. doi:10.1016/j.asr.2015.05.044

Montevideo Convention on the Rights and Duties of States, adopted in Montevideo in 1933, 165 LNTS 19; 49 Stat 3097.

Moore, W. (1813). *A Treatise on the Motion of Rockets. To which is added, An Essay on Naval Gunnery*. Military Academy at Woolwich, London: G. and S. Robinson NASA (2005-10-21). A long-overdue tribute. NASA. Retrieved from https://www.nasa.gov/centers/dryden/news/X-Press/stories/2005/102105_Wings.html

Mosher, D. (2007). *Space Station Surgery: Astronauts Mend Torn Solar Wing*. Space.com.

Moskowitz, C. (2009). *Space Station Population Hits Record High*. Retrieved January 12, 2015, from http://www.space.com/7003-space-station-population-hits-record-high.html

Motil, B., & Singh, B. (2004). NASA's microgravity fluid physics strategic research roadmap. *42nd AIAA Aerospace Sciences Meeting and Exhibit*, 123.

Munzer, S. R. (1990). *A theory of property*. Cambridge, UK: Cambridge University Press. doi:10.1017/CBO9780511609138

Musk, E. (2017). Making Humans a Multi-Planetary Species. *New Space., 5*(2), 46–61. doi:10.1089pace.2017.29009.emu

Nagai, Y. (2015). Policy Analysis: Space Programmes of Japan And India. *66th International Astronautical Congress*, 1–14.

NASA (2015a). *Capability Catalog - Black Brant XII*. NASA. Retrieved 2015-08-18

NASA Advisory Council Recommendation, Research Subcommittee of the Human Exploration and Operations Committee (HEOC-01). (n.d.). Available at https://www.nasa.gov/sites/default/files/files/NASA_Responses_Package_1.pdf

NASA awards launch services contract to SpaceX. (2008, April 22). NASA News Release. Available at: http://www.nasa.gov/home/hqnews/2008/apr/HQ_C08023_KSC_launch_services.html

NASA Media Press Kit. (2014). *Orbital-2 Mission to the International Space Station*. NASA.

NASA Office of Safety and Mission Assurance. (2017). *NASA Procedural Requirements for Limiting Orbital Debris (NPR 8715.6B)*. Washington, DC: National Aeronautics and Space Administration.

NASA Orbital Debris Program Office. (2015). Two More Collision Avoidance Maneuvers for the International Space Station. *Orbital Debris Quarterly News, 19*(4).

NASA Orbital Debris Program Office. (2018). Monthly Number of Objects in Earth Orbit by Object Type. *Orbital Debris Quarterly News, 22*(1).

NASA Policy Directive (NPD) 8020.7G.

NASA Procedural Requirements (NPR) 8020.12.

NASA. (2014, February 28). *Armstrong Fact Sheet: X-15 Hypersonic Research Program*. Retrieved from NASA: https://www.nasa.gov/centers/armstrong/news/FactSheets/FS-052-DFRC.html

NASA. (2015b, July 31). *Mercury-Redstone 3*. Retrieved from NASA: https://www.nasa.gov/mission_pages/mercury/missions/freedom7.html

NASA's Journey to Mars: Pioneering Next Steps in Space Exploration. (2015).

National Aeronautics and Space Act of 1958, As Amended. (2008). Retrieved from: https://history.nasa.gov/spaceact-legishistory.pdf

National Aeronautics and Space Administration (NASA). (2005). *NASA Facts: Mars Reconnaissance Orbiter*. Retrieved from https://mars.nasa.gov/mro/files/mro/MRO-060303.pdf

National Aeronautics and Space Administration (NASA). (2005). *NASA's Exploration Systems Architecture Study* (NASA TM No. NASA-TM-2005-214062). NASA.

National Aeronautics and Space Administration (NASA). (2015). *Life Support Baseline Values and Assumptions Document*. Houston, TX: NASA.

National Aeronautics and Space Administration. (2010). *Reference Guide to the International Space Station*. Washington, DC: NASA.

National Aeronautics and Space Administration. (2013). *Audit Report of the Office of Inspector General, July 2013, NASA's efforts to maximise research on the international space station*. NASA.

National Aeronautics and Space Administration. (2015). *ISS Reference Guide*. Author.

National Aeronautics and Space Administration. (2016). *Year Long Mission*. Retrieved June 11, 2016, from https://www.nasa.gov/1ym

National Aeronautics and Space Administration. (2018). *International Space Station Utilization Statistics*. Retrieved November 11, 2018, from https://www.nasa.gov/sites/default/files/atoms/files/iss_utilization_statistics_brochure_expeditions_0-52_final.pdf

National Aeronautics and Space Administration. Science Mission Directorate. (2010). *Gamma Rays*. Retrieved from https://science.nasa.gov/ems/12_gammarays

National Laws Governing Commercial Space Activities. Legislation, Regulation, & Enforcement, 36 Nw. J. Int'l L. & Bus. 1 (2016). (n.d.). Available at: http://scholarlycommons.law.northwestern.edu/njilb/vol36/iss1/1/

National Research Council. (1995). *Orbital Debris: A Technical Assessment*. Washington, DC: National Academies Press.

National Space Policy of the United States of America. (2010). Retrieved from https://www.nasa.gov/sites/default/files/national_space_policy_6-28-10.pdf

National Space Policy. (n.d.). Retrieved from https://space.gov.ae//Documents/PublicationPDFFiles/National_Space_Policy_English.pdf

Nazarenko, A. I., & Menshikov, I. L. (2001). Engineering model of space debris environment. In H. Sawaya-Lacoste (Ed.), *Proceedings of the Third European Conference on Space Debris*. Noordwijk, The Netherlands: European Space Agency.

Neufeld, M. J. (1995). *The Rocket and the Reich: Peenemünde and the Coming of the Ballistic Missile Era*. New York: The Free Press.

New Member States. (n.d.).

New Space Policy Directive Calls for Human Expansion across Solar System. (2017). Retrieved from https://www.nasa.gov/press-release/new-space-policy-directive-calls-for-human-expansion-across-solar-system

New Space Policy Directive Calls for Human Expansion Across Solar System. (n.d.). Retrieved from https://www.nasa.gov/press-release/new-space-policy-directive-calls-for-human-expansion-across-solar-system

New World Encyclopedia. (n.d.). Retrieved from http://www.newworldencyclopedia.org/entry/Colonialism

Newcomb, S. T. (2008). *Pagans in the Promised Land: Decoding the Doctrine of Christian Discovery*. Fulcrum Golden.

Newly, J. E., Wilson, J. W., & Townsend, L. W. (1988). Solar-Flare Shielding With Regolith at a Lunar-Base Site. Langley, VA: Academic Press.

Nimon, J. (2014). *Space Station 2024 Extension Expands Economic and Research Horizons*. Retrieved from https://www.nasa.gov/mission_pages/station/research/news/2024extension/

Nishida, S.-I., Kawamoto, S., Okawa, Y., Terui, F., & Kitamura, S. (2009). Space debris removal system using a small satellite. *Acta Astronautica*, *65*(1–2), 95–102. doi:10.1016/j.actaastro.2009.01.041

NLS II Contract, Section C, Clause 5.3; and SpaceX contract with Orbcomm, supra note 14 at p.3

NLS II contracts are different, supra note 7: Section C, Clause 1 states that The Government will only accept those launch services that successfully deliver a separated and undamaged payload to the proper orbit conditions Section C, Clause 24 explains how full success, partial success, and failure will be treated in any given LSTO

Northrop Grumman. (2018). Retrieved from Pegasus: http://www.northropgrumman.com/Capabilities/Pegasus/Documents/Pegasus_Mission_History.pdf

NTSB. (2015). *2015_spaceship2_BMG_abstract.pdf*. Retrieved from https://www.ntsb.gov/news/events/Documents/2015_spaceship2_BMG_abstract.pdf

Nyman-Metcalf, K. (1999). *Activities in space: appropriation or use?* Uppsala: Iustus Förlag.

Oduntan, G. (2015). Who owns space? US asteroid-mining act is dangerous and potentially illegal. *The Conversation*. Retrieved on 30 April 2018 from http://phys.org/news/2015-11-space-asteroid-mining-dangerous-potentially-illegal.html

Oefelein, C. J., & Yang, V. (1993). Comprehensive Review of Liquid-Propellant Combustion Instabilities in F-1 Engines. *Journal of Propulsion and Power*, *9*(5), 657–677. doi:10.2514/3.23674

On-Orbit Satellite Servicing Study: Project Report (NP-2010-08-162-GSFC). (2010). Greenbelt, MD: National Aeronautics and Space Administration.

Orbital, A. T. K. (2015). *Pegasus User's Guide*. Retrieved from orbitalatk.com

Ordyna, P. (2010). Insuring Human Space Flight: An Underwriter's Dilemma. *Journal of Space Law*, *36*, 231–251.

Osterhammerl, J. (1997). *Colonialism: A Theoretical Overview*. Markus Winer Publishers.

Outer Space Act. 1986. United Kingdom. 1986 c. 38.

Pancaldi, F., Vitetta, G., Kalbasi, R., Al-Dahir, N., Uysal, M., & Mheidat, H. (2008). Single-carrier frequency domain equalization. Signal Processing Magazine, 37–56.

Park, W. (n.d.). The Tragic Tail of Saddam Hussein's Supergun. *BBC Future*. Available at: http://www.bbc.com/future/story/20160317-the-man-who-tried-to-make-a-supergun-for-saddam-hussein

Peculjic, A. N. (2016). Identifying Elements of *Lex Mercatoria* in the Space Domain. In R. Moro-Aguilar, P.J. Blount, & T. Masson-Zwaan (Eds.), *Proceedings of the International Institute of Space Law 2015* (pp. 143-154). Hague. Eleven.

Phipps, C. R., Baker, K. L., Libby, S. B., Liedahl, D. A., Olivier, S. S., Pleasance, L. D., ... Valley, M. T. (2012). Removing orbital debris with lasers. *Advances in Space Research*, *49*(9), 1283–1300. doi:10.1016/j.asr.2012.02.003

Poulet, F., Bibring, J. P., Mustard, J. F., Gendrin, A., Mangold, N., Langevin, Y., ... Gomez, C. (2005). Phyllosilicates on Mars and implications for early martian climate. *Nature*, *438*(7068), 623–627. doi:10.1038/nature04274 PMID:16319882

Priemer, R. (1991). *Introductory Signal Processing*. Hackensack, NJ: World Scientific.

Protocol including terms, conditions and assumptions, summary balance of contributions and obligations to international space station (ISS) and resulting rights of NASA and RSA to ISS utilization accommodations and resources, and flight opportunities, done at Moscow, Jun 11 1996. (n.d.). Available at https://archive.org/stream/nsia-25923192035014/25923192035014_djvu.txt

Protocol on Environmental Protection to the Antarctic Treaty, adopted on October.4, 1991, entered into force on January 14, 1998.

Pry, P. (1999). *Black Brant XII. War scare: Russia and America on the nuclear brink*. New York: Praeger.

Purcell, E., & Morin, D. (2013). *Electricity and Magnetism* (3rd ed.). New York: Academic Press.

Quaye, C. O. (1991). *Liberation Struggles in International Law*. Temple University Press.

Quinn, J., Smith, J. J. C., Paz, A., Colaprete, A., Elphic, R., & Zacny, K. (2015). Resource Prospector. *The RESOLVE Payload*, *1863*, 2046.

Rainey, K. (2016). *Space Station Research Experiments*. Retrieved January 1, 2016, from http://www.nasa.gov/mission_pages/station/research/experiments_category/index

Rajagopalan, R. P. (2018, January 20). What's Next for India's Space Program? *The Diplomat*. Retrieved from https://thediplomat.com/2018/01/whats-next-for-indias-space-program/

Rathgeber, W., Remuss, N. K., & Schrogl, K. U. (2009). Space Security and the European Code of Conduct for Outer Space Activities. *Disarmament Forum*, *4*, 33–41.

Ratner, S., (1999). Establishing the Extraterrestrial: Criminal Jurisdiction and the International Space Station. *Boston College International and Comperative Law Review*, *22*(2).

Reals, K. (2018, March). OneWeb Looks To Disrupt Inflight Connectivity Market With LEO Network. *Aviation Week*. Retrieved from http://aviationweek.com/aircraft-interiors/oneweb-looks-disrupt-inflight-connectivity-market-leo-network

Redacted, N. L. S. II contract between NASA and Space Exploration Technologies. (n.d.). Available at: http://www.nasa.gov/centers/kennedy/pdf/541918main_nls-contract.pdf

Reed, J., Busquets, J., & White, C. (2012). Grappling System for Capturing Heavy Space Debris. In *Proceedings of the 2nd European Workshop on Active Debris Removal*. Paris, France: Centre national d'études spatiales.

Rendleman, J. (2012). *Space Traffic Management - Private Regulation?* Paper presented at AIAA SPACE 2012 Conference & Exposition.

Representative Jim Bridenstine (now NASA administrator) had proposed this: H.R. (discussion draft), F:\M14\BRIDEN\BRIDEN_160.XML, 114th Congress, 2D Session. (n.d.). Retrieved from http://www.spacepolicyonline.com/pages/images/stories/Bridenstine%20discussion%20draft%20sept%2014%202016.pdf

Restatement (Third) Of Foreign Relations Law of the United States, § 201 (1987).

Retat, I. (2012). Net capture system: a potential orbital space debris removal system. In *Proceedings of the 2nd European Workshop on Active Debris Removal*. Paris, France: Centre national d'études spatiales.

Reuter, J. L., & Reysa, R. P. (1997). International Space Station Environmental Control and Life Support System Design Overview Update. *Proceedings of the 27th International Conference on Environmental Systems.* 10.4271/972333

Roberson, L., Hintze, P., & O'Connor, G. A Conceptual Study for the Autonomous Direct Forming of Lunar Regolith into Flexlock Geomats for Lunar Surface Operations. *47th AIAA Aerospace Sciences Meeting including The New Horizons Forum and Aerospace Exposition.* 10.2514/6.2009-1013

Roesler, G., Jaffe, P., & Henshaw, G. (2017, March 8). Inside DARPA's Mission to Send a Repair Robot to Geosynchronous Orbit. *IEEE Spectrum*. Retrieved from https://www.spectrum.ieee.org

Rosenberg, S. D. Beegle, R. L., Jr., Guter, G. A., Miller, F. E., & Rothenberg, M. (1992). The onsite manufacture of propellant oxygen from lunar resources. In M. F. McKay, D. S. McKay, & M. B. Duke (Eds.), *Space Resources Materials* (3rd ed.; pp. 162–185). Academic Press. Retrieved from http://ntrs.nasa.gov/search.jsp?R=19930007699

Rosenberg, S. D., Musbah, O., & Rice, E. E. (1996). Carbothermal Reduction of Lunar Materials for Oxygen Production on the Moon: Reduction of Lunar Simulants with Carbon. In *Lunar and Planetary Science Conference XXVII* (pp. 1103–1104). Academic Press.

Rossi, A., Lewis, H., White, A., Anselmo, L., Pardini, C., Krag, H., & Bastida Virgili, B. (2016). Analysis of the consequences of fragmentations in low and geostationary orbits. *Advances in Space Research*, *57*(8), 1652–1663. doi:10.1016/j.asr.2015.05.035

Roviera, P. M., Bertrand, J., & Lardot, C. (2014). *ESA Launchers Ground Facilities: Background operational Phase and Future Developments. SpaceOps Conference*. Pasadena, CA: AIAA.

Roy, K. I., Kennedy, R. G., & Fields, D. E. (2009). Shell worlds: An approach to terraforming moons, small planets and plutoids. *J. Br. Interplanet. Soc'y*, *62*, 32–38.

Russel, K. (2017). *Restore-L Mission Passes Preliminary Design Review*. Retrieved April 30, 2018, from https://www.satellitetoday.com/government-military/2017/12/26/restore-l-mission-passes-preliminary-design-review/

Russel, K. (2018, April 4). SpaceX Lofted an Experimental Debris Removal Satellite for a UK University. *Via Satellite*. Retrieved from https://www.satellitetoday.com

Sabatier, P. (1913). *La Catalyse en chimie organique*. Academic Press.

Sagdeev, R. (n.d.). *United States-Soviet Space Cooperation during the Cold War*. Retrieved from https://www.nasa.gov/50th/50th_magazine/coldWarCoOp.html

Saleh, J. H., & Castet, J.-F. (2011). Geosynchronous communication satellites: system reliability and failures. In *Spacecraft Reliability and Multi-State Failures: A Statistical Approach*. West Sussex, UK: Wiley. doi:10.1002/9781119994077.app1

Saleh, J. H., Lamassoure, E. S., Hastings, D. E., & Newman, D. J. (2003). Flexibility and the Value of On-Orbit Servicing: New Customer-Centric Perspective. *Journal of Spacecraft and Rockets*, *40*(2), 279–291. doi:10.2514/2.3944

Salin, P.A. (n.d.). *Privatization and militarization in the space business environment*. Academic Press.

Salt, D. J. (2014). *Could a subsonic air-launched RLV enable a paradigm shift in space operations?* SpaceOps 2014 Conference, AIAA 2014-1897, Pasadena, CA.

Samplatsky, D., Grohs, K., Edeen, M., Crusan, J., & Burkey, R. (2011). Development and Integration of the Flight Sabatier Assembly on the ISS. In *41st International Conference on Environmental Systems*. Reston, VA: American Institute of Aeronautics and Astronautics. 10.2514/6.2011-5151

Sanders, G. B., Larson, W. E., Interbartolo, M. A., Mueller, R. P., & Muscatello, A. C. (2012). *Early In-Situ Resource Utilization (ISRU) Leading to Robust Sample Return and Human Exploration Missions*. Academic Press.

Sanders, G. B., Romig, K. A., Larson, W. E., Johnson, R., Rapp, D., Johnson, K. R., . . . Zubrin, R. (2005). *Results from the NASA Capability Roadmap Team for In-Situ Resource Utilization (ISRU)*. Retrieved from http://ntrs.nasa.gov/search.jsp?R=20110024178

Sanders, G. B., & Duke, M. (2005). *In-Situ Resource Utilization (ISRU)*. Capability Roadmap Progress Review.

Sands, P., Klein, P., & Bowett, D. W. (2009). *Bowett's law of international institutions*. Academic Press.

Sänger, E. (1933). *Raketenflugtechnik*. München: Academic Press.

Santarius, J. F. (2004, September). *Lunar 3He and Fusion Power.* IEEE Rock River Valley Section, Rockford, IL. Retrieved from http://fti.neep.wisc.edu/presentations/jfs_ieee0904.pdf

Saperstein, A. M. (n.d.). *"Weaponization" vs. "militarization" of space*. Retrieved from https://www.aps.org/units/fps/newsletters/2002/july/saperstein.pdf

Sariak, G. (2017). Between a Rocket and a Hard Place: Military Space Technology and Stability in International Relations. *Astropolitics*, *15*(1), 51–64. doi:10.1080/14777622.2017.1288509

Sarigul-Kijin, M., & Sarigul-Klijn, N. (2001). *A Study of Air Launch Methods for RLVs. Space Conference and Exposition*. Albuquerque, NM: AIAA.

Sarli, B., Marco Cabero, M., Lopez, A., Cardoso, J., & Jimenez, D. (2015). *South American Space Era*. 66th International Astronautical Congress - IAC 2015, Jerusalem, Israel.

Sastri, S., Duke, M. B., & Haskin, L. A. (1992). PART 3—Manufacturing and Fabrication. In M. F. McKay, D. S. McKay, & M. B. Duke (Eds.), Space Resources Materials (pp. 270–306). Academic Press.

Sauvageanu, D., & Allen, B. (1998). Launch Vehicle historical Reliability. In AIAA 1988-3979, Cleveland, OH.

Savage, J. G. (1989). *The politics of international telecommunications regulation*. Westview Pr.

Schade, C. C. (1990). Pegasus, Taurus and glimpses of the future. *Space Programs and Technologies Conference, AIAA 90-3573-CP*. 10.2514/6.1990-3573

Schade, C., & Moiser, M. (1994). The Pegasus Launch Vehicle new Capabilities and the Future. *15th International Communications Satellite Systems Conference and Exhibit*. 10.2514/6.1994-1172

Schaezler, R. N., & Cook, A. J. (2015). Report on ISS O2 Production, Gas Supply & Partial Pressure Management. *Proceedings of the 45th International Conference on Environmental Systems*.

Schaub, H., & Sternovsky, Z. (2014). Active space debris charging for contactless electrostatic disposal maneuvers. *Advances in Space Research*, *53*(1), 110–118. doi:10.1016/j.asr.2013.10.003

Schildknecht, T., Musci, R., Ploner, M., Beutler, G., Flury, W., Kuusela, J., ... de Fatima Dominguez Palmero, L. (2004). Optical observations of space debris in GEO and in highly-eccentric orbits. *Advances in Space Research*, *34*(5), 901–911. doi:10.1016/j.asr.2003.01.009

Schmidt, G. (1999). *Technik der Flüssigkeits-Raketentriebwerke*. Taufkirchen bei München: Daimler Chrysler Aerospace.

Schrogl, K.-U., Hays, P. L., Robinson, J., Moura, D., & Giannopapa, C. (Eds.). (2015). Handbook of Space Security. New York: Springer Science+Business Media. doi:10.1007/978-1-4614-2029-3

Schrogl, K.-U. (2008). Space traffic management: The new comprehensive approach for regulating the use of outer space—Results from the 2006 IAA cosmic study. *Acta Astronautica*, *62*(2-3), 272–276. doi:10.1016/j.actaastro.2007.09.001

Schrogl, K.-U., Jorgenson, C., Robinson, J., & Soucek, A. (Eds.). (2018). *Space Traffic Management – Towards a Road-map for Implementation*. International Academy of Astronautics.

Schulte, R. (2000). TUBSAT-N, an ultra low cost global communication nanosatellite system. *Air & Space Europe*, *2*(5), 80–83. doi:10.1016/S1290-0958(00)80094-5

Schwandt, C., Doughty, G. R., & Fray, D. J. (2010). The FFC-Cambridge Process for Titanium Metal Winning. *Key Engineering Materials*, *436*, 13–25. . doi:10.4028/www.scientific.net/KEM.436.13

Schwandt, C., Hamilton, J. A., Fray, D. J., & Crawford, I. A. (2012a). Oxygen from Lunar Regolith. In V. Badescu (Ed.), *Moon* (pp. 165–187). Springer. doi:10.1007/978-3-642-27969-0_7

Schwandt, C., Hamilton, J. A., Fray, D. J., & Crawford, I. A. (2012b). The production of oxygen and metal from lunar regolith. *Planetary and Space Science*, *74*(1), 49–56. doi:10.1016/j.pss.2012.06.011

Schwartz, J. (2006). A Toxic Leak Haunts the Shuttle Crew. *The New York Times*.

Scoles, S. (n.d.). *Luxembourg's Bid to Become the Silicon Valley of Space Mining*. Retrieved from https://www.wired.com/2017/01/luxembourg-setting-silicon-valley-space-mining/

Senior, C. L. (1992). Lunar oxygen production by pyrolysis. In *Space Programs and Technologies Conference*. Reston, VA: American Institute of Aeronautics and Astronautics. 10.2514/6.1992-1663

Shannon, C. E. (1949, January). Communication in the Presence of Noise. *Proceedings of the Institute of Radio Engineers*, 447-457.

Sharpe, C., & Tronchetti, F. (2015). Legal aspects of public manned spaceflight and space station operations. In *Handbook of Space Law*. Cheltenham, UK: Edward Elgar Publishing Limited. doi:10.4337/9781781000366.00020

Shaw, M. N. (1997). *International Law* (7th ed.). Cambridge University Press.

Shelton, D. (2014). International Law and 'Relative Normativity'. In M.D. Evans (Ed.), International Law (4th ed.). Oxford, UK: Oxford University Press.

Sibille, L., Sadoway, D., Sirk, A., Tripathy, P., Melendez, O., Standish, E., & … . Poizeau, S. (n.d.). Recent Advances in Scale-Up Development of Molten Regolith Electrolysis for Oxygen Production in Support of a Lunar Base. *47th AIAA Aerospace Sciences Meeting including The New Horizons Forum and Aerospace Exposition.* 10.2514/6.2009-659

Signaling strength: Russia's real Syria success is electronic warfare against the US. (n.d.). Retrieved from https://thedefensepost.com/2018/05/01/russia-syria-electronic-warfare/

Simonsen, L. C., & Nealy, J. E. (1991). *Radiation Protection for Human Missions to the Moon and Mars* (NASA Technical Paper No. 3079). Hampton, VA: NASA.

Smitham, M. C. (2012). *The Need for a Global Space-Traffic-Control Service: An Opportunity for Us Leadership. Maxwell Papers*. Air War College.

Smith, M. V. (2011). Spacepower and Warfare. *Joint Force Quarterly*, *60*, 42.

Soucek, A. (2015). Space Law Essentials (vol. 1). Linde Publishing.

Soucek, A. (2012). *International Law. In Outer Space in Society, Politics and Law*. Wien: Springer Verlag.

Space Assembly, Maintenance And Servicing (SAMS) Study, Final Report. (1986). Sunnyvale, CA: Lockheed Missiles & Space Company, Inc.

Space Generation Advisory Council. (2018). Retrieved from https://spacegeneration.org/

Space Industry Act, United Kingdom. 2018 c. 5.

Space News. (2016). Available at: http://spacenews.com/spacex-aborts-ses-9-launch/

space.com. (2006, May 29). Retrieved from Russian Submarine Launches Russian Microsatellite: https://www.space.com/2440-russian-submarine-launches-russian-microsatellite.html

Spell, T. (2011). *Estimation of the Location, Trajectory, Size, and Altitude of the "Norway Spiral" Phenomenon*. Retrieved from http://www.spellconsulting.com/reality/Norway_Spiral.html

Sputniknews. (n.d.). *Putin stresses importance of new Far East space center*. Retrieved from https://sputniknews.com/russia/20100828160373755/

Standage, T. (1998). *The Victorian Internet: The remarkable story of the telegraph and the nineteenth century's online pioneers*. Phoenix.

Stanford Encyclopedia of Philosophy. (n.d.). Retrieved from https://plato.stanford.edu/entries/colonialism/

Statement from Vice President Mike Pence on the President's Signing of Space Policy Directive. White House, May 24, 2018. (n.d.). Retrieved from https://www.whitehouse.gov/briefings-statements/statement-vice-president-mike-pence-presidents-signing-space-policy-directive-2/

Stein, R. (2007). Space Station's Damaged Panel Is Fixed. *Washington Post*.

Sternberg, D., Hilton, A., Miller, D., McCarthy, B., Jewison, C., & Roascio, D., … Saenz-Otero, A. (2016). Reconfigurable ground and flight testing facility for robotic servicing, capture, and assembly. In *Proceedings of the 2016 IEEE Aerospace Conference*. Big Sky, MT: IEEE. 10.1109/AERO.2016.7500526

Stoehr, M. (2012, December). *Getting Started With A Short Range Radio Design*. Retrieved from http://www.electronicdesign.com/communications/getting-started-short-range-radio-design

Stofan, E. R., Elachi, C., Lunine, J. I., Lorenz, R. D., Stiles, B., Mitchell, K. L., ... West, R. (2007). The lakes of Titan. *Nature*, *445*(7123), 61–64. doi:10.1038/nature05438 PMID:17203056

Subcommittee on Space. (2014). *Space Traffic Management: How to Prevent A Real Life 'Gravity,' Hearing May 9, 2014*. Retrieved from https://www.gpo.gov/fdsys/pkg/CHRG-113hhrg88144/pdf/CHRG-113hhrg88144.pdf

Sullivan, T. A., Linne, D., Bryant, L., & Kennedy, K. (1995). In-situ-produced methane and methane/carbon monoxide mixtures for return propulsion from Mars. *Journal of Propulsion and Power*, *11*(5), 1056–1062. doi:10.2514/3.23935

Sundahl, M. (2009). The Duty to Rescue Space Tourists and Return Private Spacecraft. *Journal of Space Law*, *35*, 163-200.

Sunshine, J. M., Farnham, T. L., Feaga, L. M., Groussin, O., Merlin, F., Milliken, R. E., & A'Hearn, M. F. (2009). Temporal and spatial variability of lunar hydration as observed by the Deep Impact spacecraft. *Science, 326*(5952), 565–568. doi:10.1126cience.1179788 PMID:19779149

Sutton, G. P. (2001). Rocket Propulsion Elements (7th ed.). John Wiley & Sons, Inc.

Sviatoslavsky, I. N. (1993). *The Challenge of Mining He-3 on the Lunar Surface: How All the Parts Fit Together: Prepared for Space 94, The 4th International Conference and Exposition on Engineering, Construction and Operations in Space, and The Conference and Exposition/Demonstration on Robotics for Challenging Environments, February 26 - March 3, 1994, Albuquerque NM*. Madison, WI: Academic Press.

Swiss Space Systems. (2016, Jul. 1). Retrieved from official company Link: http://www.s-3.ch

Takaya-Umehara, Y. (2014). State Responsibility and Liability for Air-Launch over the High Seas. In *Proceedings of the fifty-sixth Colloquium on the Law of Outer Space*. Eleven International Publishing.

Tate, L. (2006). The Status of the Outer Space Treaty at International Law during War and Those Measures Short of War. *Journal of Space Law*, *32*, 177–202.

Taylor, L. A., & Carrier, W. D. (1993). Oxygen Production on the Moon: an Overview and Evaluation. In J. Lewis, M. S. Matthews, & M. L. Guerrieri (Eds.), Resources of Near Earth Space (pp. 69–108). The University of Arizona Press.

Taylor, L. A., Jerde, E. A., McKay, D. S., Gibson, M. A., Knudsen, C. W., & Kanamori, H. (1993). Production of O2 on the Moon: A Lap-Top Demonstration of Ilmenite Reduction with Hydrogen. *Proceedings of the 24th Lunar and Planetary Science Conference.*

Taylor, L. A., & Meek, T. T. (2005). Microwave Sintering of Lunar Soil: Properties, Theory, and Practice. *Journal of Aerospace Engineering*, *18*(3), 188–196. doi:10.1061/(ASCE)0893-1321(2005)18:3(188)

Tchikaya, B. (1998). Le droit international des télécommunications. Presses universitaires de France.

Technology Quarterly. Space. (2016, August 27). *The Economist, 420*(9004).

Tennen, L. (2010). Towards a New Regime for Exploration of Outer Space Mineral Resources. *Nebraska Law Review, 88*(794), 802.

The 2nd International Space Exploration Forum (ISEF2). (2018, March 7). *Joint Statement.* Retrieved from http://www.mext.go.jp/isef2/pdf/1-e.pdf

The Colonization of Mars: What Legal Issues Will Arise out of a Multi-Planetary Existence? (n.d.). Retrieved from http://www.jhtl.org/blog/the-colonization-of-mars-what-legal-issues-will-arise-out-of-a-multi-planetary-existence/

The Convention on International Liability for Damage Caused by Space Objects, adopted by the General Assembly in its resolution 2777 (XXVI), 24 UST 2389, TIAS 7762, 961 UNTS 187.

The history of American space policy. (n.d.). Retrieved from https://history.nasa.gov/spdocs.html

The Permanent Court of International Justice, P.C.I.J. (ser. A) No. 10 (1927), the S.S. Lotus case (France v Turkey).

The Space Settlement Institute. (n.d.). *Lunar Land Claims Recognition.* Retrieved from http://www.space-settlement-institute.org/strategy.html

The Spaceship Company. (2018). Retrieved from Making Dreams take Flight: http://thespaceshipcompany.com/

The Tauri Group. (2015). *State of the Satellite Industry Report*. Satellite Industry Association.

The UKSA has published a list of questions applicable to launch only activities. (n.d.). Available at: https://www.gov.uk/government/uploads/system/uploads/attachment_data/file/464931/Guidance_for_applicants_-_October_2015.pdf

Third International Decade for the Eradication of Colonialism, Report of the Secretary-General

This revised Radio Regulations, complementing the Constitution and the Convention of the International Telecommunication Union, incorporating the decisions of the World Radiocommunication Conferences of 1995 (WRC-95), 1997 (WRC-97), 2000 (WRC-2000), 2003 (WRC-03), 2007 (WRC-07), 2012 (WRC-12) and 2015 (WRC-15)

Thompson, J. C. (1996). Space for rent: the international telecommunications union, space law, and orbit/spectrum leasing. *J. Air L. & Com.*, *62*, 279.

Thompson, M. (2015). *A Space Traveller's Guide to the Solar System*. Bantam Press.

Tkatchova, S. (2011). *Space-Based Technologies and Commercialized Development: Economic Implications and Benefits*. Academic Press.

Traction, B. Light and Power Company, Limited, second Phase, Judgment, ICJ reports 1970, p. 3, para 33

Treat, J., Scalamogna, A., & Conant, E. (2015). The Secret to Hubble's Success. *National Geographic*. Retrieved April 30, 2018, from http://www.nationalgeographic.com/hubble-timeline/

Treaty on Principles Governing the Activities of States in the Exploration and Use of Outer Space, Including the Moon and Other Celestial Bodies (27 January 1967) 15 U.N.T.S. 206

Treaty on Principles Governing the Activities of States in the Exploration and Use of Outer Space, Including the Moon and Other Celestial Bodies, 27 January 1967, 610 *UNTS* 205.

Treaty on Principles Governing the Activities of States in the Exploration and Use of Outer Space, including the Moon and Other Celestial Bodies, entered into force 10 October 1967, 610 UNTS 205, (Outer Space Treaty, 1967)

Treaty on principles Governing the Activities of States in the Exploration and Use of Outer Space, Including the Moon and Other Celestial Bodies, entered into force Oct. 10, 197, 18 U.N.T.S. 205.

Treaty on Principles Governing the Activities of States in the Exploration and Use of Outer Space, Including the Moon and Other Celestial Bodies, October 10, 1967, 610 U.N.T.S. 205

Treaty on Principles Governing the Activities of States in the Exploration and Use of Outer Space, including the Moon and Other Celestial Bodies. (n.d.). Retrieved from http://www.unoosa.org/pdf/gares/ARES_21_2222E.pdf

Treaty on Principles Governing the Activities of States in the Exploration and Use of Outer Space, including the Moon and other Celestial. 18 UST 2410, TIAS 6347, 610 UNTS 205.

Trentlage, C., Mindermann, P., Ben Larbi, M. K., & Stoll, E. (2016). Development and Test of an Adaptable Docking Mechanism Based on Mushroom-Shaped Adhesive Microstructures. In *Proceedings of AIAA SPACE 2016*. Reston, VA: American Institute of Aeronautics and Astronautics. 10.2514/6.2016-5486

Triggs, G. (2013). The Antarctic Treaty System: A Model of Legal Creativity and Cooperation. Washington, DC: Smithsonian Institution Scholarly Press.

Tronchetti, F. (2008). *The exploitation of natural resources of the moon and other celestial bodies: a proposal for a legal regime*. Leiden: Nijhoff.

Tronchetti, F. (2011). Regulating Sub-Orbital Flights Traffic: Using Air Traffic Control as a Model. *Proc. Int'l Inst. Space L.*, *54*, 176.

Tynan, D. (n.d.). *Galactic gold rush: the tech companies aiming to make space mining a reality*. Retrieved from https://www.theguardian.com/science/2016/dec/06/space-mining-moon-asteroids-tech-companies

U.S. Department of Energy. (2011). *Critical Materials Strategy*. Academic Press.

U.S. Office for Space Commerce. (2018). *STM Remarks from EUSpace for Business Conference*. Available at: https://www.space.commerce.gov/stm-remarks-from-euspace-for-business-conference/

U.S. Patent Law, 35 U.S.C. §§ 1 et seq., Consolidated Patent Laws (2007), (US Patent Law, Inventions in Outer Space) National and Commercial Space Programs, 51 U.S.C. Pub. L. 111–314, § 3, 124 Stat. 3328. (2010). Available at https://www.gpo.gov/fdsys/pkg/USCODE-2010-title51/pdf/USCODE-2010-title51.pdf

UK's Outer Space Act of 1986.

UN Doc. A/4141 of 14 July 1959, Part III, III.A, "Report of the Ad Hoc Committee on the Peaceful Uses of Outer Space".

UN Doc. A/AC.105/C.2/2018/DEF/L.1, "Draft report, Annex II, Report of the Chair of the Working Group on the Definition and Delimitation of Outer Space", 13 April 2018.

UN Doc. A/AC.105/C.2/L.304/Add.2, "Draft report: V. Matters relating to the definition and delimitation of outer space and the character and utilization of the geostationary orbit, including consideration of ways and means to ensure the rational and equitable use of the geostationary orbit without prejudice to the role of the International Telecommunication Union", 13 April 2018.

UN Doc. A/AC.105/C.2/L.69,

UN Doc. A/RES/72/77, "International cooperation in the peaceful uses of outer space", 14 December 2017.

UN Doc. Technical Presentations made at the Scientific and Technical Subcommittee, "The Inter-Agency Space Debris Coordination Committee (IADC) -An overview of the IADC's annual activities". (n.d.). Available at: http://www.unoosa.org/documents/pdf/copuos/stsc/2018/tech-04E.pdf

UN Doc.A/AC.105/58

UN Office for Outer Space Affairs. (n.d.). *National Space Law Collection*. Available from: http://www.unoosa.org/oosa/en/ourwork/spacelaw/nationalspacelaw/index.html.

UNGA resolution 68/74 of 11 December 2013, supra note 34

UNGA resolution 68/74 of 11 December 2013. (n.d.). Available at http://www.unoosa.org/pdf/gares/A_RES_68_074E.pdf

Union of Concerned Scientists. (2017, November). *USC Satellite Database*. Retrieved from https://www.ucsusa.org/nuclear-weapons/space-weapons/satellite-database#.WL2lyPJCjNE

Union of Concerned Scientists. (2018). *UCS Satellite Database*. Author.

United Nations Committee on the Peaceful Uses of Outer Space. (2016). Guidelines for the long-term sustainability of outer space activities: First set, UN Doc. A/71/20.

United Nations Convention on the Law of the Sea, Montego Bay, done December 10, 1982, entered into force November 16, 1994; 1833 U.N.T.S. 3.

United Nations Legal Subcommittee on the Peaceful Uses of Outer Space. Schematic overview of national regulatory frameworks for space activities, A/AC.105/C.2/2014/CRP.5 (Vienna, 2014). (n.d.). Available at: http://www.unoosa.org/pdf/limited/c2/AC105_C2_2014_CRP05E.pdf

United Nations Office for Outer Space Affairs. (2010). *Space Debris Mitigation Guidelines of the Committee on the Peaceful Uses of Outer Space*. Vienna, Austria: United Nations.

United Nations, Practice of States and international organisations in registering space objects, UN Doc. A/AC.105/C.2/L.255, 8, paras 46 & 49

United Nations. (1999). *The Space Millennium: Vienna Declaration on Space and Human Development*. Retrieved from: http://www.unoosa.org/pdf/reports/unispace/viennadeclE.pdf

UNOOSA. (2013). *The Legal Framework of the International Space Station*. Retrieved from http://www.unoosa.org/pdf/pres/lsc2013/tech-05E.pdf

UNOOSA. (n.d.). Retrieved from http://www.unoosa.org/pdf/pres/lsc2009/pres-09.pdf

US Code, 51 U.S.C. § 50904 (a)(3)-(4)

van der Pas, N., Lousada, J., Terhes, C., Bernabeu, M., & Bauer, W. (2014). Target selection and comparison of mission design for space debris removal by DLR's advanced study group. *Acta Astronautica*, *102*, 241–248. doi:10.1016/j.actaastro.2014.06.020

Van Fenema, H. P. (2005). Suborbital Flights and ICAO. *Air and Space Law*, *30*, 405.

Van Fenema, H. P. (2015). Legal aspects of launch services and space transportation. In *Handbook of Space Law* (p. 408). Edward Elgar. doi:10.4337/9781781000366.00016

Van Kesteren, M. W. (2013). *Air Launch versus Ground Launch: a Multidisciplinary Design Optimization Study of Expendable Launch Vehicles on Cost Performance*. Delft: Master of Science Thesis AE5810.

Veldhuyzen, R., & Grifoni, E. (1999). *No Exchange of Funds – The ESA Barter Agreements for the International Space Station*. Available at http://www.esa.int/esapub/bulletin/bullet99/veld99.pdf

Vielstich, W., Lamm, A., Gasteiger, H. A., & Yokokawa, H. (2010). *Handbook of Fuel Cells*. Chichester, UK: John Wiley & Sons, Ltd. doi:10.1002/9780470974001

Vienna Convention on the Law of Treaties , 23 May 1969, United Nations, Treaty Series, vol. 1155, p. 331. (n.d.). Available at: https://www.refworld.org/docid/3ae6b3a10.html

Vienna Convention on the Law of Treaties, opened for signature on 23 May 1969, 1155 *UNTS* 331.

Viikari, L. (2002). *From manganese nodules to lunar regolith: a comparative legal study of the utilization of natural resources in the deep seabed and outer space*. Rovaniemi: University of Lapland.

Villa, M. (2016). *Builders of satellites large and small must work together to benefit science*. Retrieved from: http://www.spacenewsmag.com/commentary/builders-of-satellites-large-and-small%E2%80%A8-must-work-together-to-benefit-science/

Visagie, L., Lappas, V., & Erb, S. (2015). Drag sails for space debris mitigation. *Acta Astronautica*, *109*, 65–75. doi:10.1016/j.actaastro.2014.12.013

Vlajic, N. (2010). *Analog Transmission of Digital Data course*. Stanford University. Retrieved from https://web.stanford.edu/class/ee102b/contents/DigitalModulation.pdf

Von der Dunk, F. (2009). *Space Law in the Age of the International Space Station, University of Nebraska – Lincoln: Paper 6*. Space and Telecommunications Law Program Faculty Publications.

Von der Dunk, F. (2015). International Space Law. In *Handbook of Space Law*. Cheltenham, UK: Edward Elgar Publishing. doi:10.4337/9781781000366.00011

Von der Dunk, F. G. (2007). Passing the Buck to Rogers: International Liability Issues in Private Spaceflight. *Nebraska Law Review*, *86*(2).

Von der Dunk, F. G. (2011). Space tourism, private spaceflight and the law: Key aspects. *Space Policy*, *27*(3), 146–152. doi:10.1016/j.spacepol.2011.04.015

Von der Dunk, F. G. (2015). *Legal aspects of private manned spaceflight. In Handbook of Space Law* (p. 669). Cheltenham, UK: Edgar Elgar Publishing.

Von Der Dunk, F. G., & Tronchetti, F. (Eds.). (2015). *Handbook of Space Law*. Cheltenham, UK: Edward Elgar Publisher. doi:10.4337/9781781000366

Von Karman Institute. (2016, Jul. 1). *Official project link*. Retrieved from https://www.qb50.eu

Wall, M. (2013). *Ticket Price for Private Spaceflights on Virgin Galactic's SpaceShipTwo Going Up*. Space.com.

Wall, M. (2015). Asteroid mining may be a reality by 2025. *Space News*. Retrieved on 30 April 2018 from http://www.space.com/30213-asteroid-mining-planetary-resources-2025.html

Wall, M. (2016). *Private Space Stations May Take Flight in 2020*. Available at http://www.space.com/34377-private-space-stations-may-take-flight-in-2020.html

Wall, M. (2018). *NASA Looking at Ways to Commercialize International Space Station*. Available at https://www.space.com/39880-nasa-commercialize-international-space-station.html

Walter, E. (2012). The privatisation and commercialisation of Outer Space. In C. Brünner & A. Soucek (Eds.), *Outer Space in Society, Politics and Law* (p. 500). Springer Verlag.

Waltz, D. M. (1993). *On-Orbit Servicing of Space Systems* (1st ed.). Malabar, FL: Krieger.

Wang, J. (2016, April 14). ESA head calls for 'space without borders' during first China visit. *GBTimes*. Retrieved from https://gbtimes.com/esa-head-calls-space-without-borders-during-first-china-visit

Wang, Y., Jing, X., Lv, K., Wu, B., Bai, Y., Luo, Y., ... Li, Y. (2014). During the long way to Mars: Effects of 520 days of confinement (Mars500) on the assessment of affective stimuli and stage alteration in mood and plasma hormone levels. *PLoS One*, *9*(4). PMID:24695321

Washington Post Live. (2016). NASA Administrator: We used to buy the rocket. Now we buy the service. *The Washington Post*. Available at https://www.washingtonpost.com/blogs/post-live/wp/2016/04/07/meet-nasa-administrator-charles-bolden/?utm_term=.c5f4b719f9c4

Wassenbergh, H. A. (1996). The international regulation of an equitable utilization of natural outer space resources. *Proc. Coll. L. Outer Space*, *39*, 138–143.

Weeden, B. (2016). *Insight - Small Satellites and Space Situational Awareness*. Retrieved from: https://swfound.org/news/all-news/2016/09/insight-small-satellites-and-space-situational-awareness

Weeden, B., & Samson, V. (2018, April 4). New UN Guidelines For Space Sustainability Are A Big Deal. *Breaking Defense*. Retrieved from https://www.breakingdefense.com

Weirauch, A. (2011). *F*. Houston, TX: ISS Pressurized Volume Hardware Common Interface Requirements Document.

Werkheiser, M. J., Fiske, M., Edmunson, J., & Khoshnevis, B. (2015). On The Development of Additive Construction Technologies for Application to Development of Lunar/Martian Surface Structures Using In-Situ Materials. In *AIAA SPACE 2015*. Conference and Exposition; doi:10.2514/6.2015-4451

Werner, D. (2018, March 21). ESA to investigate links between debris removal and satellite servicing. *SpaceNews*. Retrieved from http://www.spacenews.com

Wertz, J. R., & Larson, W. J. (Eds.). (1999). Space Mission Analysis and Design (3rd ed.). Microcosm Press, Kluwer Academic Publishers Incorporated.

Wheeler, J., & White, S. (2012). *Satellite launch contract: Checklist, Checklist*. Practical Law Company Ltd. Available at: www.practicallaw.com

Wheeler, J. A. (1990). *A Journey Into Gravity and Spacetime. Scientific American Library*. San Francisco, CA: W. H. Freeman.

White, A. E., & Lewis, H. G. (2014a). An adaptive strategy for active debris removal. *Advances in Space Research, 53*(8), 1195–1206. doi:10.1016/j.asr.2014.01.021

White, A. E., & Lewis, H. G. (2014b). The many futures of active debris removal. *Acta Astronautica, 95*, 189–197. doi:10.1016/j.actaastro.2013.11.009

White, R. L., White, H. M., & White, R. L. (1988). *The Law and Regulation of International Space Communication*. Artech.

Wieland, P. O. (1998). Living Together in Space: The Design and Operation of the Life Support Systems on the International Space Station (NASA TM No. 1998-206956). Huntsville, AL: NASA.

Williams, D. R. (2016a). *Earth fact sheet*. Retrieved from https://nssdc.gsfc.nasa.gov/planetary/factsheet/earthfact.html

Williams, D. R. (2016b). *Mars fact sheet*. Retrieved from https://nssdc.gsfc.nasa.gov/planetary/factsheet/marsfact.html

Williams, D., Dake, J., & Gentry, G. (2012). International Space Station Environmental Control and Life Support System Status for the Prior Year: 2010 - 2011. *Proceedings of the 42nd International Conference on Environmental Systems*. 10.2514/6.2012-3612

Williams, S. M. (1987). The law of outer space and natural resources. *Int'l & Comp. L.Q., 36*(01), 142–151. doi:10.1093/iclqaj/36.1.142

WIPO. (2004). *Intellectual Property and Space Activities*. Paper prepared by the International Bureau of WIPO. Available at http://www.wipo.int/export/sites/www/patent-law/en/developments/pdf/ip_space.pdf

Wolfrum, R. (1984). Die Internationalisierung Staatsfreier Räume. In Veröffentlichungen des Max- Planck-Instituts für ausländisches öffentliches Recht und Völkerrecht (Band 85). Academic Press.

Wright, J. (2013a). *Cupola Observational Module*. Retrieved January 12, 2015, from https://www.nasa.gov/mission_pages/station/structure/elements/cupola.html#.Vu7oYeIrKCq

Wright, J. (2013b). *Kibo Laboratory*. Retrieved January 12, 2015, from https://www.nasa.gov/mission_pages/station/structure/elements/jem.html#.Vu7ocOIrKCq

Wright, J. (2013c). *Quest Airlock*. Retrieved January 12, 2015, from https://www.nasa.gov/mission_pages/station/structure/elements/quest.html#.Vu7oaeIrKCp

Wright, J. (2013d). *Solar Arrays*. Retrieved January 12, 2015, from https://www.nasa.gov/mission_pages/station/structure/elements/solar_arrays.html#.Vu7omOIrKCp

Wright, D., Grego, L., & Gronlund, L. (2005). *The Physics of Space Security – A reference manual*. American Academy of Arts and Sciences.

Wright, D., Grego, L., & Gronlund, L. (2005). *The Physics of Space Security: A Reference Manual*. Washington, DC: American Academy of Arts and Sciences. Retrieved from http://www.amacad.org/publications/Physics_of_space_security.pdf

Xiang, M., & Tongyu, L. (2014). *The new Generation Launch Vehicles in China*. Toronto, Canada: IAC.

Yamagami, T. (n.d.). *Research on Balloons to Float over 50km Altitude*. ISAS. Available at: http://www.isas.jaxa.jp/e/special/2003/yamagami/03.shtml

Yoshida, H., Watanabe, T., Kanamori, H., Yoshida, T., Ogiwara, S., & Eguchi, K. (2000). Experimental Study on Water Production by Hydrogen Reduction of Lunar Soil Simulant in a Fixed Bed Reactor. *Second Space Resources Roundtable.*

Yulin, Z., & Zhaokui, W. (2016). Space Traffic Safety Management and Control. *IEEE Transactions on Intelligent Transportation Systems*, *17*(4), 1189–1192. doi:10.1109/TITS.2015.2494686

Zachrau, H. J. (2018, March 14). *SPHERES Tether Slosh*. Retrieved April 30, 2018, from https://www.nasa.gov/mission_pages/station/research/experiments/2553.html

Zak, A. (2015). A Rare Look at the Russian Side of the Space Station. *Air & Space Magazine Smithsonian.* Available at http://www.airspacemag.com/space/rare-look-russian-side-space-station-180956244/#rPCY1SgwIRFktBbq.99

Zelnio, R. (2005, December 5). A Model for the International Development of the Moon. *The Space Review*.

Zhai, G., Qiu, Y., Liang, B., & Li, C. (2009). System Dynamics and Feedforward Control for Tether-Net Space Robot System. *International Journal of Advanced Robotic Systems*, *6*(2), 10. doi:10.5772/6791

Zhao, Y. (2016). The Role of bilateral and multilateral agreements in international space cooperation. *Space Policy, 36*.

Zolik. (2018, April). *Europe buys from Russia 11 rocket "proton Average"*. Retrieved from http://handofmoscow.com/2018/04/12/europe-buys-from-russia-11-rockets-proton-average/

Zubrin, R. M., Muscatello, A. C., & Berggren, M. (2013). Integrated Mars In Situ Propellant Production System. *Journal of Aerospace Engineering*, *26*(1), 43–56. doi:10.1061/(ASCE)AS.1943-5525.0000201

ЦiолковскiйК. (1903). Изслѣдованiе мировыхъ пространствъ реактивными приборами.

About the Contributors

Anja Nakarada Pecujlic is a PhD candidate and assistant researcher at the University of Cologne, where she is writing her thesis on space law. In addition, since December 2017 she is the Project Manager for a European Space Agency (ESA)/Cologne University joint project: "ESA Convention Commentary". Prior to that, from 2014 to 2016, she was an External Consultant to the European Space Policy Institute in Vienna. From 2009 until 2014 she studied at the Law Faculty of the University of Vienna, where she specialised in international law, with her main focus being international space law. She subsequently continued to conduct space legal research and participated as a speaker in several international conferences. She has served as a judge in the Asia-Pacific round of the prestigious Manfred Lachs Moot Court competition since 2014. Finally, she founded "Serbian Case for Space Foundation" in Belgrade, Serbia with the aim to engage Serbia in the space domain.

Matteo Tugnoli is a Senior Research Fellow at the European Space Policy Institute (ESPI) in Vienna. Prior to joining ESPI, he worked as Trainee in the Relations with Member States Department, Director's General Cabinet, of the European Space Agency (ESA HQ) in Paris, France. He started his space career as Research Assistant at the Institute for Radioastronomy, National Institute for Astrophysics (INAF) in Bologna, Italy. Matteo has a Bachelor of Science in Astronomy and a Master of Science in Astrophysics and Cosmology, from the University of Bologna, Italy. He also has a Master in Space Policies and Institutions from the Italian Society for International Organizations (SIOI) and Italian Space Agency (ASI) in Rome, Italy.

* * *

Djordje Andrijasevic is a graduate in Economics and International Relations and has worked in the field of multilateral affairs having worked for the Austrian Foreign Ministry and later within the framework of the New Zealand United Nations Security Council tenure. He also has a Master in Space Policies and Institutions from the Italian Society for International Organizations (SIOI) and Italian Space Agency (ASI) in Rome, Italy. He has specialized several space policy issues at both the Italian as well as European Space Agency. Djordje is currently serving within the Austrian Armed Forces.

P. J. Blount (Ph.D., Global Affairs, Rutgers University, 2016; M.S., Global Affairs, Rutgers University, 2015; LL.M., Public International Law, King's College London, 2007; J.D., University of Mississippi School of Law, 2006; B.A./A.B.J., University of Georgia, 2002) is a postdoctoral researcher in the Faculty of Law, Economics, and Finance at the University of Luxembourg and an adjunct professor in the LL.M in the Air and Space Law at the University of Mississippi School of Law. He served as a Visiting Scholar at the Beijing Institute of Technology School of Law for the Fall of 2017. He has published and presented widely on the topic of space security law and has given expert testimony on space traffic management before the U.S. House of Representatives Subcommittee on Space. Blount serves as the co-editor-in-chief of the Proceedings of the IISL and was a formerly editor-in-chief of the Journal of Space Law. Additionally, he sits on the Board of Directors of the International Institute of Space Law. He is a member of the State Bar of Georgia.

Edmond Boulle is the founder of satellite imagery and LegalTech startup Orbital Witness providing services to real estate lawyers. He has 5 years experience in different branches of the space industry, having worked at the Satellite Applications Catapult in the UK and served two terms as Executive Secretary of the European Centre for Space Law based at the headquarters of the European Space Agency. He has previously written and presented papers at the International Astronautical Congress in 2014 (keynote) and 2016 and spoken at the UN COPUOUS. Edmond is a qualified barrister (England & Wales) with a Masters (LL.M) from the Institute of Air and Space Law at McGill University in Montreal.

Philip De Man is a postdoctoral research fellow with the Leuven Centre for Global Governance Studies in Leuven, Belgium, funded by the Research Foundation Flanders. Dr De Man is a lecturer in space law at the Advanced Master of Science in Space Studies of the Universities of Leuven and Ghent. He is a member of the Belgian delegation at the Legal Subcommittee of the United Nations Committee on the Peaceful Uses of Outer Space. Dr De Man is also co-chair for one of the rule-drafting committees of the McGill Manual on International Law Applicable to Military Uses of Outer Space (MILAMOS). He is also an observer with The Hague International Space Resources Governance Working Group. Among his most recent publications is the edited volume 'Commercial uses of space and space tourism: legal and policy aspects' of 2017.

Clémentine Decoopman is currently the Executive Director of the Space Generation Advisory Council (SGAC). Responsibilities of the position include (but are not limited to) Operations Management, and Events Management. The Operations Management components of this role has an emphasis on administration of the organisation and communications management, including working closely with the Executive Office on the development and facilitation of opportunities for SGAC members in the space sector. The latter includes overseeing development of the SGAC brand, web and printed content, and supporting public relations and communication efforts. The Events Management component of this role is focused on overseeing and coordinating with the SGAC event managers to ensure success of our global events, such as the Space Generation Congress, Space Generation Fusion Forum, and SGx. The SGAC Executive Director also oversees the implementation of other SGAC activities including regional and local events. Clémentine also represents the SGAC at international conferences and at the United Nations (particularly at the UN COPUOS annual sessions) and contribute to the discussions by making regular statements at the plenary sessions.

Gerold Fuchs graduated from University of Technology in Vienna, Austria with a master's degree in mechanical engineering and a specialization in aerospace engineering. Gerold has a professional background in aircraft design, certification and flight testing and has spent numerous hours in aircraft and simulators as a pilot, flight instructor and engineer. Since 2009 he is an aviation inspector for the

Austrian National Aviation Authority "Austro Control". In this job he is involved in certification as well as rulemaking tasks. For more than two years he worked as an Aviation Advisor to the Republic of Georgia on behalf of the European Commission.

Sarah Germann graduated from the University of Vienna's Law Faculty in 2016, where she specialized in international public law and international space law. After participating in the Manfred Lachs Space Law Moot Court as a student in 2013/2014, she served as tutor to the subsequent Viennese team in 2014/2015 and was invited to international conferences in Beijing and Athens as speaker. Her first academic paper was published in the Journal of Space Law in April 2017. From 2018 to 2019, Sarah was employed as Legal Officer in the Office of the Legal Adviser (Department of International Law) at the Austrian Federal Ministry for Europe, Integration and Foreign Affairs where, among other topics, she was responsible for space law matters.

Sophie Gruber is a student at the Leopold-Franzens University of Innsbruck, currently finishing her undergraduate degree in Physics. Since 2017, she is Austria´s National Point of Contact for the Space Generation Advisory Council. In 2018 she was elected as a board member of the Austrian Space Forum, where she works in the project management team for Mars analog missions and is part of the Flight Control Team. Sophie is also member of an international research team, which investigates the operational benefits of 3D printing technology in crewed planetary surface operations. She publishes and presents on analog research and planetary exploration.

Arthur K. Lacombe is a double degree student in Space Systems Engineering at ISAE-SUPAERO and in Space and Telecommunication Law at the Institute of Space and Telecommunication Law (IDEST). In 2015, he joined the SUPAERO SPACE SECTION, a student rocketry association, in which he became president for the year 2016. Meanwhile, he was the manager of the SUPAERO team in charge of the design of a supersonic rocket parts within the PERSEUS program led by the CNES (French Space Agency). The rocket was launched from Kiruna in Sweden, in April 2017. Through his studies, he worked on a measurement system functioning in harsh environment for engine applications, on embedded system design and on actual and future suborbital flight regulatory frameworks.

Arne Lahcen is Policy Officer within the Directorate of Earth Observation Programmes at the European Space Agency. Before joining ESA in 2016, he worked as Resident Fellow at the European Space Policy Institute in Vienna, Austria. His fields of research include Earth Observation governance, international cooperation schemes, innovation management, and the relationship between space and society. Prior to joining ESPI in 2011, he obtained an Advanced Master's degree in Space Studies with a specialisation in space law, policy, business and management at the Faculty of Science of the Catholic University of Leuven, Belgium. He also holds a Bachelor's and Master's degree in Social-Economic Sciences pursued at the Faculty of Applied Economics, University of Antwerp, Belgium.

Martin J. Losekamm received his M.Sc. in Applied and Engineering Physics from the Technical University of Munich (TUM) in 2016. He stayed at the TRIUMF National Laboratory at the University of British Columbia in 2013. During his M.Sc. research at the Massachusetts Institute of Technology's Space Systems Laboratory in 2015, Martin worked with the SPHERES experiment aboard the International Space Station. He took part in two stratospheric balloon campaigns as part of the German-Swedish

REXUS/BEXUS student program. Currently, Martin is a PhD student and employee at TUM, focusing on the development of radiation detectors for space applications.

Yuriy Metsker was born in 1988 in Lviv, Ukraine. In 2007 he has started to study aerospace engineering at the Technical University of Munich (TUM). In the same year, he has joined the student Scientific Workgroup for Rocketry and Spaceflight (WARR) affiliated at the TUM Chair of Astronautics. In 2014 he has started to work as a research assistant at the Chair of Turbomachinery and in parallel as a development engineer at MBDA GmbH. The main topic of his scientific work was the research on divert and attitude control systems for missile defense applications. Currently he is working as a consultant at the Air and Missile Defence Consulting GmbH.

Lauren Napier has a Bachelor of Science in Communications and a Master of Arts in International Relations. She is originally from the United States but has spent the past eight years living in Europe. She has been a member of SGAC since 2013 and served as a co-lead for the Space Law and Policy Project Group from 2015 to 2017. She was also a member of the Communication Team for the 1st European Space Generation Workshop. Most recently, Lauren was the Event Manager for the Space Generation Forum 2.0. She has been part of the SGAC delegation for the United Nations Committee on the Peaceful Uses of Outer Space and the Legal Subcommittee since 2013. Outside of SGAC Lauren has held various positions of leadership in communications, teaching, and event management. Currently Lauren is living in Vienna, Austria and is a Consultant for the Space Generation Advisory Council.

Chris Nie is a Systems Engineer working on the environmental control and life support system (ECLSS) for the Orion Multi-Purpose Crew Vehicle at Lockheed Martin Space. His career has given him the opportunity to work on GPS satellites, SmallSats, as well as NASA's Human Spaceflight Program, Orion. Chris graduated from the University of Colorado, Boulder, with a Master's Degree in Aerospace Engineering and a focus in Bioastronautics. His university career spanned work in the fields of space life science research, robotics, and CubeSats. Chris is active in the Space Generation Advisory Council (SGAC), the American Institute of Aeronautics and Astronautics (AIAA), and is an alumnus of the Students for the Exploration and Development of Space (SEDS). He is honored to have been named one of Aviation Week's Twenty20s as a rising leader in Aerospace, and to have received the President's Volunteer Service Award multiple times for work in STEM outreach.

Claas Olthoff studied aeronautics and astronautics at the Technical University of Munich (TUM), with special focus on astronautics and propulsion systems. He received a diploma in aerospace engineering from TUM in 2010 and performed the research for his master's thesis at the Massachusetts Institute of Technology (MIT). He continued his studies at the Institute of Astronautics at TUM as a Ph.D. candidate. From 2010 to 2014 he was the project manager and lead systems engineer for the "First-MOVE" CubeSat at TUM, which successfully launched in November 2013. In 2017 he earned a Ph.D. from TUM with his dissertation on the dynamic simulation of extravehicular activities. Since 2012 he is the head of the human spaceflight working group at the Institute of Astronautics. In this role he is continuing his research as a post-doctoral researcher with a focus on life support systems and spacesuits.

Philipp Reiss holds a doctoral degree (2018) and M.Sc. in aerospace engineering (2012) from the Technical University of Munich and a B.Eng. (2009) in aerospace engineering from the University of Applied Sciences Bremen. His work experience includes the participation in the European REXUS sounding rocket program and the second Joint European Partial-g Parabolic Flight Campaign. From 2012 to 2018 he was involved in several DLR/ESA/EU funded research projects on the sampling, extraction, characterisation, and utilisation of lunar resources, among them he contributed to the ESA instrument development ProSPA for the Luna-27 mission. Since 2018 he leads the lunar exploration working group at the Institute of Astronautics as a post-doctoral researcher.

Alberto Rueda Carazo is a Spanish Jurist and Legal Advisor in Outer Space Law and Public International Law. A graduate from the University of Jaén, Alberto has participated at the Manfred Lachs Moot Court Space Law competition representing his University in 2012. This experience, which helped him gaining a broader insight in the matter, led him to participate in the 2012 Summer Course on Space Law and Policy organised by the European Centre for Space Law as both, student and tutor. Alberto has served as an assistant to ESA's liaison officer in Italy, where he gave assistance in the organisation and follow-up of institutional activities linked to space issues. After that, he obtained an LL.M. degree in Public International Law at Leiden University and an LL.M degree in Space and Telecommunications Law by the Institute of Space and Telecommunications Law (IDEST) in Paris, France. Alberto is currently a legal intern for the Chief of the Committee, Policy and Legal Affairs Section at UNOOSA.

Yuri Takaya-Umehara received her B.A. in Law from the Waseda University, Japan in 1997, her LL.M. in International Space Law from the Leiden University, The Netherlands in 2001, and her Ph.D. in Public Law from the Paris Sud University, France in 2009. In 2011 she was hired as the Research Fellow by the Kobe University in Japan, lectured on space law until March of 2017. In the present, she is affiliated to the University of Tokyo in Japan as Visiting Researcher. She writes and presents widely on legal issues of military and commercial uses of outer space, including the exploitation of space resources.

Index

G

H

I

J

K

L

M

N

O

P

R

S

T

U

V

W

X

Y

Z

www.ingramcontent.com/pod-product-compliance
Lightning Source LLC
LaVergne TN
LVHW060415300926
849162LV00006B/1436